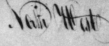

29 Light breeze from the S & E. this
Lowered after Right Whale, the
Larboard boat fastened and the Starboard
boat killed him. the
fastened to another

Saved
Lat: 58.50 N. Lo

30 Light breeze from the S & E. Lowered
after Right Whale, the Larboard boat
fastened but the line dueff

Lat: 58. 30 N.   Lost

31 Light breeze from the S. lowered after
Right Whales, the Bow boat fastened to
him & killed him he went off spouting
blood, the Starboard boat fastened to
another and the line got fouler took the
boat under water had to cut. lost all
the Craft, the Larboard boat fastened
to a calf killed him and him
alongside   Lost   Lost

Saved

ON THE NORTHWEST

ON THE

ROBERT LLOYD WEBB

UNIVERSITY OF BRITI

NORTHWEST

Commercial Whaling
IN THE
Pacific Northwest
1790–1967

LUMBIA PRESS · VANCOUVER 1988

This book has been published with the help of a grant from the Social Science Federation of Canada, using funds provided by the Social Sciences and Humanities Research Council of Canada, and a grant from the B.C. Heritage Trust.

Canadian Cataloguing in Publication Data

Webb, Robert Lloyd.
    On the Northwest : commercial whaling in the
Pacific Northwest, 1790–1967

    (University of British Columbia Press Pacific
maritime studies ; 6)
    Includes bibliographical references and index.
    ISBN 0–7748–0292–8

    1. Whaling - Northwest coast of North America -
History. 2. Whaling - British Columbia - History.
I. Title. II. Series.
SH382.7.W43 1988 338.3'7295'091643 C88–091150–6

Credits: jacket photo, "Unidentified whaleman poses heroically atop the slain leviathan," *Kendall Whaling Museum, Sharon, Massachussetts*; title-page photo, Benjamin Russell and Caleb Purrington, "Whaling on the Northwest Coast"; detail from the 1300-foot panorama *Whaling Voyage 'Round the World*, 1848, *The Whaling Museum, New Bedford, Massachusetts*; endpapers, *Historical Society of Delaware, Wilmington, Delaware*.

International Standard Book Number 0–7748–0292–8
Designed by Ron McAmmond
Printed in Canada

For
Margaret Webb, my mother,
for her constant support and confidence

CONTENTS

ILLUSTRATIONS

PREFACE

Johan Hjort is a man I would like to have known. A professor of marine biology at the University of Oslo before World War II, he carefully analysed the technological whaling of his time. He became a recognized expert in its science and was chosen to chair conferences and conventions which had belatedly convened to regulate the harvest of this valuable economic resource.

Hjort concluded, in the self-effacing tone of many of his fellow Norwegians, that the increasingly sophisticated means of hunting actually jeopardized the whaling industry because the technology offered no solution to the problem of stock replenishment. Under such telling titles as "The Story of Whaling: A Parable of Sociology," "Human Activities and the Study of Life in the Sea," and "The Optimum Catch,"[2] he explained how populations of whales, finite in number and renewable only over a long time, could not support the increased demand placed upon them by commercially motivated factory-ship fleets and coastwise chaser boats which by 1930 scoured every ocean. He showed that the harvesting of whales proceeded contrary to the expectations of agriculture and domestic husbandry, where breeding improvements lead to strains of animals and plants that reproduce in larger numbers or faster; are more impervious to disease; or thrive in wastelands formerly considered unusable. Increased and improved whaling led only to a point beyond which stocks could not replenish themselves. There was never any opportunity to improve the product or hurry its growth. "Large amounts of capital have been spent by the whalers themselves," Hjort wrote, "on experiments on new machinery for converting the flesh and bones into guano or into food for man and beast. But it is obvious that these praiseworthy efforts are not in themselves enough to establish an equilibrium between the whaling industry and the annual renewal of the stock."[3]

The subject of the present work is commercial whaling north of 49° North latitude and east of 170° West longitude, primarily in the coastal waters of Washington, British Columbia, and southeastern Alaska. Not a few maritime historians argue that the whale fishery of the Pacific Northwest has little significance, that it was a backwater both in the nineteenth century, when its returns were overshadowed by the wealth of the "South Seas" sperm-whale fishery, and also in the twentieth, when bountiful Antarctic catches dominated the business. But the whale fishery of the northeastern Pacific Ocean is a repetitive microcosm of Johan Hjort's thesis and a powerful case study of the short-term effects of intensive whaling on a local or transient whale population.

Hjort showed how the catch from a newly exploited whaling ground increased rapidly up to a turning-point, beyond which a rapid decline inevitably resulted in commercial failure. His theory can be readily demonstrated in the Pacific Northwest in the nineteenth century and again in the twentieth, both in fisheries for right whales and for the various "rorqual" (grooved-jaw) species, including the humpback, finback, and blue whales. Commercial whaling in the eastern North Pacific was supported by the carcasses of almost every species of large whale hunted anywhere on earth; every type of catching equipment from double-flued iron hand-harpoons to 90-millimetre cannons was utilized. And every commercial venture proved the soundness of Hjort's conclusions; improving technology depleted species after species until there was no more profit to be gained in the hunting of them. Then each successive industry died a malingering death.

The slow-swimming, buoyant North Pacific right whales were decimated during the first quarter-century of extensive hunting, between 1840 and 1865. By the second decade of the twentieth century, the species was described in the American press as "one of nature's rarities," whereas in truth it was commercial whaling, not natural circumstance, which had reduced it to near-extinction. Humpback whales in Pacific Northwestern waters were similarly depleted during a ten-year period beginning in 1905, and the numbers of blue and finback whales decreased significantly during the era of modern whaling in British Columbia and southeastern Alaska, as whalemen harpooned every animal they could overtake, including juveniles and lactating females.

The Northwest Coast right-whale fishery of the nineteenth century is culturally significant as well. The earliest whaleship crews who came ashore on the Northwest Coast of America left behind a scant but important record of their interaction with both native and colonial residents. In the manner of the fur traders who came before them and the better-organized mining, timber, and railroad interests which followed, the early whalemen contributed to the change which overwhelmed the Pacific Northwest and forever altered the little-populated, pristine wilderness.

Surviving journals of whaling voyages to the North Pacific frequently confuse the appellations "Northwest Coast," "Kamchatka," and "Kodiak." Before 1840 the "Northwest Coast" might even refer to Alta California and the sperm-whaling grounds offshore from San Francisco Bay and Monterey. And since the men before the mast were often uninformed of the ship's exact position, many journalkeepers made errors based on hearsay and misinformation. One such sailor, finding himself near "Trinity Island"—not so very far from Kodiak—mistakenly described the landfall to the north as the "Coast of Kamchatka," when he was in fact describing the Alaska Peninsula.[4] The boundaries of the whaling grounds were eventually delimited by the natural range of the quarry, and since the whales of choice—right whales—could be found throughout a large contiguous area adjacent to the Northwest Coast, this place soon became identified in their journals as the "North-West Coast of America" or simply, the "Nor'west" ground. During the half-century from 1840, whaleboat crews lanced the majority of the right whales to be found there, and then the coastal waters were almost abandoned by sail-whalemen. But the development of mechanized technologies, which allowed the hunters to pursue the fast and strong rorqual whales, led to the construction of mechanized shore-whaling stations from Washington State northward to the Aleutian Islands. After 1900 an entirely new industry matured and briefly flourished.

Two important aspects of whaling in the far North Pacific deserve attention but must necessarily be ignored. One is the right-whale fishery along the Russian Kamchatka peninsula. Though short-lived, its exploitation was of great economic importance and central to the removal of the right whale from the western North Pacific. The rise and decline of pelagic factory-ship whaling and the role it played after 1910 in harvesting northern whale populations are similarly set aside. It must be sufficient to note three periods of factory-ship operation in the eastern North Pacific, beginning with the arrival of the prototypical Norwegian factories, *Admiralen*, *Sommerstad*, and *Capella*, which are only briefly considered, and the *Kit*, which cruised the Arctic with Eskimo whalemen during the summers of 1912–14.

During the second period, in the mid-1930s, both Russian and Japanese factory ships made exploratory forays into the Bering Sea and along the eastern North Pacific rim. Then, after World War II, Japan and the Soviets resumed their work and subsequently maintained a presence along the edge of the continental shelf near Canada and the United States. The proximity of one such fleet to Cape Mendocino, California, inspired a confrontation in June 1975 between Greenpeace and chaser boats attached to the Soviet factory *Dalnij Vostok*[5] —an event which focused the attention of North Americans on the attrition of the large whales.

The story that is told here, of the development of whaling in the Pacific Northwest, begins to fill a remarkable void in Pacific maritime history, a virtually unknown chapter in the development of commerce in the Pacific Northwest. The seafaring history of the region has been altogether overwhelmed by the preliminary voyages of exploration, the lucrative exploits of the fur traders, and the dangerous progress of sailing cargo ships. The source of this unexplored history, and the cause of it, were the whales themselves. By a fortuitous concordance of circumstances, the coastal waters of the Pacific Northwest provide a cornucopia of food for migrating cetacean species and a comfortable and often protected home for smaller, non-migratory whales. Upwellings from a ragged sea floor constantly in ferment, combined with the salutary effect of the warm current called *Kuro Siwo*, which crosses the Pacific from Japan, nurture all manner of plant and animal life, much of which is utilized as food by the northern whales. The proximity of the best feeding areas to shore made certain whales accessible to native whale hunters, whose travel into the deepwater environment was largely restricted by the limitations of their vessels. Crews of commercial whaling ships later captured these species as well as others, such as the sperm whales and right whales, which were less likely to come within range of native harpoons.

In describing the whaling, particularly where it concerns the activities of the nineteenth-century sailors, it seems appropriate to use many of their own words and phrases. No less a maritime historian than Samuel Eliot Morison advised caution in so doing; he wrote: "There is nothing that adds so much to the charm and effectiveness of a history as good quotations from the sources, especially if the period is somewhat remote. But there is nothing so disgusting to the reader as long, tedious, broken quotations in small print, especially those in which, to make sense, the author has to interpolate words in brackets."[6]

Today, more than in Morison's time, we live in a visual world. We have less enthusiasm for the arduous and increasingly unpractised task of creating images from the inked word. But because there are so few "pictures" from the early times, so few photographs and accurate lithographs, the chapters dealing with the sail-whalemen are illuminated by the words of their journalists who reported the moment. Virtually all are quoted from entries written on the Northwest Coast. When we come to the modern whalemen of the twentieth century, then it is time to leave the quotations and look at the photographs.

While attending to the wisdom of Samuel Eliot Morison, one must also hear the American historian (and President) Theodore Roosevelt, about whose style his biographer Edmund Morris has written: "Not for him the maunderings of the 'institutional' historians, with their obsessive analyses of treatises and committee reports. He wanted his readers to smell the bitter smoke of campfires."[7] In this work I have necessarily bound myself to antique treatises and committee reports, but I have tried also to do the Roosevelt style some justice by sailing the reader from page first to page last. As for completeness, it is always true that a researcher's hundred answers raise a thousand questions, and it will be rewarding to see what questions devolve from this present work.

Robert Lloyd Webb
Sharon, Massachusetts

ACKNOWLEDGMENTS

The universal truth is: hundreds of people help assemble a research book by providing background knowledge, enthusiasm for the subject, needed resources of a financial or scholarly nature, or by assisting and encouraging, for whatever personal reasons. Published acknowledgments are never adequate—these should go back at least to Dr. Robert Johnson at Culver City High School in Culver City, California, who taught me the principles of scholarly investigation.

There is a formal beginning, however. The Trustees of the Kendall Whaling Museum in Sharon, Massachusetts, and its director, Stuart M. Frank, offered me the title of research associate—later curator of research—and thereby access to the comprehensive collections of the museum and a platform from which I could complete independent scholarship begun in Vancouver in 1979. The co-operation of everyone on the staff was invaluable, and I must particularly thank Robert H. Ellis, Jr., curator; Gare B. Reid, manager; John F. Sheldon, projects manager; and James A. Frazier, assistant curator of manuscripts.

Robin Inglis, director of the Vancouver Maritime Museum, and the museum's board of directors and staff encouraged me to develop and curate the first exhibition of west coast Canadian whaling: *West Whaling: The Whale Fishery of the Pacific Northwest*. I am particularly indebted to Leonard McCann, curator, for his enthusiastic knowledge of provincial maritime history.

Important acknowledgments are due to the museums, university libraries, research institutions, and historical societies whose staffs care for so many of the documents I studied—a list worth absorbing if you would learn anything about whaling in North America. Certain individuals assisted at particularly pressing times, giving completely of their energies to help me focus mine; by naming them I also thank their institutions:

In Canada: Judith Beattie, Hudson's Bay Company Archives, Provincial Archives of Manitoba, Winnipeg; Gabrielle Blais, Public Archives of Canada, Ottawa; Ron Denman, Museum of Northern British Columbia, Prince Rupert; Philip Goldring, Parks Canada, Ottawa; Dan Goodman, Fisheries Research Branch, Fisheries and Oceans Canada, Ottawa; Frances Gundry,

Public Archives of British Columbia, Victoria; Susan Keller, Fisheries Technology Laboratory, Vancouver; Lynn Maranda and Henry Tabbers, Vancouver Museum; Edward Mitchell, Arctic Biological Centre, St. Anne-de-Bellevue; James Wardrop, British Columbia Provincial Museum, Victoria; and Paulette Westlake, Fisheries Management Regional Library, Fisheries and Oceans Canada, Vancouver. I also received invaluable help from the Maritime Museum of British Columbia, Victoria; National Museums Canada, Ottawa; Simon Fraser University Library, Burnaby; University of British Columbia Library (Special Collections), Vancouver; the Vancouver City Archives; and the Vancouver Public Library.

In the United States: Virginia Adams, John Bockstoce, Judith Downey, and Richard Kugler, Old Dartmouth Historical Society Whaling Museum, New Bedford, Mass.; Greig Arnold and John Angus Thomas, Makah Museum, Neah Bay, Wash.; John Arrison, Hart Nautical Collection, MIT Museum, Cambridge, Mass.; Prudence Backman and Caroline Preston, Essex Institute, Salem, Mass.; Paul Cyr, New Bedford Free Public Library, New Bedford, Mass.; Robert Farwell, Cold Spring Harbor Whaling Museum, Cold Spring Harbor, N.Y.; George Finckenor, Sag Harbor Whaling and Historical Museum, Sag Harbor, N.Y.; Larry Gilmore, Columbia River Maritime Museum, Astoria, Ore.; Marion Halperin, Dukes County Historical Society, Edgartown, Mass.; Jacqueline Haring and John Welch, Nantucket Historical Association, Nantucket, Mass.; James Huelsbeck, University of Santa Clara, Calif.; Paul Johnston, Mary Malloy, and Gregor Trinkaus-Randall, Peabody Museum, Salem, Mass.; Elizabeth Knox, New London County Historical Society, New London, Conn.; John McDonough, Manuscripts Division, Library of Congress, Washington, D.C.; Paul O'Pecko, William Peterson, and Douglas Stein, Mystic Seaport Museum, Mystic, Conn.; James Owens, Federal Records Center, Waltham, Mass.; Robert Parks, Franklin D. Roosevelt Library, Hyde Park, N.Y.; Karyl Winn, University of Washington Libraries, Seattle. I also received assistance from the Baker Library, Harvard University, Cambridge, Mass.; the Huntington Library, Pasadena, Calif.; National Archives and Records Service, General Services Administration, Washington, D.C.; and the Yale University Library, New Haven, Conn..

In England: Roger Morris, National Maritime Museum, Greenwich, London; and the staff of the Public Records Office, Kew Gardens, London. In Norway: Jan Erik Ringstad of Kommandør Chr. Christensens Hvalfangstmuseum, Sandefjord. In Australia: Honore Forster, Research Institute of Pacific Studies, Australian National University, Canberra.

Special thanks are due to the Makah people of Washington State, for their welcome to the unique whaling village site at Ozette; and to its caretaker of the moment, Jim Greene, for his quiet knowledge.

Independent scholars and researchers participated "with a will": Klaus Barthelmess of Cologne, Federal Republic of Germany; Frank Clapp of Victoria; John Harland in Kelowna, B.C.; Elizabeth Ingalls, Boston, Mass.; Richard Inglis and James Haggarty, Parks Canada Anthropological Research Project, Victoria; A. G. E. Jones in Tunbridge Wells, England; Bill Merilees of Nanaimo, B.C.; A. C. S. Payton, London, England; Allan K. Reese, Sr., of North East, Penn.; Duncan Stacey, Richmond, B.C.; and Philip J. Thomas in Vancouver.

Antiquarian booksellers opened their private stacks for my research. Edward J. Lefkowicz of Fairhaven, Massachusetts, not only gave me free access to his collections, but also set up the sophisticated word-processing programme used to prepare the final drafts of the manuscript. Michael Ginsberg of Sharon, Massachusetts provided access-on-request to his extensive inventory. Steve Lunsford of Vancouver let me run free among his books and avail myself of his keen knowledge of things British Columbian; and there was important reading in the library of Dick Weatherford at Monroe, Washington.

Jill King offered research assistance and considerable support at the beginning of the project, for which I am deeply grateful. Briton C. Busch at Colgate University in Hamilton, New York, tendered invaluable editorial advice; Karen Lefkowicz provided marketing assistance. Mrs. Lewis Robbins permitted me to quote from her late husband's whaling-station correspondence. Elaine Scheier volunteered experience and skill in preparing the index; Richard Unger of the University of British Columbia carefully critiqued an early stage of the manuscript. Rick Van Krugel, radio operator for the Bull Harbour Coast Guard Station on Vancouver Island, put on his "ears" to help me find former whalemen; and Cora Veaudry, without knowing it, created a lasting impression of the whaling era while serving perhaps the world's tastiest sticky buns at her café in Coal Harbour.

Several people translated from languages that I do not read with facility: Lorraine Allen, Kerstin Gilkerson, Chris Midelfort, Nancy Elizabeth Pick, and Katrin Sermat.

To the whalemen themselves, I am without enough thanks for their willingness to share every memory with me. Their names are Richard Carruthers, of Bioproducts, Inc., Warrenton, Oregon; William Hagelund, author of a first-person narrative of whaling with the Consolidated Whaling Corporation; Lorne Hume, former general manager at Coal Harbour; John Lyon, who worked for the Western Whaling Company in the late 1950s; Jim O'Shea, a Consolidated Whaling Corporation employee during the 1920s; Harry Osselton, engineer in the S.S. *Gray*; Herb Smith, a Rose Harbour man; Jimmy Wakelen, blacksmith under the famed Larsen of Rose Harbour; and Les Wilson, a Coal Harbour employee who later managed the converted chaser boats *Pacific Challenge* and *Thorfinn*.

The descendants of whalemen also offered personal knowledge, particularly Graeme and Judy Balcom, who welcomed me into their Vancouver home; and William and Anne Lagen, who did likewise in Bellevue, Washington. Graeme Balcom is a grandson of G. W. Sprott Balcom; Bill Lagen is William Schupp's grandson. Each shared unique material concerning his grandfather's family and business activities. Joan Goddard, granddaughter of British Columbia whaling-station manager William Rolls, has researched more than her own book's worth on twentieth-century whaling in the Pacific Northwest, and contributed generously toward the completion of this one. Joshua B. Richmond opened the whaling papers of his New England shipowner-grandfather Isaac Bailey Richmond. I also received many courtesies from Alan Armour (son of David Armour, an engineer long employed in B.C. whaling); Sam Kosaka (son of Moichi Kosaka, the Japanese plant foreman at Rose Harbour from 1910 to 1919), as well as his wonderfully kind wife, Mary; and Douglas Raine (son of John Raine, Rose Harbour engineer). To the whalemen and their kin, deepest thanks of all.

A final custom in preparing acknowledgments is to place a most significant person at the close. Helen Richmond began assisting me in 1982 as a friend and fellow museum colleague. A maritime educator and editor, she soon became an essential and reliable adviser in the preparation of the manuscript. To her I owe unredeemable sums: for steering the project seaward after it had run aground; for days and weeks of careful research in musty archives; for completing the immense task of preparing the bibliography for publication; for translations from the French; for photographic and word-processing assistance; and for providing a special kind of support, that is, on account of the disassembled engine on the dining-room table.

RLW

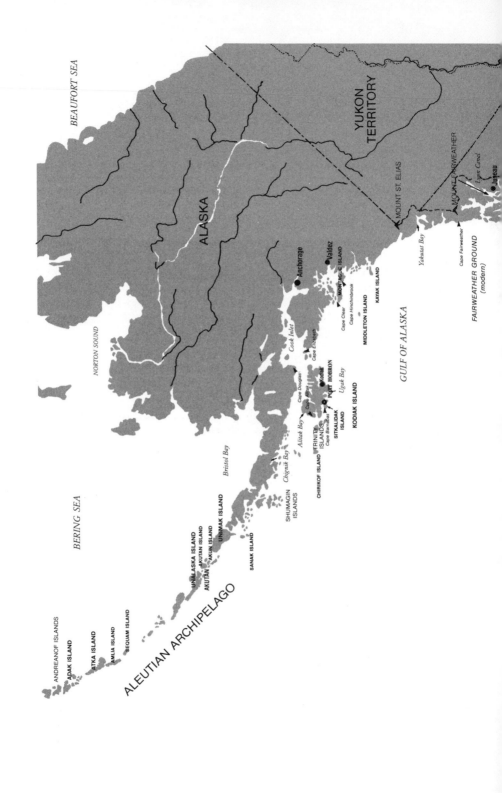

BEAUFORT SEA

YUKON
TERRITORY

ALASKA

MOUNT FAIRWEATHER

Lynn Canal

Juneau

MOUNT ST. ELIAS

Yakutat Bay

Cape Fairweather

FAIRWEATHER GROUND
(modern)

Anchorage

Valdez

MONTAGUE ISLAND

KAYAK ISLAND

Cape Clear

Cape Hinchinbrook

MIDDLETON ISLAND

Cook Inlet

Cape Elizabeth

GULF OF ALASKA

NORTON SOUND

Cape Douglas

Cape Barnabas

Cape Ikolik

Kodiak

PORT HOBRON

Ugak Bay

Afilak Bay

TRINITY
ISLANDS

SITKALIDAK
ISLAND

KODIAK ISLAND

CHIRIKOF ISLAND

Chignik Bay

SHUMAGIN
ISLANDS

Bristol Bay

SANAK ISLAND

UNIMAK ISLAND

AKUN ISLAND

AKUTAN

AKUTAN ISLAND

UNALASKA ISLAND

BERING SEA

SEGUAM ISLAND

AMLIA ISLAND

ATKA ISLAND

ADAK ISLAND

ANDREANOF ISLANDS

ALEUTIAN ARCHIPELAGO

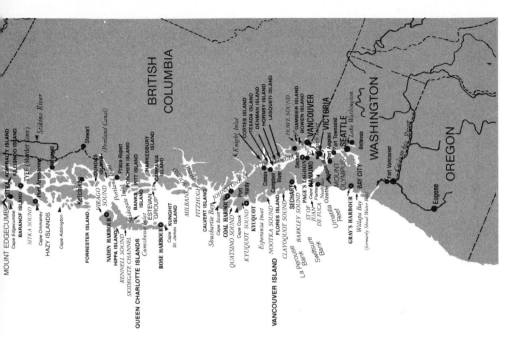

PACIFIC OCEAN

BRITISH
COLUMBIA

WASHINGTON

OREGON

MOUNT EDGECUMBE
SITKA
ADMIRALTY ISLAND
KUISNOO ISLAND
BARANOF ISLAND
TYEE (Murder Cove)
Cape Edgecumbe
SITKA SOUND
Cape Ommaney
HAZY ISLANDS
Port Armstrong
Wrangell
Sikine River
Sikine River (Mirder Cove)
Stewart
Portland Inlet (Portland Canal)
DUNDAS ISLAND
Prince Rupert
FORRESTER ISLAND
Cape Addington
Ketchikan
VIRAGO SOUND
NADEN HARBOUR
Porcher Island
HIPPA ISLAND
Skidegate Inlet
PITT ISLAND
RENNELL SOUND
Camsheva Inlet
BANKS ISLAND
ESTEVAN
PORCHER ISLAND
HAWKESBURY ISLAND
CAMPANIA ISLAND
SKIDEGATE CHANNEL
ROSE HARBOUR
KUNGHIT ISLAND
QUEEN CHARLOTTE ISLANDS
Cape St. James
MILBANK SOUND
FITZHUGH SOUND
CALVERT ISLAND
Shushartie Bay
COOL HARBOUR
JOHNSTONE
Cape Scott
QUATSINO SOUND
Port Hardy
Cape Cook
KYUQUOT SOUND
KYUQUOT
Esperanza Inlet
NOOTKA SOUND
FLORES ISLAND
VANCOUVER ISLAND
CLAYOQUOT SOUND
SECHART
BARKLEY SOUND
La Perouse Bank
Swiftsure Bank
ST. OF JUAN DE FUCA
Cape Flattery
Neah Bay
Umatilla Reef
GRAY'S HARBOUR
Willapa Bay
BAY CITY
(formerly Shoal Water Bay)

Knight Inlet
CORTES ISLAND
TEXADA ISLAND
DENMAN ISLAND
HORNBY ISLAND
LASQUETI ISLAND
HOWE SOUND
GAMBIER ISLAND
BOWEN ISLAND
STR. OF GEORGIA
Comox
Campbell River
Sechelt
PAGE'S LAGOON
NANAIMO
VANCOUVER
VICTORIA
Port Angeles
Port Townsend
SEATTLE
Bellevue
MOUNT OLYMPUS
Lake Washington
Fort Vancouver
Eugene
Port Vancouver
Columbia River

ON THE NORTHWEST

Reefed the Topsails and Furled the Top Gallant Sails and Braced up Sharp on the Wind   Stowed the Anchors on the Bows   Stowed down the Chains and Cleared up Decks   Now Hurrah For the Norwest whare the Rain And the Snow whare the Whale Fishes Blow and Knight Seldom Seen.

JOURNAL, BARK *PANTHEON*, 1 MAY 1843

I

BELOW EAGLE TOWN

*We owe much to the whale; but for it our fishers
would have clung to the shore, for nearly every fish in-
habits the river: it is the whale which emancipated them,
and led them all over the world. They were drawn by an
irresistible fascination out into the open sea, and ever
pursuing their prey, ventured from one point to another,
until without being aware of it, they found they had
passed from the Old World into the New.*

JULES MICHELET
THE SEA[1]

Captain James Cook, indefatigable Pacific explorer, steered into Nootka
Sound on the Northwest Coast of America in the spring of 1778. Already he
had traversed some unfamiliar latitudes, ostensibly lured by a parliamentary
offer of £20,000 for the first description of a Northwest Passage linking England with the rich markets of the Orient. The practical purpose of his visit
was to show the flag, to proclaim for Britain a land held in the tentative
stranglehold of the bothersome Spanish. At Friendly Cove, Cook met a native society whose artifacts suggested an unusual reliance upon the bones and
flesh of marine mammals. He arrived too late in the year to observe the
migration of gray and humpback whales across the Indian threshold, but he
and his crews examined harpoons, knives, canoes, paddles, and skin floats
more than roughly analogous to the tools employed by fellow Englishmen in
the Greenland right-whale fishery. It surely must have occurred to Cook, or
at least to some of his subalterns, that here was an issue worthy of commercial enthusiasm, an industry parallel to that already in place in Greenland.
Until they realized the value of the handsome chocolate-coloured sea otter
pelts obtained at Nootka Sound, the business of whaling may have seemed a
suitable impetus to propel British expansion into the North Pacific Ocean.
  Unbeknownst to Cook, he had arrived on the western shore of a grand,
long island and there connected with a resident maritime culture whose

ritual and corporeal replenishment depended upon the annual passage of whales and seals. On this body of land lived interrelated tribal groups — Clayoquot, Ahousat, Nitinaht, and others — who utilized whaling equipment like Cook saw to secure blubber, meat, oil, and the bones of aquatic animals. These goods filled native larders, provided material for their tools and weapons, and established a power base in trade with non-whaling cultures along the coast.

Captain Cook and his men viewed the evidence, which included "fresh skins" and pieces of whale flesh, and concluded that the people of Nootka Sound fed upon sea animals, including seals, sea otters, and whales, although they did not hunt them in all seasons. The harpoons they reported were tipped with an "oval blade of a large muscle shell, in which is the point of the instrument." This was connected to a shaft twelve to fifteen feet long which separated from the point upon impact, to float upon the water "as a buoy, when the animal darts away with the harpoon."[2] Cook must have realized that this system was equivalent but not identical to the English method, which utilized a solid harpoon with a two-barbed iron point fitted to its wooden shaft. He also saw the long, ocean-going cedar-trunk canoes which were somehow congruent with the oared whaleboats used in the Greenland service. But he missed seeing the whaling carried out.

Farther north, Cook observed whales within sight of a wild and massive snow-capped volcano which had once served as a marker for the explorer Vitus Bering and had been named by him St. Elias. Cook was well acquainted with Bering's explorations and identified Mount St. Elias on the strength of the Dane's description of it. Perhaps he also remembered that Bering's own naturalist, Georg Steller, had commented on the St. Elias whales four decades before, calling them "very numerous, not single any more, but in pairs, and travelling with and behind one another and following one another, which provoked in me the thought that this must be the time fixed for their rut."[3]

Nor was Cook the first even among his own generation of explorers to comment on them. Three years earlier, near 57° North latitude, crews belonging to the Spaniards Juan Perez and Juan Francisco de la Bodega y Quadra had determined themselves to be close in with the land by the appearance of "Orange heads" — the giant Pacific kelp — as well as by birds and many whales. Like Cook, they had previously learned the value of whale products among the native inhabitants; in 47°41′ North — modern-day Washington State — they had met nine canoes of natives who brought "fish of many sorts . . . whale, and salmon."[4]

The first French exploring expedition to arrive on the Northwest Coast, led by Jean François de la Pérouse, also commented on the large numbers of whales. La Pérouse approached the land in the same latitudes covered a decade earlier by Perez and Bodega y Quadra, and like them he wrote of the "large whales, ducks, and divers" that indicated the ships' proximity to the shore. He likewise sighted a long mountain chain covered with snow and, in among its many peaks, the "Mount S Elias of Behring, with its summit rising above the clouds."[5] La Pérouse was followed to the Northwest by Etienne Marchand, who reached Cape Edgecombe in August 1791 and subsequently visited Norfolk Sound, Sitka, portions of the Queen Charlotte Islands, Rennell Sound, and Barkley Sound. In the latter place he carefully described the whaling harpoon of the local native whalemen.[6]

In time, Cook also saw direct results of native whaling. After locating and examining a broad inlet which came to be known as "Cook's River," he sailed westward to "Oonalashka" (Unalaska) and there witnessed native hunters towing two recently killed whales.[7] Although Cook did not know it, they were whale hunters from a very different tradition, one which is said to have employed poison harpoons to kill the quarry. These Aleut whalemen were not in direct contact with the harpoon-and-chase technology at Nootka Sound. In fact, it would soon be shown that a long stretch of coastline along the Gulf of Alaska was populated by people who hunted no whales at all.

Simultaneously with Cook's explorations of the Pacific, a new whale fishery began in England, a fishery completely distinct from the familiar Greenland endeavour, and one which promised larger riches than had ever been brought back from the Arctic Circle. This grand new advance came from hunting a new sort of whale, recognized in merchant society as the spermaceti kind. In 1776, the very year Cook sailed from Plymouth, high-level negotiations were underway in London to finance and regulate this new business in the temperate waters of the Atlantic Ocean. No Englishman knew the proper method of capturing the sperm whale, but encouragements had gone abroad, to the island of Nantucket, offshore from the Massachusetts Colony. From there the spermaceti whale had first been captured for commercial gain.

The sperm whale, a deep diver and deepwater feeder, is rarely seen in the coastal waters then frequented by whalemen, but finds its way in the open ocean, feeding upon large oceanic squid at depths of perhaps a mile and more, seeking them in the compressed darkness with echolocating clicks and pings. To believe legend, the first harpooned sperm whale was brought to Nantucket Island about 1712. One of the island's whalemen was blown well out to sea and there, farther afield than ever, struck and killed one. It seemed a wrinkled, gray, toothsome monster. But when the animal's long head was opened, Nantucketers drained a "case" of liquid oil so clear and pure in quality that it could be ladled, if need be, directly into the lamps of the town's finer homes, to burn there with a nearly smokeless bright light. Sperm oil proved far superior to rendered tallow and other animal fats, including the right whale oil already used for illumination in some homes. The waxy material below the case, the true "spermaceti," could be formed into candles which replicated the white light and virtually smokeless quality of the oil.

That first lucky strike was less a discovery than advertised; the sperm whale was already known in Europe, described in detail and illustrated from stranded animals which occasionally washed ashore on the Dutch coast and elsewhere. Specimens had also come ashore on Nantucket Island long before the first one was harpooned. But the resourceful Nantucketers laid a commercial claim on the species. Soon right-whaling was given up as larger sloops and two-masted brigs were constructed for deepsea whaling. A tryworks—a brick oven structure fitted on the weather deck—made it possible to render oil from the blubber while still at sea. Sperm whale oil proved its superiority over all other kinds, and Nantucket fortunes took a quick turn toward fabulous wealth.

Spermaceti candles and sperm oil soon enough found a market in England. In the 1770s, with a colonial revolt in the offing, Parliament was persuaded to encourage Nantucketers to remove to England and establish a British sperm-whaling industry. Unstable economic conditions in the American colonies convinced a number of these experts to accept Parliament's offers. And they were scions of Yankee whaling families: Coffins, Swains, and Starbucks. Several London merchants looked upon this development with great interest, perhaps none more so than Samuel Enderby and his sons, particularly Samuel Enderby, Jr. The Enderbys invested in and encouraged the sperm-whale fishery and quickly made a reputation in it.

The new business would be prosecuted to the south of England, far from the northern habitat of the Greenland right whales. Consequently, the sperm-whalemen and their industry soon became known as the "South Sea" or "Southern" whale fishery. The term had nothing to do with the "south seas" of tropical glades and palm-frond, white sand beaches; English whaleships had not yet received legal permissions to travel beyond the Western (Atlantic) Ocean. Very soon, however, the news of Cook's voyages would encourage his countrymen to follow into the distant but lucrative waters of the Pacific.

After Cook's expedition returned—the great explorer did not return with them, having been killed by Sandwich Islanders in 1779—its findings were "fully and promptly made known."[8] The publication of the *Voyages* during the winter of 1784—85 constituted the first authorized statements in any language concerning the Northwest Coast of America. The rich profits of the fur trade were suggested, and the presence of whales and whale-hunting native peoples was noted. The former subject drew the immediate attention of a host of commercial and quasi-military venturers who expected to profit by being first in the trade; two of them, George Dixon and John Meares, soon entered into a malignant public feud over certain commercial matters, including the feasibility of a whale fishery in the North Pacific.

Dixon had served under Cook in the *Resolution* during that first examination of Nootka Sound—Cook called it King George's Sound—in 1778. Immediately after the publication of Cook's findings, a group of London merchants formed the King George's Sound Company and hired Dixon and fellow traveller Nathaniel Portlock to command two trading vessels to the Northwest Coast. In the *King George* and the *Queen Charlotte*, Dixon and Portlock traded furs there in 1786 and again in 1787. Their instructions called for the establishment of trading posts—plans never fulfilled—and similar orders were provided to Captain James Colnett in the *Prince of Wales* and the *Princess Royal*. It is clear that whaling was not to be overlooked. The company's charter called for the procurement of "whale fins" (and, presumably, oil) at King George's Sound and elsewhere along the coast.[9]

Dixon's nemesis, John Meares, has come down as one of the least engaging figures among the early British explorers in the Pacific. By all accounts an opportunist, he arrived in the Pacific Northwest from Bengal in the trading vessel *Nootka* in 1786. The *Nootka* flew the British flag, though Meares's voyage contravened the regulations of the powerful East India Company, which held a crown-granted monopoly on trade in the North Pacific; he had not bothered to obtain requisite trading licences from the company. In fact, Meares's bankroll came from employees of the East India Company itself, who had violated confidences in order to share in the rich profits from the fur trade to Canton. When the legally authorized voyage commanded by Portlock met Meares's ill-kept and illegal vessel on the Alaskan coast, he found the crew suffering from scurvy and elected to resupply them rather than seize them for operating without proper papers.

Two years later Meares returned to the Pacific Northwest, taking care this time to fly the Portuguese flag and carry a supernumerary Portuguese "captain" as further insurance. It was during this second voyage that he allegedly purchased a plot of land from Maquinna, the savvy whaleman-chief at Friendly Cove, Nootka Sound. The alleged sale—Maquinna later denied that any land had been sold—provided sufficient kindling to light the fire of war in 1789, when Spanish troops seized the property and ousted Meares and the British from Nootka Sound. Meares returned promptly to England to prod the government into making good on his land claim, and, to aid the effort, he published his advocacy of the Northwest Coast as a potential site of a whaling industry. This business, he said, would prove to be an admirable "training ground" for future Jack Tars of the Royal Navy. He further suggested that better whaling could be found farther to the north: "abundant as the whales may be in the vicinity of Nootka, they bear no comparison to the numbers seen on the Northern part of the coast: indeed the generality of these huge marine mammals delight in the frozen climates."[10]

In the northern matter he obtained incidental corroboration from Portlock, who had sailed about Cape Elizabeth at the entrance to Cook's River (Inlet) in August 1786 and found whales in vast numbers near the shore. "The land on which Cape Elizabeth is situated is an island, and in the straits formed by it and the back land there is good anchorage and shelter," he wrote. "Hereabouts would be a most desirable situation for carrying on a whale-fishery . . . there being convenient and excellent harbours quite handy for the business."[11]

Portlock's co-captain, George Dixon, disagreed. He, like many others, considered Meares's reports overblown and unvalidated by experience. Meares did exaggerate for effect; in describing the whaling implements of Maquinna's people, he noted harpoon shafts as long as twenty feet — too long to be handled easily — and whaling canoes large enough to accommodate twenty men. The usual number was eight. Even Maquinna is said to have called him "Liar Meares";[12] perhaps Dixon felt confident in challenging his suggestion that a whaling station be established at Cook's River. Dixon wrote: "should any future navigator, on the credit of your assertion, (which, begging your pardon, I scarcely think they will), go up there to catch whales, and be disappointed, for their encouragement, I can venture to affirm, they may obtain plenty of fresh salmon."[13] Meares promptly published a rebuttal, claiming that he had seen whales in Cook's River, and also off Cape Douglas, taking their course toward it. He allowed, however, as how "the Pacific Ocean and the channels of the Charlotte Isles are so abundant in these animals, that it would be an idle excursion indeed to follow them elsewhere in those latitudes."[14]

Meanwhile, James Colnett reported that he had obtained whale and dog-fish oil for his ship's lamps while at Clayoquot. His information intensified the ambitions of the restive London merchants, who had got it in their minds that the entire Pacific was full from one end to the other with the breaching and spouting of whales. And certain political factions were duly convinced that the Anglo-Spanish diplomatic dispute surrounding Nootka Sound might be the very opportunity to press suit for complete British freedom in the Spanish seas. "The [whale] fishery indeed lay near the heart of the dispute," one analyst later wrote. "The northern trade with Asia could scarcely flourish unless it was combined with whaling and sealing, and the whaling merchants were perhaps best suited to undertake the northern trade. That at least was the Government's view."[15]

The potential of a greatly expanded whale fishery attracted the attention of England's powerful and persuasive prime minister, William Pitt the Younger. Perhaps Pitt's attention was drawn by Samuel Enderby and the other London whaling merchants; whoever convinced him, Pitt's interest focused on this northern Pacific branch of the "South Sea" fishery, which suddenly promised an "immense treasure"[16] for energetic Britons. In supporting memorials from the whalemen, he earned the praise of the Duke of Leeds, who agreed with Pitt that whaling in the Pacific was "perhaps an object of full as much importance in point of national advantage as [obtaining control of] Nootka Sound itself."[17]

Pitt arranged his diplomacy to ensure that any agreement with the Spanish concerning Nootka Sound would also include British rights and prerogatives with respect to the fisheries. The Nootka Convention as finally signed on 28 October 1790 included an article which opened Spanish waters in the Pacific Northwest to British whaleships, although it precluded them from approaching nearer than ten leagues from the shore, in an effort to stave off illicit trade. This condition displeased a few short-sighted whaling merchants who felt that their captains would be unnecessarily constrained. But Pitt disagreed, as did another of his supporters, Henry Dundas. Dundas, then head of the India Board, showed that the convention gave Britain a specific and important right to fish. "The fishery had flourished with a rapidity unequalled when even cramped by restrictions," Dundas wrote. "With what rapidity, then, must it improve, when every impediment to its prosperity was removed!"[18]

Those who disagreed with their assertive young prime minister saw only that Pitt had refused a proffered Spanish apology for the military action at Nootka in order to press a much larger suit, one which brought England close to war with Spain, caused Spain to realign with France against England in the event of such a war, and allowed for the spending of some £4 million to outfit the Navy and other military branches under the guise of protecting the interests of the South Sea whale fishery. Astute political observers argued that Pitt's premise was unsound; Spain had never interrupted British subjects in their pursuit of whales, not even in the waters near their home coasts, much less at far away Nootka Sound. True, the Spanish had objected to British whaling during the resolution of the crisis, but only to add weight, they speculated, to the larger issue of territory; the whale fishery was never meant to be the object of a war. Pitt, his opponents shouted, spent the country's millions to defend a *cause célèbre* which was barely at issue, and all to aggrandize his popularity. Said one critic:

> he will boast of having strengthened the sinews of British commerce and navigation by the addition of the Southern whale fishery, with scarcely better right to boast of it, than if the Newfoundland fisheries had been acknowledged to belong to us. When he comes to Parliament . . . with a Convention from Spain in one hand, and his accounts of four millions expended in the other, he will think it a fair answer to every pos-

sible objection to say—"Look at my Convention, acknowledging your right to fish in the Spanish seas: look at the 3 per cents. at 81."—Upon some such empty, shallow title, we shall see panegyrick exhausted, extravagance itself outstripped, and new terms of adulation invented for the glorious Minister who will have done all this!—Have done what?—Who will have obtained a point which he acknowledges not to have been such an object of contest as to make *him* apprehend the least chance of a rupture with Spain when it was stated by her.[19]

Pitt and Dundas were together portrayed as masters of fraud, spending Britain's wealth needlessly. They and their new whale fishery served as subject for political satire; two surviving lithographic prints show Pitt and Dundas in a small rowboat, quite literally "fishing" for whales off the Northwest Coast of America. One, created by the well-known illustrator Isaac Cruikshank and entitled "The New South Sea Fishery or A Cheap Way to Catch Whales," depicts them just outside the ten-league marker buoy on the Northwest Coast of America, while a large ship named the *Convention* lies hove to in the background. The two fishermen have baited their lines with bags of money. An anonymous second print likewise shows Dundas and Pitt fishing; each of their moneybags contains £1 million. Pitt's pockets are crammed with tiny fish, presumably representing the unprofitable whales they intend to capture.

A caption attending Cruikshank's illustration confirms the satire. Pitt says to Dundas: "I fear Harry this Fishing will never Answer," to which Dundas replies: "Never mind that Billy the Gudgeons we have Caught in England will pay for all." The gudgeon, a tiny freshwater goby easily caught for bait, loaned its name to anyone readily duped or tricked, that is, someone who would "swallow anything."[20]

Even within England, the idea of hunting whales in the Pacific Ocean fueled heated debate throughout the closing years of the 1780s and into the early 1790s. The mighty East India Company and the South Sea Company looked askance at the whalemen, assuming that they would conduct trade without recourse to licences or permissions and so destroy the uncontested power of the crown monopolies. Nothing would convince them that the whalemen intended to harvest independent profits from the open sea, and because the focus of British maritime endeavour favoured trade, there was little leverage available to those who represented the special interests of whaling. At the time of the signing of the Nootka Convention, British whaleships were permitted to travel through only a limited area in the South Pacific Ocean; they had not yet received full diplomatic permission to engage in their business so far north in the Pacific sea as Nootka Sound.

Samuel Enderby found his way into the Pacific completely barred by the

BILLY AND HARRY FISHING FOR WHALES OFF NOOTKA SOUND.

Political caricaturists lambasted Prime Minister William Pitt and Henry Dundas when they made North Pacific fisheries rights a *cause célèbre* during the settlement of the Nootka Sound crisis. Pitt is shown fishing with money bags, each bulging with £1 million sterling. Colored engraving, published by W. Holland, London, 23 December, 1790
Kendall Whaling Museum, Sharon, Massachusetts

power of the trade monopolies, and though sperm whales had been sighted near the coast of South America and elsewhere, his crews were forbidden to go after them. There was no question of direct competition since the East India and South Sea Companies possessed no whaleships. Enderby began a long, time-consuming lobby. He earned the ear of Charles Jenkinson, Lord Hawkesbury, and in 1786 Hawkesbury, president of the newly influential Board of Trade, cleared the way for a few London whaleships to enter the South Pacific.

Parliament then introduced a bounty programme which rewarded a specific number of vessels whose masters sailed south of 7° North latitude and returned within a specified time period with not less than twenty tons of oil or spermaceti.[21] The South Sea Company was appeased by a licensing provision which gave them control over the numbers of whaleships passing through their domain in a given season. This licensing requirement also provided a revenue source. The East India Company, being the more powerful, fought a steady battle to preserve its uninterrupted rights, although it, too, eventually agreed to license a few whaleships each year. The acquiescence of the two great monopolies proved to be a "qualified but substantial victory for the whaling interest."[22]

The first permissions were narrow in scope. In the Pacific Ocean, whalemen were permitted only within fifty degrees of longitude to the west of Cape Horn, and south of the equator. In 1788 this area was enlarged westward to the 180° meridian of longitude. Similar restrictions governed travel around the Cape of Good Hope into the Indian Ocean.[23] Under the revisions of 1788, Enderby sent a 102-foot, 270-ton ship, the *Emelia*, to the South Pacific. The *Emelia* sailed on 1 September 1788 under the command of a Nantucketer, James Shields. Shields' first mate, Archieleus Hammond, was also from that island. Rounding Cape Horn, the crew of the *Emelia* on 3 March 1789 took their first sperm whale in the Pacific Ocean.

On his return to London in 1790, Shields discharged a cargo of 139 tuns of sperm oil, about 35,000 gallons.[24] Soon a small fleet of whaleships sailed from the Thames to the Pacific and quickly advanced as far westward as Sydney, Australia, through territory nominally controlled by the South Sea Company. But their path from the equator to the Northwest Coast of America was wholly blockaded by the stronger and unyielding East India Company.

Enderby's son, Samuel, Junior, wrote Pitt to express dismay over the impediment:

> The means of extending this Fishery, are by gaining as great an Extension of Latitude and Longitude round the Capes of Horn & Good Hope as is consistent with Reason or can be obtain'd from the Chartered Companies; The South Sea Company has neither Trade or Fishery, The East India Company no Fishery, yet they will not permit us to fish except in very limited Latitudes & Longitudes, whilst the Fishermen of other Nations can go without hindrance to all parts of the Ocean . . . and when the Whales are hunted away they can follow them into those Latitudes where the English Whalemen dare not go.[25]

The concern of Enderby and Sons was real. The whaleships of France and the young United States were already spreading canvas in the Pacific Ocean. The *Rebecca*, returning to Nantucket in 1793, would report some forty whaling vessels in the South Pacific during 1791–93, of which twenty-two were English, ten American, and eight French; thirty of the ships were commanded by Nantucket men. Such information, conveyed to Enderby and his colleagues, suggested that the Pacific sperm-whale fishery would be decided just as Enderby had prophesied, the others hunting freely to the northward while the English must come about or become ensnared in the political nets laid ahead of them by their own countrymen.

But the hugely profitable Northwest Coast fur trade took precedence over all, and on 20 January 1791, Hawkesbury wrote Enderby and three other leading whaling merchants—Alexander and Benjamin Champion and John St. Barbe—to inquire whether there might be some gain in combining whaling with fur on the Northwest Coast of America. The four men answered as one; it would be very much in their interest.[26] Promptly, three vessels were fitted out for such an extended voyage. It is not clear how the East India Company directors were persuaded to overlook them, since Parliament continued to prohibit whaleships north of the equator, but their names may be found buried within a long list of London whaleships prepared annually by the Board of Trade, among the sailings of 1791.[27] That their outfit lacked the boats and equipment necessary for hunting whales can scarcely be believed, since the owner of record was William Curtis, a London alderman and whaling merchant, and the overall master of the expedition was an experienced Greenland whaling master, William Brown, whose special expertise was exactly the sort required for a whale hunt in uncharted cold climates.[28]

Curtis's so-called "Butterworth Squadron," named for the large 390-ton lead ship *Butterworth*, included the smaller *Jackall* and *Prince Lee Boo*. Their avowed mission was to establish "factories" on the Northwest Coast of America where trade might be conducted; Brown had ostensibly obtained a grant from the British government to "make a settlement or rather establish a factory on some part of the [Northwest] coast."[29] Before rounding up into the Pacific, he called at Staten Land, across the Strait of Le Maire from Tierra del Fuego, and there set up a "factory" of another sort—a whaling and sealing station. This event bears further testimony to Curtis's and Brown's whaling interest and perhaps sheds some light on the type of factory they expected to establish at, or somewhere near, Nootka Sound.

The arrival of the squadron at Nootka in 1792 coincided with the visit of Captain George Vancouver to settle the final terms of the Nootka Convention with his Spanish counterpart, Bodega y Quadra. Brown's role in this meeting, if any, is not clear; it has been claimed that he carried secret orders for Captain Vancouver, but neither Brown nor his three ships are mentioned in Vancouver's personal journal of his affairs at Friendly Cove. One historian wrote: "For a time, [Brown's] vessels were anchored not far from the *Chatham* and the *Discovery*; yet Vancouver, meticulous in recording all matters of importance, never once mentions the fact. . . . Vancouver's omission has attracted the attention of historians, though a satisfactory explanation of it has not yet been found."[30]

Brown's cruise has long been explained away as a fur-trading venture. He did explore as far as the Queen Charlotte Islands, and there—Vancouver does say this—he traded for skins, whale oil, and "other marine productions."[31] It seems highly probable that the grant for developing a site on the coast was intended for both the fur trade and the whaling business, and perhaps the factory was meant to be similar to the one already built at Staten Land. If so, the matter contravened public law and the crown charter rights belonging to the East India Company. Vancouver may have felt obliged to ignore Brown; perhaps he realized that the activities of the Butterworth Squadron, if revealed, might prove an embarrassment to high government officials, including Alderman Curtis. Perhaps even Prime Minister Pitt and Henry Dundas might be implicated because of their outspoken advocacy of the new whale fishery.

The issue remained moot because Brown claimed that he found no whales. Writing to Curtis from "Whittette Bay" (Waikiki) on "Whaoo" in December 1792, he refuted John Meares's contention that whales could be had in large numbers on the Northwest Coast. He had seen none, he said, neither a right whale nor one of the spermaceti kind, in his entire voyage from Cape Horn to Nootka Sound. "Those whales Mr Mears talks of are a Small fish call'd humpback," he wrote, "and from enquiry I find that oil has been Carried to China by American vessels from this Coast but could not be sold as the Chinese have plenty of vegetable oil and cheap."[32]

Reading the interstices in Brown's report may help decipher his intentions. He wrote Curtis: "Please God all goes well I mean to send Capt Sharp in the Butterworth round Cape Horn to fill with whales and seals [in?] the latter part of next Season when we leave the coast   on board the Jackhall I mean to put all the furs and proceed myself with them to China and return home round the Cape of Good Hope where I expect to fill with Seals, Salt."[33] Brown's remarks imply that the *Butterworth* was fully outfitted as a whaleship, since there was no venue where Captain Sharp might "fill with whales"—save perhaps the fledgling station at Staten Land—unless caught by himself. And if the *Butterworth* was not so equipped, why bother transshipping the valuable and fragile cargo of furs into the smaller and subordinate *Jackall*? Captain Brown was not indisposed to vainglory; one would expect him to reserve the largest and most impressive vessel for his long trip to Canton.

Many answers were lost in the mercurial apotheosis and collapse of Brown's expedition. Soon after his arrival at Oahu he was given the rule of the island by popular will of the people—and is yet remembered for killing fellow fur trader (and former whaleman) John Kendrick while saluting Kendrick with a cannon. Presently Brown fell also, killed by Sandwich Islanders who perhaps had come to terms with his human characteristics. His untimely death robbed history of many details of the voyage of the Butterworth Squadron. The *Butterworth* reached Gravesend on 3 February 1795 with 17,500 seal skins and eighty-five tuns of "whale oil"—more than 20,000 gallons—a large amount which could have been drawn from a fair catch of eight to twelve right whales of average yield. The *Butterworth* did, in fact, begin to "fill with whales," and in one surviving document of record, when asked "Where fished?" the correspondent for the ship answered: "No. West America."[34]

In the year Brown sailed, the East India Company's objections to whaling were finally overcome. Parliament granted permission for whalemen to pass northward of the equator in the Pacific under the same bounties paid on completion of South Pacific voyages. The ruling was specifically to open a route to the Northwest Coast of America since it continued to prohibit travel west of the 180° meridian north of the Equator.[35] In 1793 the Board of Admiralty also endorsed an expedition under James Colnett, already much experienced in the North Pacific, to discover suitable reprovisioning and refreshment ports for South Sea whaleships in the Pacific Ocean. Samuel Enderby supported this effort, since he wished to know of such ports "in the south or even the north Pacific."[36] Enderby himself owned no vessel suitable for the purpose, but he convinced the Admiralty to release an aging 374-ton sloop-of-war, the *Rattler*, which had been declared unfit for further military duty. In this vessel, and properly advised as to the needs of the whale-fishers, Colnett sailed.

Although acquainted with the safe harbours of the Northwest Coast, Colnett pushed only as far as Baja California. His published report[37] eventually provided significant information to the British South Sea whalemen who sailed across the southern Pacific Ocean during the following half-century. But his failure to reach the Northwest Coast proved prophetic; for all of Pitt's assurances, British whalemen would never prosecute the extreme north. Much later, when it became economically realistic and even necessary to pursue the hunt into the North Pacific, the British southern whaling fleet was very much diminished and was not rebuilt to the purpose. The northern whales would go to others, just as Samuel Enderby had feared.

The legacy of British interest in whaling in the Pacific Northwest is largely preserved in the memorialization of its proponents: Dundas Island, Brown Passage, Pitt Island, and Hawkesbury Island lie not so far from one another on the British Columbian coastline. There is little to suggest that British whalemen went there. Whaling in the eastern North Pacific became the business of the Americans and the French, whose sailors were initially encouraged rather than prohibited, and who did, as Enderby predicted, "go without hindrance to all parts of the Ocean."

Yankees in New England faced no restraints, neither the inhibitive regulations of boards of trade nor the expense and danger of new foreign wars such as Britain soon incurred during the Napoleonic campaigns. Burdened they surely were by the cost of proclaiming national independence, but by 1785 when Cook's *Voyages* were published, the eastern American seaboard looked with fervour toward expanding world trade. With the arrival at Nootka Sound of the Boston trading vessels *Columbia Rediviva* and *Lady Washington* in September 1788, the fur trade first posited by Cook began to overpower the concept of whaling, the latter issue tabled by the staggering value of sea otter and other pelts in China. None cared to flense whales for oil as long as Indians came to the beach with furs to be traded for bits of iron and cloth.

The issue of whaling may have been ignored also because the men who developed the fur trade sailed from a port whose merchants were less interested than Nantucketers in whale oil. It was still a substantial journey, psychologically and physically, from Boston, Massachusetts to the island of Nantucket; the two communities did not share their businesses to any great degree, and the Boston men pursued trade with a passion equal to that of the Nantucketers' chase of the sperm whale. So the "Bostons" came on to Nootka Sound for furs while the Nantucketers hauled up short at the "Mocha Isles" or elsewhere on the Chile coast to pursue their own familiar work.

Bostonians were nonetheless amazed by the importance of the whale hunt in the native society along the coast of Vancouver Island. From their descriptions come some of the earliest portraits of native whaling. Robert Haswell, second mate in the *Lady Washington*, was among the first to give such an account:

> on our arrival at the place of action [Clayoquot] I found the Whale with sixteen Bladders [floats] fastened to him with harpoons   the Whale was lying at this time unmolested waiting the return of the Chief [Wickanninish] who was in pursute of another but immediately on his return he gave the order for the atack   his mandate was answered by a low but universal acclamation   the next brother to the Chief invited me into his canoe   this I rediely complied with   we were paddled up to the fish with great speed and he gave him a deadly pearce and the enormious creature instantly expired.[38]

The visitors were told that native whaling skill had been given by one or another of the pantheon of demigods who had first created and inhabited the world. By native reckoning, the first whale hunter was a giant eagle-like creature that possessed talons strong enough and wings large enough to lift a whale from the water and remove it high into the mountains to be devoured. Thunder, Haswell explained, was formed by the sound of the whales being dropped by the giant bird at its aerie.

"Thunderbird" and his kind populate the mythological stories of the coast north to south. Everywhere these giant birds resemble eagles, and everywhere they hunt the whale, usually with powerful lightning-snakes or lightning-fish which dart seaward to stun or kill. Under various guises, as *Thlu-Kluts*, *Khunnakhateth*, or *Tuta*, the Thunderbird arrived first and showed the people that the whale could be captured and utilized for food and tools.

Images of Thunderbird and its prey appeared along the length of the Northwest Coast on totem poles, tools, weapons, mortuary poles, and here, as a house frontal painting on the west coast of Vancouver Island.
Vancouver Maritime Museum

The Makah, or "Cape People"—those who yet inhabit the northwestern-most point of Washington State—claim descent from the union of animals with a star which fell from heaven. The first men were born at Cape Flattery, and from their loins sprang all the various peoples of the coast, including the Nitinaht, Clayoquot, and Makah. [39] Very soon they encountered the great being that killed whales with lightning under its wings, the great Thunderbird whose armament made the brightness in the stormy sky and whose monstrous battles with the whale set off the thunder in the clouds. Thunderbird, who feasted on the flesh of the whale and ruled the universe from the rain forests of the Olympic Peninsula to the ice-choked Bering Strait.

The origins of this whale hunter are clouded by the passage of time. Each native group tells its own story. He is often thought to have been present at the Great Flood; such a thing has always been told to Tlingit people living in the shadow of Mount Edgecombe in the Alaskan panhandle. As the great waters began to rise, consuming everything, *Khethl*—a being whose name means "Thunder-and-Lightning"—dressed in the skin of a bird and flew away to the southwest. Before leaving, he told his sister that she would nevermore see him, so the sister climbed to the top of the great mountain to watch his departure and there was swallowed into its maw. By legend, *Khethl* returns to the mountain each year. Thunder is the noise of his wings, and lightning the flashing of his eyes. His favourite food is the whale.[40]

Near the pinnacles of the highest mountains was the Thunderbird aerie. The Makah watched for him to descend from the peaks of the Olympic Mountains. Nitinaht and Clayoquot across the Strait of Juan de Fuca sought him out in the mountains of central Vancouver Island. The Haida of the Queen Charlotte Islands searched for his home too, and an old man from the village of Ninstints once described the place. He told how a man had gambled away his property and that of his family. Ashamed, he disappeared into the woods, climbed the mountain, and found the eagles wheeling about. At length he arrived at a big town, where he saw a pile of bones from animals killed and eaten. Then he saw the eagles, perched on a pole in front of the town, watching the sea below. He began to live with them and discovered that the eagles went to the sea every evening to fish. Every evening they killed one whale.[41]

Invincible as Thunderbird seemed, still he faced peril. He made enemies among the demigods, enemies like *Kwatyaht*, who dressed in a whaleskin and dove beneath the sea to grab Thunderbird and drown him.[42] And once, all the animals of the forest conspired to kill Thunderbird. They constructed an imitation whale—in the manner of the Trojan Horse—and covered the outside with pitch to make it watertight. Then they put to sea and in the ensuing battle killed several children of Thunderbird and eventually, Thunderbird himself.[43]

After all, however, Thunderbird remained immortal. The lightning and thunder belong to him yet, as well as the skill of hunting whales. And this skill, many say, was given to a few of those mortal men born of godly unions on the shores of the Northwest Coast.

> By and by he started [whaling] with a net. Then [the eagles]
> told him not to put the whole net into the sea.[44]

One young man, it is said, gave away all his salmon to the nearby eagles and himself went hungry. His ten uncles, thinking him improvident, refused to share their food and advised him to live off the food he had given to the eagles. This he would not do. The eagles rewarded his generosity with food and more food. First they brought salmon, then bits of whales, and finally entire whales, until there were ten complete whales in front of the young man's house. Then the slaves of the uncles exclaimed: "You ought to see how the one you abandoned is living. Black whales are floated ashore in front of him like driftwood."[45]

The first human to absorb the whaling power may have been Tutus. Tutus lived on Vancouver Island, and his personal spirit—his *tumanos*—told him to wait for the whales and then catch them. To do this, Tutus dressed himself as if he were the bird that caught whales. He flew from his housetop and over the sea until he found a whale. He circled time and time again. When finally he dropped down to the water, his claws dug into the soft blubber and he hauled the whale ashore in victory. Then he went home, removed his garments and took off the wings he had worn.[46] His act is repeated to this day by eagles attempting to rise from the water with a heavy fish. The eagle will flap across the surface, barely clearing the water, dragging the fish toward the shore by main muscle until, reaching the beach, it partakes of its catch.

Men and spirit-world interacting: from Tutus, some say, came the privilege of hunting the whale, ever afterward handed down among a few honoured whaling lineages. From this honour evolved the Thunderbird and lightning dances, performed before the hunt. Then, the dancers jumped and stamped around, resembling thunder, blowing a whistle and imitating the great being. In the lightning dance the leader wore a headdress consisting of a carved face with a cedar-bark hat adorned with downy eagle feathers. He, too, held a whistle, and danced with a weapon made from the skeletal bone of a whale. The dancing concluded with the leader bending downward and ever downward until his hand nearly touched the ground, his arms gradually lowering his weapon in imitation of the lightning-snakes that kill the whale.[47]

For such a dangerous trial, mortal man required the co-operation and strength of the spirit gods as well as the carefully nurtured powers of their own *tumanos*. "If a man is to do a thing that is beyond human power," they said, "he must have more than human strength for the task."[48] Preparations for whaling among the Makah, Ahousat, Clayoquot, and other West Coast peoples were stringent and severe. In the winter, canoe crews trained by eating but little. And by denying sleep; it was said that a whale hunter could remain awake four nights and days. Some said that the whale hunter could stay awake ten days.

Ritual bathing helped prepare them. The harpooner swam in a particular pond or in the surf, mimicking the actions of the swimming whale, earning the trust of the whale spirit. Ashore, herbs were rubbed on the hunter's body, and the thorny hemlock was applied until his blood flowed. Sexual abstinence strengthened his powers, too, in the weeks and perhaps months before the hunt; a violation of this rule might disturb the whale spirit and make the animal unco-operative.[49]

Ritual thus extended to the wives of whalemen. They assisted in preparing the men for the voyage, and during the long hunt they remained in their houses, speaking quietly and moving hardly at all, since it was told to them that any unnecessary activity might turn the whale away from the beach or make it wary of the approaching canoes. Then it would be woman's fault, for the men had done their best and had been defeated.[50]

There was also the solemn matter of whalemen's bones. The skulls and skeletons of dead whale hunters brought power to those living; the collection and ritual use of these bones was a clandestine activity known only to the whaling elite. And though it is not clear how extensively this ritual was practised, it is certain that the power of the bones, properly used or misused, was explained to each succeeding generation of whaling children. Skulls and bones were secretly removed from grave sites; one Makah woman went to her great-uncle's grave two years after his death and found his head gone. Speculation is that it went to the fetish-bag of another whaleman, since "the body of a successful whaler was considered an especially effective charm."[51]

Everywhere the stories. One is told on Nicagwa'ts, a Quinault whaleman from the shore south of Makah territory:

> When Nicagwa'ts trained himself for whaling he used to go out at night to bathe in the ocean and in the river. He went there where no one would see him and no one could hear him talking to himself. He would "talk to the world"... and tell "the world" what he wanted and ask it to help him...
>
> He had a skull wrapped in a blanket. He would tell his "Captain" [the skull], "You had better watch when I go to throw the harpoon at the whale. Make the whale weak—as if he were dead."
>
> Once he was "training" at night out in his canoe. He was practicing with the harpoon. He had along a skull and other bones of one of his [male] ancestors that he had dug up at night, secretly. This night... the skull took hold of the canoe and shook it. Nicagwa'ts did not look around, for if he did he would "get ghosted." He only motioned for the ghost to be still. Then he quit his "training," for he was sure of his helper.[52]

The ritual and training were strong and purposeful, for the theme of the hunt forecast untold hazards. Could not the whale return in anger to capsize the frail canoe? Could not the injured animal capture the canoe and drag it underwater to the village of whales, where the helpless men would themselves be transformed into whales? Even under the brightest light of day, away from the frightening whispers and eerie movements of the night forest, could not a running whale swim on and on, its strength enduring for days upon end? In historic times, one Ahousat whale hunt is said to have lasted two weeks; the whale was harpooned near Bartlett Island, between Vargas and Flores Islands, and then ran to Wickanninish Bay. The chase took the men so far to sea that only the snow-capped mountain peaks were visible over the horizon.[53] Nineteenth-century accounts tell of a canoe gone five days. Those men had to subsist on some blubber and roots they had brought with them from shore. They ran completely out of water and eventually had to abandon the carcass of the whale they had so bravely killed, in order to speed their passage home.[54]

Makah Indians whaling at entrance to Fuca Strait. No reliable photographic record of Makah or West Coast whale hunting exists. This illustration, by Northwest Coast surveyor Henry W. Elliott, may have been drawn from first-hand observation. It appeared in George Brown Goode's massive *Fisheries and Fish Industries of the United States* (1884–87).

So the eight men in the cedar canoe were paddling toward great danger. But they were special men, there by reason of their hereditary rights and prepared insofar as possible for the ordeal to come. Their honour was at stake, because they had inherited from their fathers and their grandfathers before them the tools and the special knowledge of the whale killing. And yet the Makah or Ahousat or Nitinaht hunter was a human being. He suffered from human weakness and was subject to human foibles. He showed himself, from time to time, to be made of material flesh. James Gilchrist Swan, a Bostonian who lived many years among the Makah near Neah Bay, recounted one such mortal:

> I was much amused by Johns moves. It seems he feels ashamed that he has not killed any whales and has concluded to go through the ceremonies to constitute him a skookum whaleman. Which ceremonies

consist of going without sleeping or eating for 6 days and nights to bathe in salt water and run on the beach to get warm. John went into the water with his accoutrements on but soon got so cold that he was glad to come and warm himself by my fire. He had gone all day without eating and I think his courage was failing him for he admitted that he thought he could not stand it more than two days and if that would not suffice to make him a whaleman he could kill sharks.[55]

Once the migrating whales had begun to arrive, the whaling lookouts would study wind and weather, awaiting a clear night when the stars would twinkle in promise of a storm-free day to follow. Near midnight, the whale-hunting crews would leave their village for whichever encampment had been selected as a base. In Makah territory this was often near Cape Flattery, close by Umatilla Reef and Swiftsure Bank where whales congregate to feed. Hunters at Neah Bay, inside the cape, kept their own vigil from Tatoosh Island. Across the water on Vancouver Island, the whaling crews also waited expectantly.

Canoes were fully prepared. Each of these had been formed from a single cedar tree, first hollowed and then widened by steam to make a thirty-five foot vessel large enough to carry eight men. Crew positions were carefully prescribed and well understood; most important was the harpooner at the bow, who usually owned boat and gear, and who had also inherited the right to be first to hurl the weapon. Behind him sat an experienced oarsman whose job it was to tie on the inflated sealskin floats and throw them overboard with the running harpoon line to hinder the whale's escape. The other paddlers waited for any command, while the steersman directed the canoe from the stern position. The Makah possessed few serviceable cedar trees for canoe construction, so they traded or purchased raw logs—or sometimes complete boats—from the cedar-rich whaling villages on Vancouver Island.[56] Whale oil collected as the result of a previous hunt might be of value in such a trade. Harpoons, lines, and other equipment were made at home.

Then they went out, perhaps silent, perhaps chanting. Some among the Nitinaht may have known special outward-bound songs.[57] Once near the whales, they may have been selective, overlooking one whale in favour of a more placid individual farther out. Or perhaps one animal was larger or in some way special; who can say what choices entered into the harpooner's final decision? But once chosen, the whale was closely approached. The men were certainly silent then, for a whale is sensitive to sounds and easily "gallied" by a paddle too roughly slapped into the water. Once alongside, the harpooner stood up and thrust his harpoon carefully yet firmly, so that its mussel-shell point entered the blubber on the top of the back and penetrated deeply through it.

There must have been great commotion then; men shouting, backing their paddles, the line-handler hurrying to attach the floats to the running line. More rope would have to be tied on and more floats attached until the whale dragged perhaps a dozen large balloons behind him. A small marker buoy, floating on the surface, gave away the animal's whereabouts. Then the whale hunters would begin to pray to the whale, talk to it, perhaps even sing to it, begging its spirit to turn toward the shore, to close in with the village where men and women stood ready to give it praise. It was known that the animal must be flattered and cajoled, its spirit guided to the beach by the struggle of human desire against animal will, by overcoming the whale's strength and purpose with the power of the whalemen's carefully guarded *tumanos*.

Often several canoes would join in chasing a single whale, adding gradually to the hindrance of harpoons, lines, and floats. As the injured animal swam at the surface, near exhaustion, smaller harpoons might be darted into its head, with floats attached to support its great weight at death. Then the whale would be lanced many times over with a bone-tipped hand-lance. When it was dead, the "diver"—a crewman specially honoured with this cold and risky task—would jump overboard and cut holes in the jaws through which a line might be passed. This line held the mouth closed against the sea and served also as a tow-rope.

Then would be the time for victory songs, calling for extra strength for the heavy row homeward. Details of these songs differed from place to place; many spoke or sang directly to the whale, to its spirit who kept his dwelling at the bottom of the sea and only dressed in the whale's suit to come to the surface. It was hoped that the spirit would be flattered by the praise, and come willingly to the village to share in the ceremonies attending the flensing of its blubber and the distribution of its wealth.

The carcass was inclined to sink after death, and only the rapid and successful work of the diver prevented this from happening. The swell of the sea might impede the rowing, carrying the whale backwards a half mile for every mile gained, until each man reached his own personal exhaustion. Expediency permitted the hunters to beach their prize on territory not their own; one Makah harpooner told how his crew towed a whale to Vancouver Island—it was nearer than their village beach on the opposite shore of the Strait of Juan de Fuca—and there cut up the carcass. The crew gave some of the whale to the local people and carried the rest away.[58]

Once safely home, the whale was dragged as far as practicable up the beach on the high tide. As the water receded, the entire community hastily removed the blubber in square slabs perhaps two or three feet to the side. This was easily accomplished since the beaches were gradually and gently sloped. The operation began with the ceremonial removal of the "saddle," a special

Among artifacts recovered from the buried whaling village of Ozette, Washington, was this three-foot wide fetish representing the "saddle," the most important ceremonial piece cut from the whale. This unique wooden object is adorned with 700 sea-otter teeth.
Makah Museum, Neah Bay, Washington

portion of the back withheld for ritual treatment in celebration of the whale-spirit. Reporters have described several variations in this event, but in most instances the piece was adorned with the feathers and down of eagles and temporarily set aside in a place of respect. The bulk of the oil was stored in large boxes or in bags made from the intestines of whales or the stomachs of seals. After completing the flensing, every Makah would scour himself clean with beach sand and salt water. In later years, James Swan reported that the Makah at Neah Bay considered themselves cleaner than the white men they met along the waterfront at Port Townsend.[59]

For three or four nights, the chief and perhaps his wife conducted the prescribed rituals, singing songs of praise to the whale-spirit and drinking water brought in ceremonial pails. Haswell reported in the late 1780s that it was still customary for the people at Nootka Sound to sacrifice a slave over the first whale killed in a season[60] and that a piece of that whale—no doubt the saddle—was adorned with eagle feathers to honour Thunderbird, who gave them the whaling power.

At the conclusion of the ceremonial period, the saddle was cooked and presented to guests who appeared for the final celebration. Its tip was removed and awarded to the harpooner; the rest was cut up for distribution.[61] The many bones received no special treatment. They were piled onto the midden in heaps, and there remained to be discovered and analysed by scholars. "Evidently," T. T. Waterman wrote, "the courtesies extended to the taboo-piece or saddle fulfilled all the necessities."[62]

A month might be required to prepare the oil and dry the meat. The rendered blubber was roasted and stored in baskets. Dried whale meat and fish were often served together, oil providing the lubrication to complement their dryness. Whale and fish oils were everywhere useful for cooking and preserving and constituted a source of wealth redeemable outside the whaling territory. Whale oil was highly valued, even far inland, and a regular trading cycle developed among the Makah, the Nitinaht, and others both northward and to the south. The trade worked both ways; Makah purchased oil from Nitinaht and also traded "in a single season, it is said, as much as 30,000 gallons."[63]

After the feasting, harpoons and lances would be repaired. A whaling harpoon that had proven itself successful acquired added value; an old and often-patched weapon therefore became more valuable than an untried new harpoon. Even today it is a far different matter to touch such a harpoon than to describe its construction, and yet again different to witness the residue of its power imbedded in the porous skeletal bones of whales. Some years ago, during archaeological excavations south of Cape Flattery at the Makah whaling village of Ozette, several whale scapula bones were uncovered which display a perfect inlay work, a tiny, jagged edge of mussel shell shining silver against dull bone, the remains of a harpoon point imbedded by the force of the harpooner's arm and cleanly shattered. Such an artifact reaffirms the great power of the whaleman's *tumanos*, and the great readiness of the man himself.

But for plain description of the harpoon, Waterman's serves well for accuracy and clarity:

> The harpoon-head is made up of several elements cunningly joined together. The foundation consists of a pair of pointed barbs, made of elk-antler or bone, which fit neatly together.... The end of the whale-sinew lanyard is attached by unlaying the strands and making them fast around the barbs; and then winding the whole with cord and strips of cherry-bark.... A socket is fashioned in the rear of the head, into which the point of the wooden harpoon-shaft exactly fits. The blade, or cutting edge, was formerly made of a large mussel shell ... but in recent years of copper, or steel saw-blade.... This blade, whatever the material, is cut into the shape ... and inserted between two barbs, the ends of which are cut away to accommodate it. A "fat" knot of spruce is set in front of a fire, and the gum which melts out ... is kneaded and spread

smoothly over the harpoon-head . . . so that it may cut its way deeper when the quarry is struck.[64]

The mussel shells selected for harpoon points were very large specimens, perhaps twelve inches long. These were obtained from offshore rocks where the pounding of the surf promotes growth. The Makah utilized Umatilla Reef for this harvest, and the shells were brought back and sharpened on sandstone before being dried on a housetop for as long as two years. Occasionally they were soaked in oil to prevent them from becoming brittle.[65] The fragile nature of the shell-points actually improved a hunter's chances, for seemingly frail natural points penetrate farther than the more solid but less-sharp substances, and only break after they have accomplished their mission. Waterman wrote: "The mussel shell harpoon blade inflicted a deeper wound . . . than the more modern steel blade which has replaced it, even though the shell blade may often have been shattered in the process. The shattering . . . does not come until after the impact is completed, and the wound inflicted."[66] The proof of Waterman's assertion are the scapula bones with their thin inlays of shell-point, as if the most skilled artisan had carefully hollowed out the place and cut and sanded to make a perfect fit.

The entire harpoon assembly fastened to a long shaft, made of two or three pieces of yew wood scarfed together near the middle with matching diagonal cuts. The total length varied from about fourteen to perhaps eighteen feet. Unlike the harpoons used by British and American whalemen, the shaft was designed to separate from the point, and could be retrieved immediately after being hurled. It was necessary to make such shafts from very straight wood, which the West Coast hunters went to great pains to locate. The entirety was connected to a whaling line made of sinew, served with a string of nettle-fibre and covered with a wrapping of this string, or perhaps bark.

The harpoon was thrown from very close range. One whale hunter said the distance could be three feet or even less. There is no doubt that the canoe came directly alongside the quarry; after the harpoon was thrown, it sometimes happened that the sealskin floats thrown overboard by the line-handler served as "bumpers" between the whale and the canoe, so close had the men paddled to the animal's massive back.[67]

Archaeologists and other students of native culture remain divided over the importance of whaling within the context of Northwest Coast life. They disagree also over the species of whales most preferred by these whalemen. The California gray whale has long been considered the most hunted species, but newly-analysed evidence suggests that the humpback was equally favoured. Studies of bones recovered from the Ozette excavations show an equal proportion of gray and humpback remains, although the identifiable sample included only 25 per cent of the total recovered.[68] Most West Coast

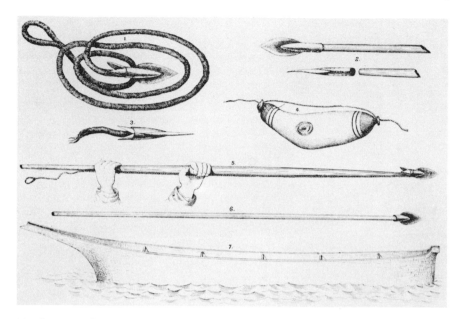

North West Indians Whaling Canoe and Whaling Implements. Charles M. Scammon's 1874 illustration shows the harpoons, sealskin floats, and other gear employed by Makah whalemen.

and Makah depictions of whales include a distinctive, triangulated dorsal fin and obvious flippers—diagnostics of the humpback species that are completely absent from gray whales. John Meares observed that the whale hunters at Nootka Sound preferred "those small ones with hunches on their backs, as being the most easy to kill,"[69] implying that the whales in question were humpbacks, which often show a visible hump or hunch just by the dorsal fin and which also hump their bodies in a characteristic manner when diving.

Far to the north, near the top curve of the Gulf of Alaska, lived other native whalemen whose hunting techniques were vastly different. These were the whalemen whom Cook saw at Unalaska. This group hunted with slate-pointed harpoons probably soaked in a natural poison, aconite, derived from plant material fermented in water. From Kodiak Island and elsewhere in the far North Pacific, the Koniag whalemen hunted in two-man skin boats. Once harpooned, the whale would be left in the sea to "go to sleep" for a period of three days; it was hoped that favourable winds and currents would carry the dead body to some landfall near the home village.[70]

Between this northern whaling area and the southern whaling at Vancouver Island existed a cultural milieu in which both whaling and the eating of whale meat was abjured. The killing of whales by poisoned harpoons may account for their strict taboo against eating the whale's flesh, since it is possible that some of the poisoned whales drifted to the south. Many Tlingit, who formerly inhabited much of the coast of the Alaskan panhandle, refused this source of food and ritual entirely, holding the meat to be unclean.

Indiens Kodiak Chassant la Baleine: Amérique Russe. Native whalemen of Kodiak Island utilized small boats, and are thought to have added aconite poison to their harpoon points. Rather than remaining "fast" to the whale, they waited for the poison to take effect. This illustration was published in Duflot de Mofras's *Exploration*, published in Paris in 1844.
Kendall Whaling Museum, Sharon, Massachusetts

Russians who began the colonization of the coast on Kodiak Island in 1783 made note of the many whales to be found there but expended no effort on a whale fishery. Their economic interests, like those of Britain and the United States, centred on the fabulously valuable sea-otter pelts and other furs. When a Russian exploring party led by the hired Englishman Joseph Billings reached Kodiak Island in 1790, "whales in amazing numbers" were found in the vicinity,[71] but no hunt was undertaken. Gawrila Sarytschew, the major-general of the expedition, wrote how whales swam around the ship and "perpetually occasioned, by their violent lashing of the waves, a report very similar to that from the discharge of a cannon."[72] Sarytschew also reported the use of whale oil by the local natives, and theorized that commercial whaling might prove profitable. "The most important commerce," he wrote, "might be carried on with train-oil and whalebone [baleen], whenever proper arrangements could be made for catching the whales which frequent these seas in vast numbers."[73]

Native whaling was no small issue. Much later, in 1831, naturalist Ferdinand von Wrangell calculated that 118 whales had been struck that year by natives he visited in the Gulf of Alaska, and of these, 43 had been recovered for their use.[74] But for complex bureaucratic and economic reasons the far northern whale fishery was never prosecuted by the Russians; the Russian American Company—which after 1799 enjoyed a trade monopoly in these lands—was unable or unwilling to retrain the Koniags and Aleuts to use modern whaling equipment. Even as late as 1858, Baron von Kittlitz could say that the traditional poison-harpoon method of whaling had continued unchanged. Whatever whaling was carried out by the Russians was done only to sustain the local residents.[75] The market in sea-otter pelts and other furs proved so lucrative that there was little impetus to invest in commercial whaling.

There were whales in great numbers—right whales, gray whales, humpback whales—yet from the several nations claiming territory in the Pacific Northwest there were few, if any, whalemen arriving to kill them. The great distance to such far waters put these animals out of easy reach. Besides, the more valuable sperm whales were to be had in temperate and tropical seas closer to home. Who need travel farther for less? The early reports certainly made no mention of spermaceti whales; almost surely, the indigenous whales of the Northwest Coast of America were not these. No, it would be better to remain in the South Pacific, in places where profit was virtually guaranteed and the waters better known, than to venture northward on a cold, uncertain speculation.

The sperm-whale fishery had grown rapidly, particularly in the young American states, and it continued to grow during the period of economic stabilization that followed the Napoleonic wars and the Anglo-American war of 1812–14. The hopes and dreams of this increasing enterprise were clearly revealed in the christened names of whaleships sailing from New England ports. The limitless expanse of their owners' ambition could be observed as early as 1789, when the whaleships *Africa* and *Asia* cleared on voyages to the South Atlantic. In 1805 the ship *Chile* of Nantucket made its first voyage, and it was followed to sea, as dream and reality coincided, by the *South America* in 1816, the *Peru* —and the *Peruvian* —in 1818, the *Japan* in 1822, the *Oregon* in 1826, the *Albion* in 1828, and the *Hoogley* in 1835, these last two vessels bearing, respectively, the name given by Sir Francis Drake to California in 1579 and a phonetic spelling of the wide river Houghli that flows past Calcutta on its way to the Bay of Bengal. Finally, in 1850, the barks *Anadir* and *Arctic* would mark the farthest passage of the dream and the last great ambition of American whalemen in the northern seas.

The Pacific sperm-whale fishery quickly expanded beyond the limits of the Chile and Peru grounds, to New Zealand and Australia, northward to the "Line," and beyond to the Galápagos Islands and the coast of Mexico. The call went out for more vessels and more crew to man them, and overnight — almost literally so — the whaling changed from a closely knit family business to an employer-employee industry that would hire any man fit and willing to go to the latest-discovered haunts of the whale. The fleet, once belonging almost entirely to Nantucket, soon included ships from New Bedford and a dozen Massachusetts villages; Poughkeepsie, Sag Harbor, and Cold Spring Harbor, New York; New London, Stonington, and Mystic, Connecticut; Bristol, Providence, and Warren, Rhode Island; and other seaports from Wiscasset, Maine to Wilmington, Delaware. The number of available berths quickly exceeded the number of competent seamen available to fill them, so recruiters began to travel inland to encourage the most adventurous young men to leave their monotonous farm lives and come away to sea. Most of these had never seen the ocean, and they believed tales of incredibly rich voyages that made every man-jack wealthy. They were told how each sailor earned a percentage — called a "lay" — of the oil, and that such percentages, when tallied on the docks at New Bedford, often made the captain his fortune from a one-tenth lay, and left enough to satisfy the rest.

The recruiters did not say that the earnings of the owners and the ship's costs were taken off the top, nor that the shipboard "advances" in cash, clothing, and tobacco might eat away the miniscule 1/150 or 1/190 lay paid to a "green hand" and leave him penniless after perhaps three years of hard service. These young men were conned and lied to, but the lies were supported by facts. There had been, and continued to be, impressive returns from the whaling grounds, and such returns required no exaggeration in the retelling. So these hopeful young men, many of them second and third sons with little prospect at home, signed on and went to sea in an unimproved condition befitting their rank, into the unforgiving ocean on a voyage of two to four years around the globe.[76]

Others signed on in the Azores or the Cape Verde Islands, and still more were shipped in the South Pacific. The Azorean and the South Pacific islander were as anxious as the Yankee farm lad to go "see the world." Family connections in the forecastle dissolved as Amerindians, Cape Verdeans, Azoreans, and South Sea islanders joined ship. A "mixed crew" was no oddity by 1830, and whereas the eighteenth-century Yankee whaleship had been manned by the hardy Quaker youth of Nantucket — perhaps two or three of the same surname in a ship — crew lists now included Romontars, Zafouks, Sylvias, and the ubiquitous "Kanakas" — David, George, John, Levy, and Spooner[77] — among the Coffins, Folgers, Howlands, Starbucks, and Tabers. It was no fiction that Herman Melville chose for his three boatsteerers of the *Pe-*

quod a Yankee, an American Indian, and a Pacific islander. By the time *Moby-Dick* appeared in 1851, that sort of interracial mix was virtually always the case in fact.

Kanakas were the native Sandwich Islanders, a name given them by themselves. That distinctive appellation was loosely appropriated by shipmasters because the Hawaiian names were too much to be rolled easily off a Yankee tongue. It was simpler to style a man "Jack Kanaka" or "John Maui" than to imitate his four-syllable, vowel-laden moniker. The first American whaleships to call at the Sandwich Islands, the scorbutic *Balaena* of New Bedford and the *Equator* of Nantucket, reached Kealekekua Bay, Hawaii, in September 1819 and seem to have instituted among whalemen the practice of hiring Hawaiians. This was not a new idea; captains in the China trade had taken Kanakas to the Northwest Coast before 1800, and even William Brown is said to have taken some Sandwich Islanders there in the ships of the Butterworth Squadron.[78] But the first to go in a whaleship may have been the two taken off by Captain Edmund Gardner when he left the islands in 1819. Following a process which would persist more than half a century, he styled them "Joe Bal" and "Jack Ena," their new surnames together forming the name of Gardner's ship.[79]

Had the Sandwich Islands been inaccessible, by virtue of disagreeable winds and currents, or by the unfriendly disposition of the natives, or by being unpopulated and barren, the development of the whaling business in the North Pacific must have been very differently organized. As it happened, Honolulu, Lahaina, and to a lesser extent, Kealekekua Bay and Hilo, became welcome stops for whalemen. English whaling and sealing crews knew of these places at least by 1810, as their Southern fishery made its delayed but insistent push north of the equator. That year, the *Duke of Portland*, a "South Sea Whaler," called at the Sandwich Islands to obtain refreshment before the homeward-bound journey.[80] But the visit of the *Balaena* and the *Equator* in 1819 proved the true harbinger, placing the island residents on notice that the flood tide from Cape Horn had begun. Within fifty years, western "civilization" would forever alter the complexion of the faces and souls of Hawaii. Missionaries and fornicators abound everywhere; there was no shortage of either, and the infection of commerce proved incurable.

In 1820 the American whaleship *Maro* of Nantucket arrived in Honolulu Harbor, with information, probably obtained from the fur-trade master Jonathan Winship of Boston, that large numbers of sperm whales could be found off the uncharted coast of Japan. After recruiting supplies, the *Maro* set out on an experimental cruise northward and westward. The whales were indeed there, and the crew shared them with only one other whaleship, the *Syren* of London, whose captain had independently acted upon another shipmaster's account of sperm whales observed near Japan in 1806.[81] The *Maro* brought back 2,425 barrels of sperm oil on 10 March 1822; the *Syren* arrived in London a month later with 346 tons—about 2,650 American barrels—in casks. Immediately after these two successful cruises the Japan Ground became the most important whaling area in the North Pacific.

Because the new ground could be reached handily from the Sandwich Islands, the ports of Lahaina and Honolulu quickly gained in importance. Summer cruises to the waters near Japan provided an impetus for the establishment of a regular seasonal rotation of whaling that brought whaleships to the Sandwich Islands twice annually. Masters found it convenient to proceed northwestward in the summer, reprovision at Honolulu or Lahaina in the fall, and then whale the southern-hemisphere summer in the South Pacific, perhaps as far away as Australia, before returning to the Sandwich Isles again in the spring.

And so they came. Almost all of them were three-masted, square-rigged vessels, stoutly constructed and round-bottomed to withstand the rigours of long voyaging. They could be easily identified by their long, narrow rowing boats slung on davits overside. Inevitably there were spares, keel-up on racks above the deck, for the owners knew all too well that one or another of the whaleboats would be destroyed in the flash of a flipper or flukes disappearing downward.

On closer inspection, the whaleship might also be recognized by the brick tryworks standing between the foremast and the mainmast and topped with two or three huge iron cauldrons called "trypots." In these pots the whale blubber would be rendered to produce the oil that constituted the major article of commerce. Such "trying-out" made the whaling life one of the most hellish of seagoing occupations. The business of whaling demanded that the men first "cut-in" the blubber—that is, strip it off the carcass—and then melt it to liquid oil in the smoking, blazing "works." Slippery and hazardous, the operation could take two to four days. Ship and crew could only flense and try-out one whale at a time, because the carcass would be tied up alongside while the blubber was removed; most crews ignored any opportunity to take more whales until the work was finished.

During this period of intense labour the men worked six-hour watches around the clock. At night, flames and smoke from the tryworks created the ghastly spectre of a ship afire, and many a merchant captain, obeying the first rule of the sea, steered well off course to assist brother mariners in peril, only to find a busy blubber-hunter engaged in its oily calling. The smoking tryworks, wrote Thomas Pennant in 1792, made "the darkness visible, and the night hideous."[82]

If the sweating, greasy crew derived any pleasure from their handiwork—unless it be in counting the anticipated profits—it was in the preparation of doughnuts. Whale oil proved as suitable as lard for this purpose, and these

A barefooted sailor tends the trypots aboard an American whaleship. The blubber scraps have been rendered for oil, and will later be used to fuel the fire in the tryworks.
Photograph by William Tripp, Kendall Whaling Museum, Sharon, Massachusetts

delicacies, deep-fried in fresh boiling oil, were so much appreciated that they were often withheld by the captain as a reward: "at 7 PM boats got fast to a whale   at 9 got him to the ship. Men all singing and bawling Doughnuts Doughnuts tomorrow, as this will certainly make us 1,000 barrels and it is a custom among the whalemen a bache of doughnuts to every thousand."[83]

Frederick Crapser, aboard the bark *Pantheon* in 1843, elaborated more fully on the scene:

> Old man gave the Crew a barel of flour (its the custom with all whalers to give their Crew a Barel of flour for every 1000 Barrels of Oil taken) and they had a regular built spree frying Dough nuts Doughboys and Every thing else in the shape of eatables cooked in the Hot oil   at first some of them that had never seen anything cooked in hot [whale] oil before would not eat any of them but they soon came too their appetites and the way they made their jaws wag wasent by no means lazy.[84]

So it was, in such sailers as these, that the world's merchants eventually sent men to hunt the whales in the Pacific Northwest. By the time they arrived, they were fifty years behind the fur traders, and their voyages were supported by the efforts of the sperm-whalemen, who had opened the Sandwich Islands as a reprovisioning place in the midst of the vast North Pacific. Notices of whales on the Northwest Coast of America had been largely ignored, but by 1835 the southern grounds had begun to show signs of depletion. Only then, more than a half-century after Captain Cook arrived at Nootka Sound, was the whaling industry sufficiently curious to pay any heed.

2

CRUISING OVER THE GROUND

*But this much let me say: that Right Whaling on
the Nor'-West Coast, in chill and dismal fogs, the sullen
inert monsters rafting the sea all round like Hartz forest
logs on the Rhine, and submitting to the harpoon like
half-stunned bullocks to the Knife; this horrid and in-
decent Right whaling, I say, compared to a spirited hunt
for the gentlemanly Cachalot in southern and more genial
seas, is as the butchery of the white bears upon blank
Greenland icebergs to zebra hunting in Caffraria, where
the lively quarry bounds before you through leafy glades.*

HERMAN MELVILLE
MARDI (1849)[1]

It is so writ, that Captain Barzillai Folger in the ship *Ganges* of Nantucket
did take the first whales from the Northwest Coast of America, in the year
1835. This honour was accepted by the citizens of Nantucket, who were
probably unaware of the abortive British efforts of the 1790s and un-
concerned by more recent exploits of others than themselves. But one can
hardly confirm a "first," particularly since the Northwest Coast had been
chock-a-block with traders, explorers and naval expeditions since the publication of Cook's *Voyages*. Had not Captain Sharp claimed the Northwest
Coast as his fishing ground when he returned to London in 1795 in the *Butterworth*?

The Butterworth Squadron provides the earliest whaling enigma; afterwards there are others. The ship *Eleanora*, for example, sailed from Providence, Rhode Island on 11 September 1802 and arrived sometime later at
Rio short of food and water. Since Rio was then closed to American vessels,
the master explained to curious Brazilian officials that he intended to round
Cape Horn and hunt whales and seals on the Northwest Coast of America. In
November 1805 the captain of the brig *Minerva* said the same thing when he
called there to repair damages sustained during a circuitous passage from
Norfolk, Virginia, via Nantes, France, to London and thence toward the
Horn.[2]

If one wished to name the first commercial whaleman on the Northwest Coast, another contender could be Benjamin Worth of Nantucket. Worth, one of the island's most revered whaling masters, wrote in his dotage that he had been twenty-nine years a shipmaster during forty-one years at sea between 1783 and 1824 and had "traversed the west coasts of North and South America. . . . to 59 degrees N, on the N.W. Coast, and up Christian Sound to Lynn Canal." Worth also said he had "assisted in obtaining 20,000 barrels of oil."[3] Unfortunately, his voyages were not sufficiently well documented; he may have been on the Northwest Coast in command of a fur-trader. And the Nantucket *Inquirer*, under the date of 28 February 1825, reported only that Worth had sailed as far north as the Columbia River.[4]

These uncertainties aside, one thing is perfectly clear: Barzillai Folger was neither the first, nor even the first of his own generation, to hunt whales on the Northwest Coast of America. In 1834 a British "South Seas" whaleship had cruised there without result, and during the following summer at least one French crew successfully hunted right whales above 49° North in the eastern Pacific. At that time, Folger was still on Nantucket Island. He did not take the oath of command until 10 October 1835, and only on the twenty-sixth did he sail the ninety-one-foot *Ganges* from Edgartown, on nearby Martha's Vineyard, bound southward for the Pacific.[5]

His voyage was predicated on discovering a new sperm-whaling ground to fill a burgeoning demand caused in part by an increasing need for illumination and industrial lubrication. Sperm oil was sought not only to light a growing number of homes, but also for the gross lubrication of the new machinery of the Industrial Age. As the numbers of machines increased, so did the call for the oil; the annual consumption of sperm oil by a single cotton mill in Lowell, Massachusetts in 1851 was calculated to be 6,772 gallons — about 220 barrels[6] — roughly the amount obtainable from three large sperm whales. And there were many machines, and more under construction, at Lowell, Boston, Providence, New York; and at London, Paris, and Bremen. Whaling merchants remained alert for news of newly-discovered whaling grounds and listened with care to the advice of merchant captains and other seafarers who encountered whales while pursuing their maritime occupations.

Among this class of consultant was Jonathan Winship of Boston — actually Brighton — Massachusetts, who in 1820 had directed Nantucketers to the rich sperm-whale grounds off Japan. Winship wrote to Frederick Sanford and others on Nantucket to say that he had seen sperm whales on the Northwest Coast of America, near Kodiak Island in the Gulf of Alaska, while on passage to the Sandwich Islands.[7] His advice about Japan had been proven correct; perhaps this information was also worthwhile?

"Being convinced that there was something to be made out of this," Sanford later wrote, "we fitted out the ship Ganges and sent her, in the summer of 1835, to the locality mentioned."[8] But when the *Ganges* finally reached the Northwest Coast, Folger sailed back and forth and found not the object of his search: the men at masthead raised "plenty Right whales but no Sperm whales, as Capt Winship of Boston supposed."[9]

"In his first report which we received from there," Sanford later explained, "Captain Folger stated that he had seen nothing but right whales. As whale oil and bone were of little value at that period, and as he did not care to lower his boats for them, he had, after taking 300 barrels, dropped down the coast to Pudder Bay, California."[10]

Though much is made of Folger's "first," it is little understood that he was disappointed to find only right whales. Island pride may have forged his disdain — Nantucketers hunted the sperm whale exclusively — but the problem was economic; if he filled his ship with the relatively worthless "whale oil" he would have neither casks nor stowage space when he reached the sperm-whaling grounds. And the *Ganges* was a relatively small ship even by standards of the day, measuring only 265 tons.[11] So Folger ignored this untapped population of right whales and turned southward toward Rudder Bay, a sperm-whalemen's rendezvous at the southwestern tip of Cape St. Lucas, Baja California. In early December 1836 he had 500 barrels of oil aboard, and by the time he reached Honolulu the following May he had casked 200 more. He then sailed to Japan where he filled up his ship, or very nearly so, before returning to Edgartown on 10 May 1839 after forty-two and a half months at sea.[12]

Those who owned the *Ganges* may have sensed they were creating history by sending a whaleship to the Northwest Coast. Their excitement would have been quelled at once had Folger found the spent but not yet rusty harpoons of the French *baleiniers* who had already taken right whales there, and perhaps even more so had they learned of the Englishman who cruised there unsuccessfully during 1834. This man was Thomas Richard Stavers, an experienced Pacific whaling master, his vessel the London whaleship *Tuscan*, which had been his charge for most, if not all, of the previous ten years.[13] Stavers received instructions from the shipowners to determine the feasibility of conducting a whale fishery in the far North Pacific; the *Tuscan* was fitted out for an extended voyage and carried several supernumeraries, including the naturalist Frederick Debell Bennett, who later chronicled the expedition.[14]

Captain Stavers was instructed to cruise as far as 55° North latitude in search of sperm whales. Accordingly, in the early summer of 1834, the *Tuscan* arrived in the vicinity of Cape St. James at the southern tip of the Queen Charlotte Islands.[15] Fogbound there, his men saw none of the species for which they had come. If they cared about or saw right whales, Bennett's account makes no mention of it, although he reported that the "smaller cetaceans" were seen. The crew undoubtedly formed their own opinion about whale-fishing on the ill-tempered North Pacific, and Stavers decided—or perhaps he was persuaded—that the *Tuscan* had gone far enough:

> we had advanced sufficiently to the north to experience its dis-advantages, without any compensating appearances of the object of our pursuit . . . The sentiments of the crew, also, being averse to con-verting a southern into a northern whale fishery—the inclemency of the weather leading to the conclusion that the peculiar operations of south-ern whaling could scarcely be conducted in a more northern latitude, be-ing barely practicable in this—Captain Stavers determined upon returning . . . to the southward.[16]

Just as Folger would later sail for California, Stavers put away to other seas where he could be certain of obtaining sperm oil. Captain Abraham Gardner of the American whaleship *Canton* spoke him in March 1836 at Tahiti, and by then the *Tuscan* carried 1,700 barrels of oil on board.[17] Perhaps Captain Gardner heard an earful about the impossibility of whaling on the Northwest Coast.

Elsewhere in the Pacific, the news was much better. A month before Gardner spoke to Stavers at Tahiti, another "gam" had taken place close by the Mocha Isles, near Valparaiso, Chile. Ebenezer Stetson of the New Bed-ford whaleship *Endeavour* had spent an hour aboard a French whaling vessel recently arrived from the northward, and the word he brought back caused something of a sensation. The Frenchman was "late from the NW Coast 17 months out with 1500 bbls. oil." Stetson, or perhaps his mate, thought enough of this to enter the sailing co-ordinates into the logbook of the *Endeavour*, in a margin never otherwise breached: "Whaling on the N.W. Coast from the Lat of 47 to 51 N, Long 176 to 180 West from April to No-vember is reported by the French ship."[18]

The French captain who gave so unstintingly of this information was Nar-cisse Chaudière, and his ship—unfortunately for a later generation of historians—was named the *Gange*. Chaudière had just extended French whal-ing experience north of 49° in the eastern North Pacific; his report to Stetson may have been the first announcement of success there.[19] Stavers had missed the whales, and Barzillai Folger, finally at sea in the *Ganges*, had only just got to Cape Horn. It is even possible that Folger learned the news as he hurried across the Chile grounds.

Chaudière's voyage would not have been possible just a few years before. The French sperm-whale fishery begun by Nantucketers in the late 1780s had not survived the Napoleonic upheaval. In 1815 French whaling was moribund. The government instituted a subsidy in 1816 in the form of a premium awarded to returning whaleships, but these initial prizes had little effect in motivating merchants to return to whaling. During the following decade the amounts were steadily increased, so that by 1829 the owner of a homeward-bound whaleship could expect a reward of ninety francs per ton, provided the ship's crew consisted entirely of French nationals. Whaleships employing American or other foreign-born men also earned a bounty, though it was reduced to less than half the usual amount.[20]

In 1832 the premiums were again improved, and the larger part of the bounty could now be claimed upon the ship's departure. The increased prize and the advance payment stimulated a new interest among investors, and a supplementary bonus awarded on Pacific sperm-whaling voyages drew particular notice. These combined factors encouraged the Havre-based merchant C. A. Gaudin to send several whaleships immediately the bounties of 1832 were enacted; his *Constance* cleared out that year under Narcisse Chaudière, and Chaudière sailed all the way to Monterey, the territorial capital of Alta California, in search of sperm whales. Seven were taken during June 1833 between 29° and 33° North, and right whales were also seen, though the captain chose to ignore them.[21]

Chaudière—whose name may be translated to mean a large cauldron for cooking seafood—earned himself another voyage. Just four weeks after returning to Havre in August 1834 he again set sail for the Pacific, this time in Gaudin's *Gange*. His new command was large. At a time when French students of whaling recommended vessels of 350 to 400 tons, the *Gange* measured 573. Chaudière sailed with a very full complement of thirty-eight men,[22] and, given the size of vessel and crew, one wonders if Gaudin instructed his captain to push farther than before. *Le Capitaine* no doubt seemed the right man for the job. He was in the prime of life, about forty years old, described as a large man with stiff black hair. He seems to have been a careful sailor, and perhaps a man of some humour, for he is said to have habitually scratched his nose and smiled when he told jokes.[23]

In the *Gange*, Chaudière sailed farther north than he had in the *Constance*. He came up from the Chile grounds and reached 40° North by 30 May 1835, remaining well to sea in 152° West longitude, perhaps in order to escape prevailing northerly winds which in California make an upcoast sailing passage a lengthy beat to windward. Five days into June he lowered for right whales, but the boats could not reach them. After that he had success, despite the fogs which constantly hampered visibility. The right whales his men killed were fat and large, the blubber thick, and the baleen plates very long.[24] Continuing to the north, the *Gange* reached 48°39' North, 158° West —not far south of the latitude of Cape Flattery—on 9 June.[25] No further positions were obtained due to the poor weather, but Chaudière believed that he attained 51° North.[26]

Early in August Chaudière could have substantiated Jonathan Winship's

observations; his crew lowered for sperm whales in the vicinity of 42–45°
North but could not make fast. In his farthest northward journey, however,
he did succeed in taking seven right whales, which made about one-third of
the total cargo of 2,393 barrels of oil carried home to France on 18 April
1837.[27] After turning south, the *Gange* met another whaleship, a coun-
tryman, whose crew may also have reached the latitudes of the Northwest
Coast. This was the *Rhône*, with twenty-seven men under the command of
Joseph Marie Letellier. They had been whaling without success. The two
crews met in 42° North, parted, and never afterwards saw each other. Letellier set his course for Chile, and reached Paita on 24 September. Chaudière
soon followed, calling at San Francisco and Monterey late in August, but by
the time he reached the Chile grounds, Letellier's *Rhône* had been wrecked in
the Chonos Archipelago, not far from Lemus Island.[28]

Perhaps — just perhaps — the seven right whales taken by the *Gange* on the
northernmost leg of its long journey were the first ever taken by the crew of a
commercial whaleship in the Pacific Northwest. In any event, they were signal enough, and soon the whalemen of the United States, France, and several
other nations set out to collect their shares of these grand animals from the
northern seas:

> the [North Pacific right] whales average from about 125 bbls. each, the
> male or bull whale making from 60 to 100 bbls. and the female or cow
> from 100 to 200 bbls. The latter is the most I have seen one make
> though I have heard of them taken that made 250 bbls and one that
> made 290. The bone [baleen] will go about 1000 lbs. to a 100 bbls. oil
> and it is much longer than the South Sea [right whale] bone, the number
> of nabs to a full grown head about 200 and varying in length from one to
> eleven foot.[29]

Some of the Northwest Coast right whales may have been even larger;
Samuel Stafford, whaling in the bark *Laurens* of Sag Harbor, described one
taken in the summer of 1846 which made 300 barrels.[30] Frederick Crapser,
the young sailor from the *Pantheon*, believed them to be bigger than that: "it
is 10 times the largest livving thing that i ever saw in my life," he wrote;
"the Elaphant is no more to be compared to a Norwest whale than a Calf is to
the Elaphant   the whale that we just cut in was 97 feet in length and 84 feet
in Surcumferance and he made 220 bls of oil."[31]

To catch such giants, the whalemen employed the same technology as the
sperm-whalemen and the North Atlantic right-whalemen before them. They
rowed their shallow whaleboats alongside the living animal, thrust a harpoon
through the blubber of its back, and pulled away with a will to await
whatever might happen next. If they could hold on to the animal long

Several men stand on the cutting stage, and another on the whale's back, to begin re-moving the blubber from a large right whale. The photograph was taken about 1900.
Kendall Whaling Museum, Sharon, Massachusetts

enough to tire it, they would try to thrust the killing lance into its body and turn it "fin out." Otherwise they risked a dunking or worse, for the whale could easily upset the boat. Author Gustave Duboc's epic portrayal of the first North Pacific right whale taken by the crew of the *Gange* serves well enough to describe all the whaling that would very soon come to pass:

After a chase of two or three hours, a whaleboat managed to approach a
whale; at the signal from the officer, the harpooner left his oar, seized his
harpoon, turned it and lanced vigorously at the animal; the iron entered
deeply, but the boat was suddenly smashed by a blow of the tail; the harpooner was thrown . . . in the air, also one of the oarsmen, who fell
back down and injured himself critically on the wreckage. One of the
other boats . . . advanced boldly, and soon we saw it fly like an arrow, it was fast [to the whale], and the whale dashed along with it in its
flight. . . . But this frightening chase slackened little by little; the
other boats . . . surrounded it, harassed it and wounded it with the
lance; it received a mortal blow . . . but night arrived, and all disappeared from sight. Our eyes, always fixed toward the point on the
horizon where we had lost them from view, soon distinguished a light: it
was the light of a lantern from one of the boats. At the same time the six
men who remained on board [the *Gange*] . . . steered toward the
light. . . . Finally, at midnight, we suddenly heard the men of our
four boats meet . . . a half-mile from us . . . a quarter of an hour
later they were tied up alongside, as well as a huge dead whale, which
our poor oarsmen had towed for two hours.[32]

That first whale gave 200 barrels of oil, according to Duboc, and now it
seems like a fit beginning — even if improved by the author's literary skill —
to the right-whale hunt in the northeastern Pacific.

It took some time for the news to drift through the whaling fleet, though
the notices were apparently passed from ship to ship. In the final pages of a
journal kept aboard the whaleship *American* of Nantucket — which returned
home in 1841 — someone entered the co-ordinates of the northern ground for
future reference. Clearly, the news brought by Narcisse Chaudière and later
by Barzillai Folger had begun to be heard: "Right Whales Lati 56:00 N Long
145 to 147 W and to the West of that But None to the East   Ship Mary
Fisher."[33]

It took longer still for the information to reach the offices of whaling merchants in New England and France, upon whose instructions the shipmasters
set sail. So the Northwest Coast seems to have been visited by only a few
whaling vessels between 1835 and 1840, perhaps those whose owners allowed their captains to follow new leads based on the latest news. The rush to
the Pacific Northwest may have been forestalled also by the unappealing reports of the earliest captains to investigate there. Both Folger and Stavers
were conspicuous on the Pacific sperm-whaling grounds, and their experiences with the foul weather of the Northwest Coast cannot have been encouraging to others. Even the successful Chaudière had to admit that the
northern weather was terrible; he had encountered nineteen days of fog and

rain in June 1835, ten more like days in July, and eight additional up until 16 August, when he turned the *Gange* toward Chile. "The northwest coast cruising ground," Captain Sanford later wrote, "was first visited ... by two or three of the Chili whalers who saw, indeed, numerous whales, but gave it as their opinion that the fishery could never be prosecuted there with any success, by reason of constant and dense fogs."[34]

A Northwest Coast whaling cruise during the late 1830s was therefore liable to be a lonely pastime. A crew might pass an entire season without sighting another vessel; so the rediscovery of these early voyages is now made difficult by the absence of corroboration from the logbooks of other ships similarly employed and by the dearth of arrival records at various Pacific ports of call. Captain Sanford—who owned a part of the 1835–39 voyage of the *Ganges*—did keep a careful ledger of whaling departures from Nantucket during this period, and there, under the date of 1835 he entered marginalia which shed a small glimmer:

> Ganges was the first ship that went in N.W. Coast Whaling 1835–36 Saw millions of right whales. "Villi D'Lion" of Havre next ship, Elbe of Poughkeepsie 3d Ship and filled with Whale oil, completed her voyage & return to her home port 1840   1800 bbls W[hale oil] & 1200 bbls. Sperm oil. Cap[tain] Waterman reported great numbers of Whales.[35]

The history of the French whaleship *Ville de Lyon* has vanished into the mists of the North Pacific. Perhaps Sanford erred on the name; he may have meant the *Liancourt*, which made early voyages to the North Pacific. Or perhaps he thought of Captain Largeteau in the *Ville de Bordeaux*, who explored and hunted whales in the Sea of Japan and along the Kamchatka peninsula during 1837–38.[36] The identity of this visitor remains another of the mysteries. But of the Poughkeepsie ship *Elbe* and "Cap" Waterman, there is much to say.

Poughkeepsie, New York is not a likely place for a whaling town, though it is not far from Hudson, whose Nantucket-born families sent ships to the South Pacific sperm-whale fishery before 1800. It lies along the shore of the Hudson River, some sixty-five miles from the egress of that great waterway into New York Harbor. Despite its upriver location, the Hudson's great depth allowed whaleships to come up to the Poughkeepsie docks, and there, in 1832, money was assembled to form the Poughkeepsie Whaling Company and, in the following year, the Dutchess Whaling Company.[37] The *Elbe* sailed in 1833 on a four-year voyage for the Poughkeepsie Whaling Company under Captain Whipple. Whipple hunted sperm whales as far north as "St. Clements Island"[38]—San Clemente Island—off the coast of California. While cruising the sperm-whale grounds off Baja California, he spoke at least twice with Thomas Stavers in the *Tuscan*, not long after Stavers had brought his ship back from 50° North.[39] Whipple did not attempt to duplicate Stavers's voyage, but the next captain of the *Elbe* did.

As Whipple sailed homeward, the national economic depression of 1837 forced the closure of many American businesses. Whaling was hard hit; the price of a gallon of right-whale oil plummeted from about fifty cents to twenty-eight cents.[40] The *Elbe* returned in May of that year, and a few days later it was sold at auction at the wharf in Poughkeepsie, "as she lies discharged for a whaling voyage to the Pacific."[41] Purchased by the rival—and more solvent—Dutchess Whaling Company, the *Elbe* was soon dispatched to sea again, under the command of Charles C. Waterman. He declared for the South Atlantic but soon entered into a long voyage to the Northwest Coast of America, returning in 1840 with a good cargo of 1,850 barrels of whale oil and 850 barrels of sperm oil.[42] It is not known if his adventurous journey was influenced by Whipple's reports from California, but Waterman was certainly among the first to take advantage of the new whaling grounds, as noted in Captain Sanford's memorandum.

It was clear to virtually all that the business of whaling, if it were to continue to provide large cargoes of oil, must seek out the whale in its farthest haunts. Long-exploited whaling grounds now showed signs of depletion; while the numbers of vessels in the fishery were increasing yearly, the catch from the South Pacific had begun to decline. One whaleman with a flair for journalism later explained how "two or three" whaleships had gone to the North Pacific after the southern whaling season of 1836 and had found both whales and continual fogs. The next year several more of the fleet went farther north and west and found better weather. For the three following years, he said, there were few ships on the new cruising ground, but "upon the almost entire failure of the Southern fishery [about 1840], the right whalemen were forced to turn their faces hither, and the North West now became a very El Dorado to this extensive branch of our marine."[43]

Learning the identity of these ships is problematic, but there are some clues. The first is a curious sidelight to Frederick Debell Bennett's narrative of the 1834 exploration by the *Tuscan*: a whaleship was seen trying-out blubber north of 40° in the longitude of 142° West. This vessel was far to the east of the Japan ground and well north of the cruising area sometimes frequented by whalemen near the Sandwich Islands. But there it was, as Bennett puts it, a "South-Seaman which had lately been successful."[44] Bennett did not record its name, but the following day the *Thomas Williams* of Stonington, Connecticut, was spoken. Perhaps these aggressive Stonington men, who had been instrumental in opening the South Atlantic sealing industry some years earlier, now had troubled themselves to explore the furthermost frontier of whaling in the Pacific.

The ship *Gratitude* of New Bedford may have been an early arrival on the Northwest Coast, and was a frequent visitor to the North Pacific. Watercolor, attributed to Benjamin Russell, circa 1848.
The Whaling Museum, New Bedford, Massachusetts

Then there is the matter of the whaleship *Vineyard* from Edgartown, Massachusetts, which almost surely made a season's cruise on the Northwest Coast in 1837. George Pelly, agent for the Hudson's Bay Company in Honolulu, reported to London that the French warship *Nereide* had encountered the *Vineyard* "on the N W Coast" on 7 July and that the American was bound "to the Northward."[45] Another curiosity is the ship *Gratitude* of New Bedford, which returned home in 1839 with 2,500 barrels of whale oil, 300 barrels of sperm oil, and a staggering 34,000 pounds of baleen. This enormous cargo, but not its source, was duly entered on the inward manifest submitted to the New Bedford Custom House. And while it is true that the crew of the *Gratitude* might have caught their right whales in any of a number of seas from Patagonia to the coast of Zanzibar, the large catch suggests an untapped field, a sea of right whales not much "gallied" by the approaching whaleboats. To add spice to the mystery, the inward manifest also declares an enigmatic cargo of "763 hides,"[46] suggesting perhaps that the master and crew wandered up the North American coast to Alta California and encountered the bustling cowhide trade, which one year after their return would be the subject of Richard Henry Dana's *Two Years Before the Mast*.

One positive identification may be made, not surprisingly a New Bedford whaleship which hunted successfully on the Northwest Coast during the spring of 1838. Captain Joshua Bunker departed Massachusetts in the *Timoleon* during mid-November 1835, just a few weeks after Barzillai Folger cleared Edgartown in the *Ganges*. But Bunker's itinerary did not include the Northwest Coast until thirty months later. On 3 May 1838 he left Oahu "to cruise," and on 29 May the *Timoleon* crossed 49° North latitude in 146° West longitude. Bunker continued to 54° North without a sighting, but on 3 June he came upon the right whales. His men lowered boats twice that day and at least three times the next, and in the course of those pursuits struck six whales. One was killed and brought to the ship's side. Eventually, twelve whales were harpooned north of 49°, and two successfully brought to the ship. Then the fog rolled in, and Bunker, being unfamiliar with the neighbourhood, beat a premature retreat southward, recrossing 49° North during 19 June and thereafter passing down the coast to the sperm-whaling anchorage at Bahia de Todos Santos, Baja California.[47]

Almost certainly some of the early arrivals of 1837–40 were French; sadly, the French whaling campaigns are equally difficult to isolate. Captain Letellier, after losing the *Rhône* in 1834, returned to the North Pacific in the *Ajax*; in 1838 he cruised in the vicinity of 42–45° North, 145–155° West, though his search once more proved fruitless. The *Gange* also returned, this time under Pierre Francis Phobin, who examined regions near the Kamchatka coast. And the *Cachalot* of Havre cruised in 41° North, 147–170° West as early as April 1840.[48]

Beginning in 1839, New England shipowners sampled their first oil from the Northwest Coast. Folger returned the *Ganges* on 10 May that year, the *Vineyard* arrived early in 1840, and in Poughkeepsie, Captain Waterman brought home the *Elbe*. The effect of the news in the managerial offices was immediate. Waterman, for example, took his knowledge to New Bedford and returned to sea in six months' time as master of that port's large 470-ton whaleship *Braganza*. In it, he would make a fabulous voyage to the Northwest Coast and bring home 3,600 barrels of whale oil, 400 barrels of sperm oil, and a tremendous lot of 42,000 pounds of baleen in February 1843.[49]

Despite a half-century of trade and exploration, there was little consensus among whalemen concerning the precise co-ordinates of the "Northwest Coast of America." Some navigators considered the whole of the country north of Spanish Mexico as the Northwest Coast, though the portion which is now Alta California was often referred to instead as "New Albion." At least as late as 1837 some whaling masters referred to Alta California as the "North West Coast."[50] Nor did they know exactly where in this vast tract of ocean whales were most likely to be found. Consequently, shipowners did not immediately think of the "Northwest Coast" as a *destination* for their whaling vessels. In 1839, for example, the port of New London, Connecticut sent twenty-one whaleships, most declaring for New Zealand or the South Seas or the Indian Ocean. Yet some of these ships, on arrival at the Sandwich Islands, revised their itineraries to include a cruise northward.

These New London whalemen were not as concerned as Nantucketers with the extra value of sperm oil. They paid some heed to a rising market for baleen, which provided a flexible stiffener in the wasp-waisted corsets and widely-flared skirts just coming into fashion among Victorian women on both sides of the Atlantic. "Whalebone" also gave service in the manufacture of other flexible products, such as buggy whips, umbrella ribbing, and collar stays. And if they did not yet have them, cadets studying war at West Point—among them Ulysses S. Grant and Robert E. Lee—would soon engage in hand-to-hand combat drills with flexible bayonets made from the filter plates of whales.[51]

A brief look at the development of the market for baleen will indicate the extent of change: in 1841 the average of baleen prices in New England was 19 2/3 cents per pound, but this rose to 23 cents the following year, then to 35 3/4 in 1843, and to 40 cents in 1844.[52] Consequently, the Connecticut fleet and also those from the outports of Long Island, New York—in fact, ships and barks from most of the whaling villages of New England—soon looked with interest upon the Northwest Coast right whales. Even New Bedford, by 1830 the premier whaling port, investigated; the *Timoleon* had cruised there in 1838, and on 29 June 1840 their ship *Logan* took a northern right whale in 45°12' North, 146°51' West and then cruised as far as 56°50' to gather eight more.[53]

The New London ship *Phoenix* also went northward in 1840. The crew saw

New Bedford artist Benjamin Russell created this popular and accurate print, *A Ship on the North-West Coast Cutting in Her Last Right Whale*, after he visited the Northwest Coast and Nootka Sound aboard the whaleship *Kutusoff* in 1843. The large upper jaw and baleen of a right whale are being hoisted aboard by a crew working at the windlass forward. *Kendall Whaling Museum, Sharon, Massachusetts*

one right whale in 40° North, and from 1 June until 16 July five right whales were taken, though the ship never worked north of 48°.[54] Another New London ship, the *Superior*, reached the latitude of the Queen Charlotte Islands during the summer of 1841. The crew of this fortunate ship struck no fewer than fifty-eight right whales between 11 May and 19 August while cruising in 51–54° North, 148–154° West. They saw whales almost every day, and in their enthusiasm—dare say greed—they lowered boats even when a carcass was already lashed alongside and the tryworks set to be lit.

The *Superior*'s men brought twenty-six whales to the ship and cut them in to make a remarkable whaling season on the Northwest Coast.[55] Theirs was a one-season voyage. The *Superior* had sailed 29 September 1840; by 3 July 1842 Captain Albert McLane and his crew were home again, and with an impressive cargo: 2,000 barrels of whale oil, a few hundred barrels of sperm oil, and a very satisfactory 20,000 pounds of baleen.[56] McLane abstracted the catch for the ship's major investors, Noyes Billings and William W. Billings, who then, belatedly, held in their hands a North Pacific destination to which they could send other ships. Outward-bound masters working for N. and W. W. Billings received new instructions:

The ship Robt Bowne made your command being now ready for sea you will embrace the first favorable opportunity to proceed on your whaling voyage to the North West Coast. After you pass Cape Horn we should advise you to cruise on the coast of Chili—or go up to the Line & cruise [there] for Sperm whales taking care to leave in sufficient time to recruit at the Marquese Islands or at Mowee Sandwich Islands & be on the North West Coast by the 1 May. The cruising ground on the North West coast for right whales is all North of 40° & from one continent to the other . . . The Superior last voyage steered *North* from the Sandwich Islands & found whales in great abundance in lat 48 to 53° & we have no doubt the whales are to be found all over the Northern Ocean.[57]

"When you leave the coast," Billings added cautiously, "you must be guided by your own judgment what course to pursue—we think it always an object to avoid the Sandwich Islands as they are expensive & the crew likely to desert—& probably it will be best for you to come to the N West Coast to recruit say at St. Francisco which is a good port & supplies readily obtained."[58] With these instructions in hand, Captain Fitch set sail in the *Robert Bowne*, and made a good voyage. On his return in February 1845 he brought 4,600 barrels of whale oil—a huge cargo—plus 200 barrels of sperm oil and yet another staggering lot of baleen totalling 40,000 pounds.[59]

Others who worked for N. and W. W. Billings were encouraged to go even farther north to obtain choice cargoes. Captain Alexander Hart, upon replacing McLane in the *Superior*, was advised to redouble the ship's previous effort by returning to the North Pacific and, if unsuccessful there, sailing "if necessary up North of Berings Straits or inside of the Islands of Kamskatchka."[60] New London merchants as surely as any other—and more surely than most—took advantage of the new whaling. In 1841 two-thirds of their departing whaling fleet, twenty of twenty-nine vessels, declared for the Pacific. By 1843, half of the port's thirty outward-bound masters would identify the Northwest Coast as their specific destination.[61]

According to Alexander Starbuck's massive compilation of American whaling voyages up to 1876, the first whaleships to declare for the North-west Coast sailed in 1841; one each from Sag Harbor, New York; Edgartown, Massachusetts; and Newark, New Jersey—this latter an unusual departure since few whaling voyages were prosecuted from the mid-Atlantic states. All three succeeded: the *John Wells* of Newark returned 1,950 barrels of whale oil and 650 of sperm oil; the *Daniel Webster* of Sag Harbor brought in 3,300 barrels of whale oil and 33,000 pounds of baleen; and the *York* of Edgartown collected 4,200 barrels of whale, 400 barrels of sperm oil, and 30,000 pounds of baleen.[62]

It is clear, however, that accounting lagged behind the reality of whaling in the Pacific. Other records reveal that the summer of 1841 constituted the first important "season" on the Northwest Coast whaling ground. It may be difficult now to determine how many whaleships visited the far North Pacific in that year, but at least eleven of them called at Lahaina at the conclusion of the summer to reprovision. Their names are culled from a list of whaleships arriving at that place which was published in 1842 in *The Sailor's Magazine*; there it was reported that the eleven ships brought 15,650 newly casked bar-rels of oil—a most impressive average of 1,423 barrels per vessel. Some, like the *Superior*, took sufficient whales to allow them to return home after just a single cruise.

Their names represent the breadth of the American whaling industry and a French ship for good measure: from New Bedford the *India*, *Milton*, *Orozimbo*, and *Roman*; from Dartmouth the *South Carolina*; from Salem the *Mount Wollaston*; from Warren the *Phillip Tabb*; from Stonington the *Mer-cury*; from Hudson the *Beaver*; and from Poughkeepsie the aforementioned *Elbe*. To complete the census more fully, this group must be augmented at least by the *Sapphire* of Salem and Charles Waterman's *Braganza* of New Bed-ford, both of which called at Monterey instead of Lahaina; McLane's *Superior* of New London and the *Magnolia* of New Bedford, whose captains sailed directly home; a Boston ship called the *Fama*; and the *Parachute* of New Bed-ford. This last-named vessel perhaps went farther north than anyone else; its Captain Wilcox claimed to have obtained 56° North, 140–152° West, within thirty miles of the southeastern corner of Kodiak Island in the Gulf of Alaska.[63]

The French whaleship that called at Lahaina from the Northwest Coast was the *John Cockerill* of Havre, under Captain Richard Walsh — sometimes Walch — a New Yorker long in the French whale fishery. It was Walsh's second trip in command of the *Cockerill* for Duroselle et Cie. Later, in the fall of 1842, he would enter Honolulu Harbour and announce to the French consul that he had found, to the north, whales large enough to give 140 to 200 barrels each. The consul, relaying this news to the minister of foreign affairs, would note fervently that *"la pêche est bien plus avantageuse et plus abondante dans le Nord-Ouest et en vue de la côte du Kodiak."*[64] Walsh confirmed for the French mercantile the voyages of Chaudière, Largeteau, and Letellier and vindicated its interest in the North Pacific; the Northwest Coast was, indeed, home to large whales and source of large profits for the inquisitive *baleiniers*. Nor was Walsh the only master of a French whaleship in the northeastern Pacific in 1841. Captain Malherbe of the Havre ship *Elisa* came to anchor at Monterey in the fall of the year and pronounced himself the first French whaleman to reach 54° North latitude.[65]

Even with this much traffic, the Northwest Coast remained a solitary place. The men in the *Magnolia*, for example, spoke only three vessels between June and August 1841. One was the *Braganza*, and Captain Waterman could boast that he had taken 1,850 barrels in just eight months.[66] McLane in the *Superior* spoke only the *Elbe* and the *Braganza*. Other ships spoke three, or four, or five, during perhaps two months or more of whaling.

They found few fellow humans, but a plenitude of whales. Captain David Barnard and the crew of the *Magnolia* took nine large right whales, averaging 112 barrels each, during their two-month stay north of the forty-ninth parallel.[67] The *Orozimbo*'s people likewise found fine whaling. They crossed 49° on 20 June and took their first right whale on the twenty-ninth. During the following ten days they struck five and killed three. By 29 July they had advanced to within sixty miles of Cape St. James, very close to the spot where Captain Stavers had to wear ship seven years before. The men at masthead raised spouts, and there were the very whales Stavers and Folger had come to find but had not: sperm whales. Some of the crew were incapacitated by scurvy but the rest managed to lower the *Orozimbo*'s boats. Before nightfall five sperm whales had been struck and three of them killed; the crew cut-in and continued the cruise despite the advance of the scurvy.[68]

Perhaps the sperm whales were considered a good omen. But there were no fresh provisions on board or ships in sight to provide assistance. The illness intensified. On 20 July the blacksmith was stricken; others suffered in the forecastle. Three days later the logkeeper noted six men sick "and the rest complaining,"[69] and soon Captain Bartlett could do little else but turn southward for help. At the mouth of the Columbia River they were met by the schooner *Flying Fish* belonging to Lieutenant Charles Wilkes's United States Exploring Expedition. Not ten days earlier, Wilkes had suffered the loss of the sloop-of-war *Peacock* on the Columbia River bar, and the commander of the *Flying Fish* was obliged to fire a shot across thé whaleship's bows as a signal to heave-to, since the *Orozimbo* was entering in the track of the shipwrecked vessel.[70]

The *Orozimbo* spent one day "laying off and on in with the squadron."[71] Lieutenant Wilkes supplied its suffering crew with preserved meats and antiscorbutics; then the whaleship continued southward.[72] At San Francisco Captain Bartlett obtained additional relief from William Dane Phelps, master of the trading vessel *Alert*, who sent "all the fruits & vegetables I could find on board."[73]

The advance of unarmed whaling ships into uncharted seas provided a purposeful justification for exploration. Few examinations of the North Pacific had been conducted since the early years of the century, and now, with a growing pattern of development in the American west and with the proliferation of missionaries and traders in the Pacific islands, it seemed appropriate to investigate the mercantile, naval, and political potential of the Pacific basin. Consequently, both the French and the Americans sent expeditions to the North Pacific, in part to update charts and to identify repair and reprovisioning ports for whalemen, but also to demonstrate national interest in this new international arena. Each nation's newest explorers sought, among their other tasks, to identify the boundaries of the whaling grounds.

Sir George Simpson, late overseas governor of the crown-chartered Hudson's Bay Company, made some effort on behalf of British interests to determine its extent while inspecting Company properties during 1841–42. He received his best information from a Dutch-born American sea captain, Cornelius Hoyer, who told Simpson that the whaling on the Northwest Coast encompassed 52–57° North latitude by 144–152° West longitude. Hoyer had practical knowledge; he had recently served as master of the Boston ship *Fama*, which arrived in California from the United States on 6 April 1839 and had been "awhaling ever since."[74] And his advice was also current; the *Fama* had been on the Northwest Coast at least as far as 53°30' North during June 1841.[75]

The French government, whose whale fishery had grown considerably under the benefits of the improved bounty system, now used whaling as the guise to mount a major exhibition of naval and diplomatic might in the Pacific. The voyage of Admiral Abel du Petit-Thouars in the frigate *Vénus* (1836–39) proposed to benefit their whalemen in the Pacific, but its larger mission was to lay a claim for France in the territories of the Pacific basin. The admiral was instructed to identify harbours where French whaleships and merchant vessels might obtain provisions and repairs and to determine if there was a need—a great need was surely anticipated—for naval protection of French merchantmen.

Officers of the expedition were instructed to inform themselves as completely as possible of the direction taken by whalemen, the circumstances which determined their choice of certain waters, their chances for success, their most frequent ports of call, and finally, all circumstances of importance to the naval vessels which would be entrusted with the protection of the industry. It was also hoped that du Petit-Thouars's presence would assist in quelling the revolts of whaling crews, "who did not feel confined to their duties by any coercive force," and whose actions led "to the ruin of expeditions and often even to the loss of ships."[76]

Du Petit-Thouars sailed as far north as the Kamchatka peninsula and then crossed eastward to meet the American coastline in 45° North latitude before returning to France. Even as his reports were being made known, another French observer, Eugène Duflot de Mofras, visited California and the Pacific Northwest, and he confirmed what Cornelius Hoyer had told Sir George Simpson about the North Pacific whaling. Duflot's account, published in 1844, belatedly encouraged the French whale-fishers to take up the Northwest Coast ground.

> our ships should be urged to explore regions not as yet visited, situated north of the Equator and along the coast of Mexico, the Red Sea [Gulf of California], the two Californias, the Northwest Coast, with its many archipelagoes, and the Aleutian Islands, the waters, briefly speaking, between the equinoctial line and 60° north latitude and from 90° to 180° west longitude, Paris time. . . .
>
> Notwithstanding the dense fogs, the heavy seas, and the northwest winds that blow almost without respite, whaling is conceded to be comparatively simple in these waters during the summer season, especially in the many bays that are sheltered by the thousands of small islands along the coast.[77]

After considerable delay, the Americans confirmed the United States Exploring Expedition and its commander, Lieutenant Wilkes. Wilkes, too, thoroughly investigated the North Pacific on behalf of the whaling industry and delineated no fewer than fifteen Pacific whaling grounds in the expedition's report, which appeared in 1845. He clearly separated California whaling from the Northwest Coast and noted that the favourite North Pacific ground "is between the fiftieth and fifty-fifth parallel, where vast numbers [of whales] have been recently taken in June and July, of great size."[78]

The efforts of governments to identify the location of the whales were redoubled by shipowners and shipmasters, whose immediate economic gain rested on the accuracy of such information. The co-ordinates they obtained were culled from the logbooks and journals of their own captains just returned from sea; it was hoped that the whales would remain in or return to these places. After the Northwest Coast whaling ground became more commonly known, masters at sea could boast of their successes there: "spoke . . . the [*William Hamilton*] of newbedford 1800 Wright 300 Sperm She went on the nor west last season [1843] and the season was 2/3 out and she got 1800 barrels."[79]

Contrary to the advice of N. and W. W. Billings, it had begun to appear that right whales were not found uniformly across the ocean, but rather were concentrated near shore on both sides of the Pacific. Whalemen had seen right whales from the Bonin Islands northward along the Russian Kamchatka peninsula as well as on the Northwest Coast of America. The Northwest Coast population was confirmed by Charles Wetherby Gelett, master of the *India* during 1841, who hunted whales between the Queen Charlotte Islands and "Kodiack" Island.[80]

In the early twentieth century, an ambitious zoologist named Charles Haskins Townsend began a study — later entrusted to underlings — to chart the distribution of whales based on worldwide catches by whalemen of the nineteenth century. Townsend and his colleagues sifted through logbook accounts of 1,665 voyages in 774 vessels, marking the location of each whale taken by species and date. These data were then organized and charted.[81] The findings verified that the southernmost promontory of the Queen Charlotte Islands — Cape St. James — marked the operational southeastern corner of the whaling ground; only a few whales were taken farther to the south. Most of the right-whaling took place in the Gulf of Alaska, from the north end of the Queen Charlottes around the curve of the shore to Kodiak Island and beyond to the Aleutian archipelago.

In 1841, the first important year there, a considerable amount of whaling was conducted between 50° and 54°30'; even in later seasons a few right whales continued to be taken south of 50°. But the right-whaling soon shifted to the upper curve of the Gulf of Alaska, excluding waters south of 54°. This became so much the case that whalemen began to refer to the northeastern Pacific as the "Kodiak" — or "Kadiak," or "Coodiac" — Ground. Charles Melville Scammon, the foremost nineteenth-century authority on North Pacific whales, placed the Kodiak Ground from Vancouver Island northwestward to the Aleutians and from the coast westward to longitude 150° West.[82]

The geometric progress of this new fishery may be observed in a table published in the *Shipping & Commercial List* of New York in 1848. By that account, only two whaleships visited the Northwest Coast in 1839, and but three in 1840. Twenty arrived in 1841 — probably very near the true total — then 29 in 1842. In the summer of 1843, the *Shipping & Commercial List* accounted for 108.[83] The bulk of this traffic arrived at the Sandwich Is-

Packing Whalebone. Whalemen customarily cleaned and bundled the long plates of baleen at the conclusion of a cruise or during the homeward-bound voyage. Ink drawing, by "Porte Crayon" [David Hunter Strother], dated 21 May 1859. *Kendall Whaling Museum, Sharon, Massachusetts*

lands direct from Cape Horn or else returning from winter cruises in the South Pacific. In the spring of 1843, 104 American whaleships called at Lahaina alone — 36,279 tons of shipping carrying 2,961 sailors representing thirteen New England ports. [84] The Hawaiian government quickly realized the value of such commerce and made various exceptions in favour of the whalemen, including the reduction or abolition of port dues and pilot charges and the granting of permission for crews to sell $200 in goods without payment of import duties. Of all Sandwich Islands ports, Lahaina was most preferred by whaling masters. Goods were less costly than at Honolulu, and the open roadstead meant that the vessels could come up to anchor without paying for a pilot. Ship captains endeavoured to stay in the complicated harbour of Honolulu only long enough to complete such business as could not be conducted elsewhere; then their vessel would be moved off to the anchorage at Maui.

For the "green hands," Lahaina or Honolulu was perhaps the first place they had been away from home, and the first time on shore since leaving it. The islands were spiced with the lush promises of paradise that the crew dreamed about during months of hardship rounding Cape Horn and whaling. But some observers thought it the very devil of a place.

"The truth is," one visitor wrote, "the whole nation is rotten with licentiousness. Men hire out their wives & daughters without the least scruple, for the sake of money. It is computed by Dr. — — of Lahaina that at that port during the whaling season there are upwards of 400 instances of intercourse with sailors daily. [One] establishment at that place is a perfect sink of iniquity. They are accustomed to have dances of naked girls for the entertainment of their customers the whalemen."[85]

A stopover at the Sandwich Islands included the loading of fresh water, potatoes, pumpkins, cabbages, sweet potatoes, yams, bananas, and firewood; also painting such parts of the ship as needed it, repairing any damages incurred at sea and, glory be, liberty on shore. It also included attempts at desertion, suicides, and even efforts to set fire to or otherwise damage the ship in order to forestall its departure from Eden. It became a serious offense to jump ship; escape was virtually impossible since a reward was paid to informants, and there were few living in the Sandwich Islands who wished to encourage the settlement of chaff from the harbour. Arrested sailors were taken to the fort, and their captains were obliged to fetch them and pay a fine. At Lahaina it became illegal for sailors to be ashore "after the beating of the drum," or at Honolulu "after the ringing of the bell." Nighthawks were apprehended and fined two dollars.[86]

The delights of the shore would sorely tempt any sailor, after so long a time at sea. One whaleman, visiting Lahaina in 1848, wrote home to say that he had enjoyed "a fine cruise about the town," during the course of which he saw an old man "who is said to have seen Capt Cook who was killed on Oyhee while on a voyage of discovery."[87] Another, a teenager in a New London vessel, sent home the latest news about the successes of the arrivals:

> Here I am once more as well as can be expected, fatting on salt horse and rusty pork. With Islands every way I can look, and in a harbour protected by them along with about twenty seven vessels beside our own all

bound for the North West, and are stopping here for a recruit   I will
here give you a list of them as it might be of some interest to those who
have got friends in them   I will limit it to those who belong to New
London Gen¹ Williams[,] Holt hails 1000. bbls Palladium, McLane had
400 [barrels right whale]. 300 sperm but sold part & sent home the rest.
Chelsea, Potts 56 sperm. Columbus, Crocker, [40?] Clematis 600 bls
her Captain and one man were killed by a whale off St. Pauls
[Kamchatka]. The Friends 900 and Nantasket 600 left here a few days
before we arrived.[88]

Most whaleships calling at the Sandwich Islands hailed from New England
ports, and a tenth as many more were French. The British whaling fleet had
been driven off the sea by the repeal in 1824 of bounties for their South Sea
whalemen. That repeal, one observer noted, led to a reduction of British
whaling from 322 ships and 12,788 men in 1821 to but 85 ships and 3,008
men in 1841. And though he hoped that the British industry might be
revived — he suggested the establishment of settlements on Vancouver Island
or in the Oregon territory for this purpose[89] — few of the British whaling vessels made their way to the Sandwich Islands after 1840. Fewer still carried on
toward the North American coast, a sad fact lamented by a correspondent in
the Boston *Atlas*, who wrote under the date of 16 July 1844: "The success of
the Yankees in [the Northwest] fishery is a matter of great astonishment to
the Englishmen, and their zeal and enterprize are certainly deserving of high
praise."[90]

The Americans meanwhile had grown an industry correspondingly large as
the British fishery had diminished. In 1835, the year Captain Folger took
the *Ganges* from Edgartown for the Pacific Northwest, the annual product of
the Yankee whaling fleet exceeded $6 million for the first time, based on
5,181,529 gallons of sperm oil and 3,950,289 gallons of whale oil.[91] There
was little competition against the Americans in the Pacific whaling. But if
one were to encounter a foreign flag, it would most likely belong to the
French.

The expansion of French interest in whaling followed the bounty revisions
of 1832, and was demonstrated by the rising number of French-flag
whaleships at sea. The law of 1832 established a prize of 70 francs per ton at
departure and 50 francs per ton on return. And though this bounty was subsequently reduced each succeeding year — until the payments of 1841 were
34 and 23 francs, respectively — the French merchants found sufficient profit
to continue whaling. Realizing that support of the premiums was essential to
French participation in North Pacific whaling, the amounts were slightly increased in 1841 and stabilized for a period of ten years, through 31 December 1850. French ships manned by all-French crews would receive 40 francs

Navire Baleinier. In *Moby-Dick*, Herman Melville gave his opinion that the French provided "the only finished sketches at all capable of conveying the real spirit of the whale hunt." Morel-Fatio's rendition shows a typical French *baleinier* of the type used in the Pacific Northwest during the 1840s.

Kendall Whaling Museum, Sharon, Massachusetts

per ton outbound and 27 francs per ton on return; those with mixed crews would be paid 29 francs out, 14.5 francs in. A special sperm-whale bounty applied to all vessels remaining at sea longer than thirty months, provided each cruised north of 28° North in the Pacific; this additional payment amounted to 20 francs on each 200 pounds of oil and spermaceti taken prior to New Year's Day 1846, and 15 francs per 200 pounds thereafter, until expiry.[92]

The 1832 regulations favoured ships with all-French crews by reducing the bounty payable on vessels engaging foreigners. It also required that masters and officers be French by birth or by naturalization. A number of American-born captains who had earlier expatriated themselves acquired naturalization papers in order to retain their berths. Shipowners objected to the requirement, since most of the experienced officers and harpooners were Americans. One merchant of Havre, Jeremiah Winslow, was particularly well acquainted with Yankee skill; he had come from New Bedford in 1817 to begin in the French fishery. He avoided the law through a loophole which denied the bounty to owners sending more than five ships to sea. Winslow outfitted more than this number, and though he lost the bounty, he expected to profit by employing experienced Americans on the whaling grounds.[93]

Encouraged by their government's confidence, the French actively participated in the whale fishery on both sides of the North Pacific basin. In the Sandwich Islands in the summer of 1842 it was supposed that thirty or more French whaleships would be arriving, "owing to the late discovery of right whales on the northwest coast."[94] Among them proved to be the *Liancourt*, which reached the coast of Alaska; the *Nancy*, whose crew took nine whales on the Northwest Coast; the *Faune*, which also took nine; the *Nil*; the *Adele*;[95] the *Elisa*; and the *Angelina*. The latter two crews found themselves busy, by account; Malherbe of the *Elisa* reported striking forty-four whales between 4 April and 30 August, while the *Angelina*'s men successfully killed fourteen of twenty struck.[96]

The crew of the New Bedford whaleship *Orozimbo* spoke also the "Magret Champlain" of Havre on the ground on 15 June 1842, which vessel reported 800 barrels of whale oil and 1,100 barrels of sperm oil.[97] And there were other French crews hunting right whales on the Kamchatka coast. Although sending scarcely one-tenth as many ships as the Americans, the French nevertheless mustered the second-largest whaling fleet in the Pacific, and by 1842 sent much of it to the North Pacific.

A Yankee sailor lad would have felt right at home aboard a French ship. They were modelled on American lines, equipped similarly to their New England counterparts, and manned by the *harponneurs*, the *forgeron*, *cuisinier*, and, of course, the *quatre hommes par pirogue* — the four hapless and nameless *matelots* who would row their harpooner and boat's officer to an everlasting

glory. And, just as in American whaleships, many of the young Frenchmen
were novices, newly recruited and never before at sea. The whaleboats in
which they had to go were virtually identical to the New England model,
though they were constructed in France by well-known boatbuilders such as
Leford, Deliquaire, Fossat, and Thaler.[98]

American whalemen who met with French ships at sea did not always attempt to enter their exotic names in the log. Often, their presence is noted
simply as "a French ship from Havre." French mates often wrote only "un
baleinier Américain." But the relatively large number of Americans, or descendants of Americans, living aboard the tricolor ships provided a base for
communication. During the 1840s, for example, the *Pallas* sailed under
Henry Thayer, a native of Taunton, Massachusetts. He had shipped out to
Havre in 1824 as a boy of twenty, whaling in the ship *Bourbon* and eventually
taking out French citizenship in order to matriculate to master. Another
American, Asa Bullard Casper, was master of the *Jacques Lafitte* in 1846–48.
Casper, then on his fifteenth whaling voyage—twelfth as master—spent virtually his entire adult life in French whaleships. His two daughters both
married French whaling captains.[99]

The camaraderie between French and American whalemen must have been
intense; they came to the far north for the same reason, met with some of the
same success, and suffered many of the same disappointments. If successful,
they were at least occasionally wildly and fully successful, and well repaid for
their trouble. If not, then they were all, Americans and Frenchmen alike, a
very long time at sea.

> The [French] captain came on board and stopped an hour. As I was not
> up I did not see him but heard the conversation and a short sketch of his
> life. He had been two voyages previous to this then got married and
> thought he had sufficient [money] to remain at home. Kept his house
> and carriage and lived in great style.
>
> At the expiration of 4 years had two babies added which circum-
> stances caused them to have more servants and expenses enlarged. He
> had heard much of the NW whaling and thought he could make another
> voyage in a very short time. So he told his wife how easy he could catch
> the whale and so soon be back she consented for him to come. At this
> time he had been gone 2 years and lacked 1000 [barrels] of being full
> and said if I ever get home I will stay there.[100]

Nor were the Americans and French the only nations looking with a
speculative eye toward the Northwest Coast. Surviving logbooks and journals record a League of Nations afloat in the northeastern Pacific throughout
the 1840s and 1850s, each ship seeking to bring home a full cargo of oil and

the utilitarian plates of baleen. In the recruitment ports, particularly in the Sandwich Islands, Yankees and French anchored alongside Canadian, Dutch, Germanic, Prussian and Norwegian whaleships. Among the 161 whaleships which called at Honolulu in 1845 were 19 French, six from Bremen, three from New Brunswick, two Danish, and one each Hanoverian, Norwegian, Prussian, and Hamburger.[101]

Many of these ships, but not all, went to the Northwest Coast in hopes of reaping a summer harvest that was depicted by the press in glowing terms: "During the year 1844," the *Sheet Anchor* announced, "there were 400 sail reported at the Islands, manned by 13,200 seamen, and valued at $23,374,000. The great increase of this branch of shipping . . . is owing to the opening of the new fishing ground on the N.W. Coast."[102] Most of these non-American vessels employed Americans in their crews, and all used the same technology; some were sent to New England ports to outfit before coming to the Pacific. The Bremen ship *Alexander Barclay* was one of these. It cleared Bremen in October 1839 and proceeded to Edgartown, Massachusetts to ship men and crew before sailing for whales in 1840.

A significant business network linked New England and Bremen. Bremen merchants handled much of the importation of whale oil into Europe — 34,825 barrels of whale oil from the United States in 1843 alone.[103] The major whaling firm, Christian A. Heineken and Company, maintained an office in Baltimore, Maryland.[104] Heineken's best-known vessel, the 315-ton *Averick Heineken*, made six whaling voyages for the company between 1837 and 1855. It also sailed with American crew; during the voyage of 1840–42 Captain J. A. Schneider shipped seven men from New Bedford and one from Philadelphia, and these were clearly the experienced whalemen on board, since they signed on lays of 1/29 to 1/100, whereas none of the Germans before the mast was paid greater than a 1/100 lay. On the subsequent voyage Captain Schneider did very well on the Northwest Coast, and returned home 15 April 1845 with a total catch of 3,800 *tonnen* of oil and a very respectable 38,000 pounds of baleen.[105]

The Northwest became a beacon for the Bremen whaleships; in the spring of 1844 no fewer than five recruited at Maui for cruises there: the *Sophie*, under command of Sir George Simpson's informant, Cornelius Hoyer; the *Gustave*, Norton; the *Europa*, Fitch; the *Patriot*, Cranston; and the *Mozart*, Fisher.[106] The Germanic whaling merchants exceeded all but the Americans and the French in prosecuting the North Pacific fishery, and they carried it on longer than most, at least until 1868, when the Bremen ship *Eagle* hunted on the Kodiak Ground and the *Count Bismarck* returned from the western Arctic.

On 4 August 1846, the *Patricet* of Bremen was spoken by the crew of the whaleship *Montpelier* in 50°30' North; fourteen days later in 49°30' North the crew of the *Edward* spoke the "oehite" — *Otaheite* — of Bremen.[107] Another Bremen whaleship, the *Clementine*, was described by an American sailor as "a large splendid ship for N West whaling 700 Tons carries 6000 barrels."[108] Its crew apparently had a bad time on the Northwest in 1844, since the captain stopped at Acapulco, Mexico after leaving the northern grounds "to obtain officers, and the complete fitting for the whale fishery in the Indian ocean and on the N.W. Coast of America," and reported four men killed by whales.[109] The hard lesson was also reiterated to Captain Hoyer and the crew of the *Sophie*. When spoken by the American whaleship *Columbus* on the Northwest Coast in late July 1844, the Bremeners had been out twenty-six months and had lost their regular captain, who died at the Sandwich Islands, and seven men. Six died from dysentery and one was "Drownd by a Whale this Season on the North West."[110]

"She Is a large splendid ship for N West Whaleing," wrote journalkeeper John Francis Akin of the Bremen whaleship *Clementine*. Several whaleships from Bremen cruised in the Pacific Northwest during the 1840s.
Ship Virginia *of New Bedford, John Francis Akin, keeper, January* 1845, *Kendall Whaling Museum, Sharon, Massachusetts*

Other Americans commanded Bremen ships on the Northwest Coast. The *Gustave* was sailed to that whaling ground by Clement Norton, a native of Martha's Vineyard, Massachusetts. The whaling merchant Frederick C. Sanford considered Norton one of the two or three most successful Yankee whaling masters of the nineteenth century. According to Sanford, Norton's career began on the *Apollo* in 1816 and ended with his death in Honolulu, "after obtaining a magnificent voyage upon the Northwest Coast" which sent the *Gustave* home with 4,500 barrels of oil.[111]

Other nations and districts sent one whaling ship here, one there. A single Prussian whaleship, the *Rica*, also employed American tools, and men obtained in New Bedford.[112] The *Zuid Pool* of Amsterdam, Captain Braun, was hailed by the crew of the *William Hamilton* in May 1846, in 52°45' North, 156°60' West.[113] Braun reported his ship eight months out and "clean" — that is, without any oil. A Norwegian, the *17th of May*, called at Honolulu in October 1844, ten months out with 50 barrels of sperm oil and 800 of whale oil on board. A Norwegian bark was spoken on the Northwest Coast by the American whaleship *Eugene* the following 30 July,[114] and it seems probable that this unidentified vessel was the *17th of May*.

The Danish brought the *Concordia* to the Pacific in 1842 under Captain Soldering. He arrived in Honolulu in the fall of that year with 1,750 barrels of whale oil after a year at sea. All save 100 barrels were obtained between May and September 1842 in the North Pacific. Soldering afterward cruised on the Kamchatka ground in another Danish whaleship, the *Neptune*, and he may have been among the first to take a bowhead whale in the northwestern Pacific, in 1845. A third Danish whaleship, the *Copenhagen*, also hunted whales near the Kamchatka coast.[115]

Hobart Town, the capital of Van Dieman's Land, sent several whaleships in the 1840s, the most "visible" being the 300-ton *Eamont* — sometimes recorded as *Earmont* or *Ermont* — which was seen whaling in 50°58' North, 150°15' West on 12 July 1846.[116] The *Maria Orr*, a Hobart-built whaleship of 289 tons, is known to have reached Kodiak Island before being shipwrecked in Recherche Bay, Van Dieman's Land in 1846.[117]

Maritime Canada also expressed an interest in whaling, particularly New Brunswick, whose merchants perpetuated a traditional relationship with New England which had begun in the eighteenth century. Then, a small number of Nantucketers had elected to establish the whaling industry in New Brunswick and Nova Scotia rather than removing to London or Dunkirk.[118] Between 1833 and 1850, however, local merchants established their own companies, and Saint John hulls soon scoured every ocean of the world. The majority were owned and operated by the Saint John Mechanics' Whale-Fishing Company, formed in 1836, or by Charles Stewart.

Like crews from other nations, Saint John whalemen continued to rely upon American expertise and equipment. In addition, some of their vessels were insured in part by American underwriters. The 418-ton *Java*, for example, was protected by a $36,000 policy split into four unequal parts among Boston and New Bedford insurance companies.[119] The interlocking nature of New Brunswick-New England whaling ventures was soured somewhat by the imposition in England of a duty on whale oil entering from the United States. New Brunswick vessels could import freely and this disparity rankled Americans who watched their whaling men and equipment put to sea in these foreign-registered ships.[120]

The Saint John Mechanics' Whale-Fishing Company remained active only until 1846, despite the promise of company directors that the ships could fill with oil on the Northwest Coast. But the dissolution of the New Brunswick companies did not deter Canadians from entering the business; they continued to arrive in southeastern Massachusetts aboard coastwise vessels or overland and were welcomed in the forecastles of outward-bound American whaleships, particularly in the 1840s when the expansion of the fishery had robbed New England of every man with whaling skills.

> among [the crew] was . . . a Canadian Frenchman from Montreal, a young man of perhaps five-and-twenty, and who, when he first came among us, spoke not a word of English, but in which, before six months had passed, he was tolerably proficient. . . . He had come from Canada with his brother, both designing to seek employment in the United States, but missing the route of their proposed destination, they found themselves at one of our seaport towns, where some prowling land-shark, by holding out brilliant inducements, set at naught the resolution of the unsophisticated Gauls, and in perhaps an evil hour, they were persuaded to ship.[121]

Nor were these "unsophisticated Gauls" the only gullibles persuaded to go whaling by the roving "land-sharks" and boarding-house runners. The expansion of the whaling grounds left shipowners in need of crew, and the romantic young men who signed on were convinced of becoming rich while enjoying the salutary effects of the trade winds and the scenic views of tropical paradises; not to mention the unspeakable, nay, unthinkable kindnesses of the island maidens who were as yet unaccustomed to the corset and frock. But those naive young men — Canadian and American alike — whose fortunes took them to the Northwest Coast of America found the service very different. The realities of whaling in that cold clime quickly smothered their initial expectations. "If any man wants to be cured of his romance," one returning whaleman wrote, "let him go to sea. If he is laboring under severe and frequent attacks, let him go in a whale ship, and if the disease is deeply seated and given up by all authorized physicians as incurable, then let him go to the North West Coast."[122]

The change of weather upon leaving the Sandwich Islands offered the first clue that northern whaling would be very much different from sperm-whaling "on the Line." "After we left Mowee," Sir George Simpson wrote of his trip to Sitka, "every hour added to the severity of the cold . . . we had certainly no great satisfaction in seeing how gradually the thermometer fell, rarely indeed did it stand for two days at one point, but continued to fall lower & lower."[123] Perhaps the whalemen knew — certainly after the 1841 season they must have all known — that the evanescent days of sunlight and calm would be inextricably mixed with rain and fog and dark shadows to endanger the lives of men fast to running whales. Accounts of the Northwest Coast cruises are saturated with weather, often "rugged" with rain, sleet, fog, and snow, punctuated by living gales from the Arctic North and every other point on the compass. "Thick," they wrote. "Thick and rainy." "Thick, wet & disagreeable." All danger and claustrophobia, the men sealed below decks until called to stand a clammy watch in a condensed, ethereal world, neither lookouts nor officers knowing when they might find the shore, another ship, or an uncountable multitude of right whales disporting in the fog, taunting the suddenly impotent boatsteerers.

"All these 24 hours fresh Winds from the Northward and thick foggy weather," wrote the logkeeper aboard the ship *Robin Hood* of Mystic, Connecticut. "Steering SE by Compass; Saw nothing this day in Shape of Whales, So Ends."[124] A taciturn Sag Harbor whaleman wrote: "Some foggy." Two degrees of latitude to the north, the fog cleared off. "Heading along WNW by the wind," he wrote then, "fog some thin."[125] Coming with the fog, and sometimes apart from it, were the "strong breezes," gales, and "near hurricanes" which drove the seas to frothy crests and necessarily inhibited the whaling. The phrase "Too rugged to lower" pervades the journals; too rugged to chase whales, too rugged to start the tryworks, too rugged to make sail. And if a whale was somehow taken in such dismal conditions, there was afterward the obligation to cut-in and try-out before the carcass spoiled. In the deep Pacific swell, the rolling of the whale against the fluke chain and the hull often caused its body to part company with the ship. Then the effort and danger invested in its capture was lost as the carcass floated away into the mist to the scavengers that awaited its coming.

Often a hunt which began in poor weather ended with the whale swimming briskly to windward, in the direction of travel forbidden to square-rigged ships. The boat's crew would find themselves on a "Nantucket sleighride" toward oblivion, well out of sight of either land or ship. Then they would have to cut the line and let the whale go, and with that cut went all their work and all their gear, too. An enthusiastic crew taken too far to windward could find themselves wholly separated from their ship and compelled to spend a cold, dangerous night at sea.

Thus was Captain Cushman of the ship *Morea* rescued by the New London whaleship *Dromo* after losing sight of his own command while whaling in fog. He and his boat's crew were forced to remain two full weeks on board before they again met their own ship. Similarly, a boat's crew from the ship *Gratitude*, towed into the fog by a whale, returned to the obscured silhouette of their ship only to find the *Phocion* whaling where the *Gratitude* had been.[126]

Rising winds, fog, or the coming of night forced more than one boatsteerer to cut the line when pulled too far to windward on the Northwest Coast. In some cases whales near to exhaustion had to be abandoned, so that the boats could return to the whaleship in safety. Oil on canvas, by J. S. Ryder.
Kendall Whaling Museum, Sharon, Massachusetts

Whaleboat crews separated from their ship were sometimes assisted by the firing of a gun, as during the cruise of the ship *Edward* on the Northwest Coast in 1848:

> the whale spouted thick blood   it shut in thick fog at 6 P.M. lost sight of the Ship and being 10 miles to the windward was obliged to cut and look for the Ship with the reports of Muskits discharged on board   at 7:30 P.M. we found her.[127]

The extreme danger of being pulled too far to windward, with little chance of an immediate rescue, gave rise to a stanza of sailor poetry which no doubt provided instruction as well as melancholy entertainment for those young hands about to sail for the Northwest:

> In these Dreary seas midst fogs & frost
> Poore Whalemen oft in the boats are lost
> Unsheltered from the biteing blast
> With frozen limbs they breath their last.[128]

Nor was the ship itself immune from danger. In heavy fog the possibility of collision with another whaleship increased yearly as the Northwest Coast ground became more and more frequented. In such fog, life closed up, sounds became muffled, enlarged and then diminished, their direction suddenly uncertain. The lookout at the bow would strain his senses to hear the eerie blow and whistling explosion of whales spouting. Then, they were less the mortal animals of the hunt than ghostly, invisible monsters. Superstition haunted the forecastle then. Where are the whales? And where the land, and its jagged, dark promontories?

As the whales became fewer in number and more afraid of the men in the boats, captains were forced to remain longer in the north; first into August, and then into September, in hopes of retrieving just one whale more. Those who remained too long received curt eviction notices, delivered without warning or grace. In 1843, James A. Norton in the ship *Callao* stretched his good fortune until 13 September, but that day the usual gale grew into a near hurricane. Strong winds continued throughout the night and by morning the seas ran green over the lee rail. The crew hastily secured the decks; the barometer had fallen to 28 6/10 inches of mercury, with the swell running very high. That was enough. Captain Norton set his course by the wind for the Sandwich Islands.[129]

All too infrequently the weather turned favourable. Then the sun shone upon drying decks and dry clothing, and the sea calmed to a gentle roll. At such a time as this one whaleman was moved to write: "This has been the most pleasant day that we have had this season. The sun has been in sight 12 hours which is a remarkable occurance in this region." Then he added sourly, "Friday night and all day yesterday it rained like Jehu."[130] Sometimes, the night gave of its beauty: "At Midnight clear Starlight the only Night since we came on to the NW that we have seen the Stars or so fine a Night   saw the Auroraborales verey Bright between twelve and one this morning."[131]

The clearing skies of the Pacific Northwest sometimes exposed a view of land. In 1847 the crew of the ship *Citizen* of Sag Harbor killed several whales in sight of Vancouver Island bearing away northeast. Nor was the *Citizen* a solitary whaleship on this serene blue sea; the following day it was joined by the *Brooklyn* and the *Vesper*, both of New London.[132] Other logbooks and journals likewise comment on the proximity of the land. The rising mountains and evergreen forest provided a scent and a view long to be remembered.

> saw Queen Charlbotte's Island bearing E dis 40 miles saw humbacks no right whales.
> [ship *Armata*, 24 August 1847]

> At 3 p.m. raised The land. At 5 tacked Off leaving Queen Sharlotts Island Baring E 30 miles headding SSW several humpbacks in sight.
> [ship *Golconda*, 15 August 1852]

> made Charlottes Island bearing NE stood in 10 Miles saw nothing.
> [ship *Splendid*, 1 August 1855]

> night time took in Sail luft too [luffed to] at sun rise made all [sail] steered off ENE saw the Iland of Green Charlottes I.S. and 2 sails.
> [ship *Java*, 4 August 1855][133]

So they came. To the "Queen Charlbottes" and the "Green Charlottes," and even further, to Mount Edgecombe, St. Elias, Kodiak Island, the Trinity Islands, and finally to Shumagin Island and the Fox—Aleutian— Islands. Just before mid-century they pushed through the narrow Aleutian passes into Bristol Bay and eventually into the narrow jaws of Bering Strait and beyond.

At night, if not too close to the land, captains would let their ships drift, setting a lookout but leaving the other men to sleep. At first light everyone was rousted out to prepare for the chase. The hunt was the same as in every other sea. Masthead sings out, the captain decides to lower, boats hit the water, the men give a long pull, the boatsteerer stands up, darts his harpoon, and then it is "Stern All!" to escape the anticipated slap of flukes and flippers. Then the line singing through the chock in the bow, the boatsteerer

Whaleman John Martin illustrated his personal journal with watercolour sketches depicting the perils of right-whaling on the Northwest Coast. Martin went to the North Pacific aboard the ship *Lucy Ann* of Wilmington, Delaware, during the voyage of 1841–44.
Kendall Whaling Museum, Sharon, Massachusetts

and his officer exchanging places, perhaps a "Nantucket sleighride," and eventually, the lancing, the whale's "chimney on fire," the death flurry, and the contest is over.

Otherwise, a quick dive, a flipper crushing a man's neck or swamping a boat, an imperfect fastening by a single iron barb—there were many ways to lose the whale. In truth, the harpoon was but a fishhook, the boat and men its rod and reel, and the "fish" would have to be coaxed, goaded, played, flirted with, and exhausted before any bit of the killing could begin.

The common ways to lose a whale were these: by the harpoon breaking, or by a line being accidentally cut with a second harpoon; by a harpoon "drawing" from soft blubber; by a deep dive, necessitating a quick cut to prevent the swamping of the whaleboat; by a fluke or flipper knocking a boat to pieces or capsizing it. The success rate among whaleboat crews on the Northwest Coast during the 1840s was dismal. Even the most experienced men brought back fewer than 50 per cent of the whales they struck with their harpoons; in some ships barely 20 per cent of the whales struck were killed; those crews losing four of five whales. At first, when the whales were ignorant of whaleboats, a great many were hastily harpooned that afterwards escaped to die. In later years, as they came to distrust the boats, when they were less likely to lie inert like Melville's "Hartz Mountain logs" upon the sea, the harpooner had fewer opportunities to properly position himself for the thrust, and his misplaced thrusts contributed to the wounding of whales.

Killing a whale was no guarantee that it could be converted into oil. Even the normally buoyant right whales would occasionally sink, to the dismay of the sailors who had risked so much in the capture. One sunken Northwest right whale prompted a whaleman to write: "After turning up fin out just to pay us for 4 hours hard pulling besides the whet jackets down he went like so much lead taking with him 4 harpoons and a lance. This is what I call taking oil and stowing it down in a hurry."[134]

The New London whaleship *Superior* struck no fewer than fifty-eight right whales on the Northwest Coast during the Garden-of-Eden summer of 1841, but recovered only twenty-six; five additional whales were killed but sank. The ship *Magnet* of Warren, Rhode Island in 1844 struck twenty-six but saved only thirteen; two others sank. In 1848, by which time the whales were extremely wary, the ship *Julian* managed to strike only a dozen from the first week of May until 15 August. Of those, only three were saved.[135] These were an average of the lot. The armament simply, disshearteningly, failed to meet the demands placed upon it, and much was wasted:

> At 2 P.M. raised a right whale   lowered and struck by the Starbourd boat   took the line   got fast again and fought him until 7 P.M.   cut and came on board tired and wet. this is what we called a N.W. horse.
> [ship *South Carolina*, 20 May 1843]

> Noct Boat Steerer over board and hurt him badly in Consiquence had to cut the line.
> [ship *Cabinet*, 22 June 1844]

> The whale dodged around a spell and then set off to windward with two boats tailed on to the fast and ran like a steam Boat several miles when the line parted and lost him and I suppose he is going yet.
> [ship *William Hamilton*, 3 August 1846][136]

And every vessel the same, some worse than others. In the summer of 1852, the crew of the ship *Golconda* remained on the Northwest Coast until 2 September, lowering seventy times for whales and realizing a return of only eight animals. Forty-nine attempts ended in total failure, the oarsmen not able to bring the boatsteerer close enough to the gallied animals. During twenty-one other attempts only twelve whales were killed, and four of these sank irretrievably. The men lost fifteen more to causes beyond their control: three by irons drawing, four by irons breaking, another four escaping when the twisted strands of the whaling line parted. They intentionally cut from two whales, and accidently separated themselves from a third when a second harpoon severed the fast line. In another case a loose whale stove a boat and forced a cancellation of the hunt.[137]

This poor showing was not in the least unusual. In 1845 the crew of the ship *Hibernia* struck forty-six whales and lost all but thirteen owing to "bad irons."[138] This particular happenstance reached the pages of the *Whalemen's Shipping List and Merchants' Transcript*, and subsequent commercial pressure apparently forced something of a retraction. In a following edition, the editor noted that "recent letters received by [the *Hibernia*'s] owners impete the loss not to any defect in the irons, but to the roughness of the weather which prevented them from being firmly fixed." The harpoons, the editor concluded, were "from a manufactory deservedly of high repute."[139]

The double-barbed iron harpoon seemed incapable of halting the mighty ocean-dragons. In the closing years of the 1840s a black New Bedford blacksmith named Lewis Temple designed an improved harpoon, one with a single barb that toggled sideways in the blubber, perpendicular to the shaft. The open T-shape held more firmly in the whale than the two solid barbs. This design quickly supplanted the old, but even Temple's improvement fell short of perfection, and came too late for use during the halcyon years on the Northwest Coast.

No one knows how many of these injured Pacific right whales died of their wounds, but almost every logbook from the Northwest Coast reports the discovery of a carcass or two, some bearing in their flesh the identifiable harpoons from the fatal struggle. If the whale had been recently killed, then perhaps the blubber could be salvaged to the benefit of the finders. Otherwise, the body would be foul and "blasted," inflated beyond its normal girth by the mordant gases of decomposition, and would have to be left behind at a total loss to everyone.

"This day at 11 AM raised a Dead Whale," one logbook records. "[Run] for him but he was to far gone for our use   cut out an Iron marked S.S.A. N° 19 it was a New Bedford harpoon and we supposed it to belong to the ship South America or Sally Ann of New Bedford."[140] Occasionally a ship would reclaim its own harpoons from a whale previously struck and lost, but such recoveries were rare. The spoils usually went to the sea birds or scavenger fish or else to some other whaleship. To its salvagers went all the oil and bone, though the men were honour-bound to return the "trade"—harpoons, lances, and lines—to the rightful owners, if encountered.

From a modern perspective the tragedy of such failure lies in the high numbers of wounded animals left to swim the sea with soft iron in their flesh. The twelve whales so "tagged" by the *Golconda*'s crew in 1852, and the thirty-three marked by the men of the *Hibernia* in 1845 may be multiplied by the fifty or a hundred whaleships annually visiting the Northwest Coast throughout the 1840s, and by a lesser but important number sailing there until early in the twentieth century. Then one begins to consider the possibility of very great carnage among the whale tribe.

"The havoc they make of whales is intense," one writer noted at midcentury. "I have heard of one ship that sunk twenty-six whales after she had killed them; of another that killed nine before she saved one; of another that killed six in one day, and all of them sunk."[141]

> The idea sometimes advanced by captains of ships, who ought to know better, that "there are now as many whales as ever there were, that they had only been driven from particular grounds, Ec." is preposterous in the extreme. A simple calculation will shew the utter fallacy of such an assertion. There were upon the North West and Kamschatka, last season [1844], 300 ships. Each of these struck, and captured, or so badly wounded that they afterwards died, we will say 40 whales, and I knew one to have struck 75 whales in taking a thousand barrels. . . . Now this gives 12,000 whales killed in about four months. Many of these—the larger half—were cows with calf, and hence we may safely assert, the number of whales necessary to be born and arrive to maturity, in order to make up for this sweeping destruction among them, must be not less than 18,000. Suppose the increase is adequate to this, and from the time this "mighty monarch of the main" was first established in his domain, up to the first commencement of the fishery (less than 200 years ago) they had gone on multiplying their numbers in the same ratio, making allowance for the gradual increase of the whole number, verily the oceans would have swarmed with them, and the "highway of nations" would have been blocked up with their huge bodies.[142]

If the toll on whales was higher than many cared to believe, the toll on men working before the mast was in some ways completely unbelievable. The Northwest Coast whaling grounds toughened men beyond sensitivity, sometimes to the point of uncontrollability. Whaling masters and mates were more often feared for their highhanded punishment and their unreasonable use of authority and power than praised for their compassion. Complaints from the foremast hands were customarily met with harsh discipline and almost always with complete indifference, whether such complaint grew out of boredom, fear, or some precisely accurate objection. There was very little occasion for joy, unless the ship was full and bound home. The captain was frequently considered a monster and his chief mate a bully or worse. The smallest disagreement, a point of order over division of authority between officers, a perceived breach of protocol—in the cramped circumstances of a fogbound whaleship on the Northwest Coast these became issues of contention, disturbance, uproar, everything upwelling and none satisfied.

as I cut the Iron out The End of the pole flew arand & hit him over his eye & Barked the Skin & hee swoar out with menney oaths that i had knocked is Eye out & i told him to git out the [cutting] Stage if hee was hurt & goo below & hee swore hee whoulde & stay thair. I then toled him to do so & that i had no further order for him to put in faurse [force] as an officer on Board of the Ship under my command for I do not Consider him a fit man to have Enney atharety over men or to hold Enney offic on Ship Board as i have put up with so much of his disrespect that i will not have Enney Thing to do with him as a mate.[143]

The business of whaling was in lowering the frail whaleboats, rowing, darting, paying out line, lancing, pulling a long oar back to the ship, then thankless hours of stripping off blubber and feeding it into the smoking heat of the tryworks from a pitching, slippery deck, with a collection of razor-sharp implements lying all around. Little wonder mortal men felt disagreeable and ill-used. And such disagreements sometimes led to mutiny.

Mutineers and deserters usually carried out their plans in the southern seas, at some port or roadstead of the South Pacific where they anticipated a pleasant expatriation. But mutinies began to occur on vessels bound north, and they can not all be placed at the doorstep of the severe climate on the right-whaling grounds. Before anything was known of Northwest Coast whaling, there had been threats. Captain Stavers may have had some trouble with his crew during the 1834 reconnaissance; certainly in 1840 both the American whaleship *Phoenix* and the French *Cachalot* had their run-in with potential mutiny. The *Phoenix*'s sailors "locked the ship in irons" for a half-hour one day on the Northwest Coast, refusing to do duty, and the situation aboard the French ship was even more severe. Its crew repeatedly threatened mutiny, charging Captain Frederic Mauger with drunkenness and maltreatment.[144]

Crew dissatisfactions seemed to increase each year, but never so severely as in 1846. That was an uncertain year, for the Northwest Coast ground had begun to show sure signs of overfishing. Seamen in the Pacific anticipated a war between England and the United States over the Oregon Territory, and war began between the States and Mexico over Texas and the Southwest. In 1846 also the weather proved particularly unpredictable, and this added to the whalemen's woes.

In several cases, mutiny resulted from unusual circumstances. The death of the captain of the bark *Fame* of New London on the Northwest Coast occasioned an early return to port in that discordant year of 1846; on arrival at the Sandwich Islands, about half the crew refused to go back north.[145] On 14 July 1846 the bark *American* of Sag Harbor likewise returned to Honolulu Harbour, only three months and three days out. The horrified crew told how, while cruising in 52°30' North, 155° West, Captain William Pierson, his boatsteerer, and two seamen had been drowned "by a whale running over the boat." After that, the crew refused to lower for whales. The mate returned the bark to Lahaina, and there eight men were put in irons by the American consul for refusing duty. New men signed to take their places and the *American* sailed again.[146]

A bona fide mutiny disrupted the northward passage of the *Meteor* of Mystic that same year. While chasing "blackfish"—pilot whales—on 22 May, an altercation took place between the second mate and one of his boat's crew, a man shipped at Lahaina. Captain Francis Lester signalled the boats to return to the ship, and it was clear that there was trouble in the second officer's boat. The Captain allegedly told his first mate, "Prepare, we may have difficulty." They both went below and loaded firearms. When the second mate came aboard, he, too, went below, and some of the crew took that opportunity to secure the companionway and the hatches, trapping all three officers in the cabin. The foremast hands claimed that they were "in danger of their lives" and refused to let the captain and his mates on deck unless they surrendered their guns.

This they would not do. The mutineers placed the cooper in charge of the vessel and set sail for California, where they hoped to have their situation adjudicated by the American consul. Unfortunately, none knew how to navigate, so for three days the ship sailed more or less aimlessly to the southward. The officers probably expected to be murdered if they surrendered their pistols, and the men on deck likewise feared reprisals. They did send food to their prisoners, passing it over the taffrail and through the cabin windows.

At the end of the third day the mutineers spoke the whaleship *Midas* of New Bedford, which had been severely damaged in a gale en route north. Captain Jacob Davis came aboard the troubled *Meteor* and made arrangements to remove Captain Lester and the beleaguered second officer. The mutineers agreed to allow their own first mate to take command of the *Meteor* and return it to the Sandwich Islands. Lester and his two mates then signed a pledge, promising to discharge any men who wished it and allow them to take their gear off at Honolulu without hindrance. After signing, Captain Lester and the second mate climbed through the cabin windows and dropped into the *Midas*'s whaleboat, which was brought under the counter for the purpose. The firearms were discharged, and the first mate was allowed to return to the weather deck and assume command. Both ships returned to Honolulu.[147]

North Pacific gales that year damaged several whaleships. Perhaps worst abused was the *Luminary*, which shipped a tremendous sea while hove-to in

33° North. The rush of water tore the capstan from the deck, destroyed every whaleboat and boat-davit save one, ripped the wheel from the steering gear and brought down the main-spencer sail and all its rigging complete. Six men went overboard, including the cooper and a boatsteerer. The second officer died later that day of his injuries. A half-dozen men, hauling on tackles, were required to right the helm and get the vessel once again before the wind, and even so they could steer but poorly without the wheel.[148]

Two whaleships were lost in the north that summer. The *Konohasset* of Sag Harbor sank off Kamchatka, and the *Baltic* of Fairhaven went on Bering's Island. The *Peruvian* grounded on Atka Island in the Aleutians but got off again. These two losses, storm damage to the *Midas* and another whaleship, the *Albion*, and the mutinies aboard the *Fame* and the *Meteor* were all reported in a single issue of *The Friend*, a maritime newspaper offering a strange admixture of religion, temperance, and shipping news. What a sense of evil foreboding must have grown up among the readership in New England.

A captain's illness or death might occasion a return to port, but for the rest it was business as usual, struggling along with the dubious balms to be found in the small shipboard medicine chest. Some men sickened on the whaling grounds, from scurvy, smallpox, typhus, consumption, venereal disease, and infection from injuries. Even when they died, the whaling continued.

> at 4 PM the Capt calld the Officers and Boatsteerers in to the Cabin to lay the Mate out   The Ships Company then went into the Cabin to take a Last looke of thair deceased Ship mate   he was then sewed up in [his] Blankets   at 5 P.M. whales close to the Ship   Lower the Lar and Waste boats after them but returned without being able to Strike   at 7 P.M. Brought the Body of the Mate on deck and Commited it to the Deepe after Capt Fish red the funeral Servis.[149]

The medicine chest contained generic remedies administered from a book of instructions according to the symptoms proclaimed by the patient. These were transitional medicines, somewhere between folk cures and pharmacological compounds, dispensed at a time when purgation was the best response to illness. There were cathartics and emetics, epispastics, diaphoretics and escharotics, astringents, and, of course, narcotics. The old medicine chest included calomel to move the bowels, laudanum and paregoric to slow them, Peruvian bark and antimony for fevers, ipecac to induce vomiting, and mercurial ointments to heal sores. Every small injury betokened the possibility of infection, and though sailors in general were healthy, still there was potential for decaying death. One whaleman wrote:

I was oblidged to heave up this Evening on account of my hands, whale oil is poison to my flesh and my hands is one running sore. . . . I am a mere skeleton of what I was 4 months ago and God knows I shall be glad when this cruise is ended.[150]

The names of those lost were recorded faithfully in *The Friend* or the *Whalemen's Shipping List*; dead from illness, dead after falling from aloft, dead by drowning, dead by whale. Their ages were thirteen to fifty, most of them just eighteen, twenty, twenty-two. Some were from Sag Harbor and New Bedford and Stonington; some from farms in Pennsylvania and New York. Others came from the Sandwich Islands, the Cape Verde Islands, the Azores, and other places where whaling masters called to fill their complement. Some were black men and some Pacific islanders, while others were the sons of Yankee whaling fortunes.

Thomas Joseph, "a Portuguese belonging to Fayal," was drowned en route from the Northwest Coast. Likewise died "William Maui," a Sandwich Island native. Died a hero, "on the N. West, Aug. 14, 1846, Antonio Sylva, of ship Saratoga, Capt. John Smith, New Bedford. The boat had been badly stoven, and most of the men thrown overboard, but Antonio kept his place, and the whale then ran directly over the broken boat, after which poor Antonio was seen no more."[151]

John T. Perkins, aged 23 years, of Norwich, Connecticut, belonging to the ship Tiger, Capt. Brewster, while off in the boat after whales, June 15, 1846, in lat 55 N long 145 W, was struck by the flukes of a whale on the back and neck, and never spoke afterwards. The boat was badly stove, so that the body sunk before assistance came.[152]

"Henry Weber, (a Prusian by birth,) was killed by a whale on the N. West Coast, on the 1st of July last." And: "At sea, killed by a whale on N W Coast, Aug 5th, '47, John G. W. Wate, aged 26; boatsteerer of ship Nimrod, and formerly of Ledyard, Conn."[153]

THOMAS HILL, seaman, on board the Bremen whaleship Sophie, was killed by a whale, July 4th, in N lat 55 and W long 154. This unfortunate young man is known to have been of highly respectable connexions in Canada, North America. At the time he left home his father, Col. Hill, resided at Rosebank Cottage, New Market, and a brother, Rev. George Hill, at Tecumseh, Canada, North America. He came to the Sandwich Islands on board the whaleship Ann Mary Ann, having sailed in her from Sag Harbor, Long Island. . . . This notice may meet the observation of his friends, if editors in Canada will take the trouble to copy the same.[154]

Nor was Thomas Hill the only Canadian lost in the Northwest Coast whale fishery, for there is, among the cenotaphs standing sentry in the famous Seamen's Bethel in New Bedford, a dedication to two seamen—one a Canadian—lost from a Northwest Coast voyage of the whaleship *Magnolia*:

TO THE MEMORY OF
JOHN W. SAMSON
Son of Capt. John D. and
Rachel H. Samson of New-Bedford,
who died of consumption
at sea April 8, 1848,
aged 18 years.
AND
CHARLES E. FITZGERALD
of Kingston, upper Canada,
who was drowned on the North West
July 8, 1847, aged 28 years
ERECTED
By the Capt. Officers, & Crew
of the Ship Magnolia, as
a tribute of respect to their
deceased Brothers.[155]

The *Magnolia* was hardly a stranger to death on the Northwest Coast, having given up a man there six years before, in July 1841. That had been a fine time for Captain Barnard and his crew. Several whales had come their way, and the weather continued to hold, even at 57° North, 139° West. On 17 July all hands were busily engaged in boiling the previous catch; the North American shoreline was seventy miles distant and the sea calm. Perhaps it was one of those Northwest days described five years afterward by a captain's wife: "Have never seen so fine a day since we have been here . . . the sea very smooth and calm   So calm were the waves that its appearance was more like glass   The sun shining in all her glory and beauty and as far as the eye could reach not a [colour] to be seen save of a beautiful blue."[156]

It was certainly a rare and welcome day in the fogbound north, and the *Magnolia*'s men busied themselves rendering oil and stowing it below. But the respite was brief, for their luck began to change; perhaps they overextended themselves, confident in their success and comfortable in dry clothing. The next day they lowered and made fast to a right whale, but had to cut from it after it stove one of the boats. Two days later they lowered again and a boat's crew went on and fastened, but the whale flashed its flukes and once more a boat was stove. Along with the whale went the harpoon and the long length of line from the tub, probably looped and kinked in the bottom of the foundering boat. In the confusion of men scrambling for safety, escaping overboard, trying to swim — in that moment Mr. Luce, the third mate, became entangled in the line and was whisked away on the instant: "we saw him no more," Captain Barnard wrote, "which was a sad event to ourselves   he has left an amiable wife to mourn his loss nor will he soon be forgotten in our circle here."[157]

Next day the rain returned. All certainty had been drained, all the joy gone from the whaling. Perhaps the men had built dreams in the fine weather. Then they set out after another whale, to bring them closer to a full ship and homeward-bound, and their card house tumbled in a rushing line. Lost a man, lost a boat, the weather all turned to cloud and rain.

> SAD DISASTER. — We have just received an interesting communication from a Mr. Jameson, in regard to the death of two young men belonging to the Superior of New London. The facts are these: — a boat-steerer, George, (colored) of New London, and Uriah Coffin, of Martha's Vineyard, were drowned during the last season on the N. W. under these circumstances. George is supposed to have been taken out of the boat by the line, and to have drawn Coffin with him. The boat to which they belonged was obliged to cut from the whale. Soon after another boat fastened to the same whale and was drawn rapidly through the water, and the crew as they looked over the sides of the boat saw the bodies of their shipmates dragging under the water. One had the line several times around his legs, the other had grasped the line with the convulsive grasp of death. The whale was killed and sunk. The bodies of the unfortunate young men were not recovered.[158]

There the apotheosis of the common whaleman; the struggle, the loss, the death vindicated, and finally the corpses all, whale and men, sunk beyond the scrutiny of more interrogative minds. Ahab's Parsee, "Lashed round and round to the fish's back; pinioned in the turns upon turns in which, during the past night, the whale had reeled the involutions of the lines around him";[159] Herman Melville's masterpiece was yet five years away, while reports such as these, championed by a strictly pious press, percolated through the forecastles of greasy whaleships from the coast of New Zealand to Atka Island.

There were other portents. Aboard the *Golconda*, on the Northwest in

1852, the crew added to its usual complement of gale-rent sails and poor weather the various problems of mechanical and structural failure. On 26 April, "the Eigh[t] day Chronometer stoped and Refused duty," and on 12 May, with gales again hampering the ship's progress, the hull "sprun a leaking" and the men had to pump. It was duly noted in the log that "the ship has leaked Some ever seance we left home and Now grows worse." Nevertheless, the men carried on, "Cruising Over The ground"[160] long enough to make seventy-odd strikes on whales, turning south only when the foul weather of September forced the captain to wear 'round at last.

Of course, most wooden vessels leak, and whaleships commonly leaked more than that. One hundred strokes per hour on the pump-windlass, two hundred, one thousand. More than one whaling vessel, outward bound from Lahaina or Honolulu, put back owing to severe leakage. The bark *Levant* of Wareham, Massachusetts returned in June 1844 "leaking 2000 strokes per hour."[161] Toward the close of the 1844 summer, the *Magnet* was driven from the Northwest Coast by the same cause. "Our ship leakes about 200 Strokes per hour & Increasing daily," the master wrote, "so I have Concluded to Leave the Whales & go to the Sandwich Islands & Endeavour to Stop the Same   It cheats us out of A full cargo of oil But I hope the Lord will Preserve us to Reach home & Meet our dear friends In health."[162]

Faith took hold among the officers and crews of some whaleships departing for the Northwest Coast, though others were hellships, where neither a Bible was read nor a Sunday given over to any cause save the oily job of whaling. The most reverent of Christian missionaries who invaded the Pacific in search of converts counted themselves among a tiny minority conscripted into the holy militia of the Lord. Some of these zealots troubled to bring their dogmatic gear into the forecastles of the whaling ships, which by reputation were infested with debauchers and drunks, petty thieves and murderers and other criminals of the worst and most sordid order. Missionary efforts to convert these heathens to the glories of God became almost fanatical. Never mind that the whaleman's opportunity to live as a landsman came only during twice-yearly visits to island ports where there was an ample supply of grog and plenty of women; who save monks could resist the allure of either after six to eight months in a sea filled with human death and animal carnage? Yet *The Friend* and the *Sailor's Magazine*, both of them careful admixtures of sentimental piety and shipping news, harangued these sailors, telling them that they drank too much, debauched too much, and, what is more, instilled their evils upon the peaceable and childlike flesh of the Pacific islanders.

If pure religion failed to take hold, the concept of temperance did succeed to some degree in certain whaleships. Officers and men alike foreswore the use of alcohol as one of the sins most likely to interfere with a sailor's long

The baleen of a right whale provided an unusual backdrop for this deck portrait of American whalemen.
Pardon B. Gifford Collection, Kendall Whaling Museum, Sharon, Massachusetts

life. The concept was linked inextricably with pious Christian religion; temperance and Christ were practically one and the same, and the tract newspapers rejoiced at every conversion to the temperance pledge. Missionaries kept their tallies and published them as proof that the army of the Lord was daily increased by these volunteers from the whaling fleet. The name of Captain Charles Weatherby Gelett was burnished frequently in the annals of the *Sailor's Magazine* and elsewhere for his adherence to temperance principles. Formerly master of the *India*, which made an early voyage to the Northwest Coast in 1841, Gelett's 1844 cruise in the ship *Uncas* smacked more of an ecclesiastical adventure into the Upper Realms than a greasy voyage for blubber in the fogbound North Pacific. Witness testimony of the Reverend Samuel Chenery Damon, writing from Honolulu:

> During the whaling season of 1844, on the North West Coast, the ship's company of this vessel experienced a most remarkable refreshening of the Holy Spirit. At least one half of the crew professed to have passed from death to life. The fact became known among the whaling fleet. I felt extremely anxious to have the ship visit this harbor; and yet I feared the temptations of "Oahu" would prove too potent for the faith of the young converts. Imagine, however, my joy, when I tell you that the vessel came in and remained in the Inner harbor six weeks; and not a man, to my knowledge, broke his temperance pledge. You may ask, "How many of the ship's company had signed?" I answer, "All, to a man; from the captain to the cook." The piety of nearly all who had professed a change of heart, stood the trials of the port equally well.[163]

By the year of our Lord eighteen hundred and forty-seven, fears began to be entertained, among both the pious and the not-so, that the whaling men might have done too good a job on the Northwest Coast. The number of whaleships sailing there had more than doubled in just four years, but the average catch begun to fall. In 1843, 108 vessels reported 145,692 barrels of oil, an average of 1,349 barrels per ship. The average in 1844 reached 1,528 barrels. But in 1845, a huge fleet of 263 ships and barks returned only 953 barrels each, causing one astute whaleman to speculate that

> the poor whale is doomed to utter extermination, or at least, so near to it that too few will remain to tempt the cupidity of man. . . . In view of the destruction which has been made among them since the time when five and six months were sufficient to complete a cargo upon Brazil, or fifteen upon Peru, never looking to the Northward of 18 degrees South, for Sperm Whales, we may safely suppose another century will witness the entire abandonment of the fishery, from a scarcity of whales.[164]

In 1846, the busiest year in American whaling history, 292 whaleships hunted over the Northwest Coast of America and returned an average of only 896 barrels each. Voyages grew longer, prompting ever more letters of explanation from masters who could not complete their ships' cargoes: "I have made my second cruise on the Northwest Coast and have been unsuccessful," one captain wrote in the fall of 1847. "I only got 900 barrels this season. I have now got 2,450 of whale, 50 of sperm. I am going to stay another season."[165]

After 1847, many of the ships were withdrawn from the whaling service and chartered for freighting. A writer in the *New London News* observed that whale oil was worth only 38 cents per gallon—about half the price of sperm oil—and that ships now remained at sea just as long for this income as they formerly did for a cargo of sperm oil; he concluded therefore that unless "the price of whale oil should advance, or new cruising ground be discovered . . . the number of vessels in the whaling service must be diminished."[166]

Or a new cruising ground be discovered: in July 1848 a Sag Harbor whaling master, Thomas Welcome Roys, sailed the bark *Superior* farther than anyone else in pursuit of whales. Against the strenuous objections of his crew, Roys followed five-knot northerly currents that took the bark through the narrow Bering Strait. Almost immediately he came upon pods of "gentle" large whales which proved to be the economically profitable bowhead species, a relative of the right whale. Roys filled the *Superior* and by the end of January 1849 he was off Cape Horn, "full and bound home."[167]

Roys earned credit, though he never claimed it,[168] for discovering the most profitable baleen-whale fishery of all, and for catalyzing a rush to the western Arctic which resembled the race for gold that began the same year in California. In 1849 a flotilla of 154 whaleships followed his lead into the Bering Sea and harvested 206,850 gallons of oil and an unbelievable 2.5 million pounds of baleen, some of it as long as thirteen feet. The average take per ship was 1,334 barrels, equalling the early output from the Northwest Coast ground. The number of ships departing New England for the "Northwest Coast" collapsed to none in just one season. Starbuck's 1876 compilation showed 59 vessels departing for the Northwest in 1847, and a further 61 sailing there in 1848. In 1849 the number fell to just 20; the premier whaling port of New Bedford instantly abandoned the Northwest Coast as a destination. Every face looked beyond the Bering Strait to the newest whaling ground; suddenly the Gulf of Alaska had been eclipsed.

The salient point is that the men who financed whaling voyages realized there would be a few more years of profit, and the early arrivals would take the lion's share as they had done on Peru, on Japan, and on the Northwest Coast. The reports from the Bering Sea were very satisfying: whales were reputed to be fat and sluggish and full of oil. "We doubt if so much oil was ever taken in [one season], by the same number of ships and attended with so few casualties," one report boasted. "In fact, a cruise in the Arctic Ocean has got to be but a Summer pastime."[169]

It is a parable of the whale fishery, now well understood, that each new whaling area yields up the bulk of its harvest in the first few years of whaling. Increased use only depletes the natural resources and leads to a failure of the ground. This proved to be as true in the western Arctic as it had been on the Northwest Coast. The Arctic whalemen hunted farther and longer and, eventually, the bowheads were extirpated like their more southerly cousins.

After the first two hunting seasons in the Arctic, *The Friend* left off its sermonizing long enough to publish a lengthy letter received from the Anadir Sea. Under the date of "The second Year of Trouble," it is the only known communiqué from the newly discovered leviathan:

MR. EDITOR, — In behalf of my species, allow an inhabitant of this sea, to make an appeal through your columns to the friends of the whale in general. A few of the knowing old inhabitants of this sea have recently held a meeting to consult respecting our safety, and in some way or other, if possible, to avert the doom that seems to await all of the whale *Genus* throughout the world, including the Sperm, Right, and Polar whales. Although our situation, and that of our neighbors in the Arctic, is remote from our enemy's country, yet we have been knowing to the progress of affairs in the Japan and Ochotsk seas, the Atlantic and Indian oceans, and all the other "whaling grounds." We have imagined that we were safe in these cold regions; but no; within these last two years a furious attack has been made upon us, an attack more deadly and bloody, than any of our race ever experienced in any part of the world. I scorn to speak of the cruelty that has been practised by our blood-thirsty enemies, armed with harpoon and lance; no age or sex has been spared. Multitudes of our species (the Polar), have been murdered in "cold" blood. Our enemies have wondered at our mild and inoffensive conduct; we have heard them cry, "there she blows," and our hearts have quailed as we saw their glittering steel reflecting the sun beams, and realized that in a few moments our life-blood oozing out, would discolor the briny deep in which we have gambolled for scores of years. We have never been trained to contend with a race of warriors, who sail in large three-masted vessels, on the sterns of which we have read "New Bedford," "Sag Harbor," and "New London." ... We have heard of the desperate encounters between these whale-killing monsters and our brethren the Right whales on the North-west coast. Some from that quarter have taken shelter in the quiet bays of our sea, others of the

spermaciti species from Japan, have also visited us and reported their
battles and disasters; they have told us it is no use to contend with the
Nortons, the Tabers, the Coffins, the Coxs, the Smiths, the Halseys,
and the other families of whale-killers. We Polar whales are a quiet inoffensive race, desirous of life and peace, but, alas, we fear our doom is
sealed; we have heard the threat that in one season more we shall all be
"cut up," and "tried out." Is there no redress? I write in behalf of my
butchered and dying species. I appeal to the friends of the whole race of
whales. Must we all be murdered in cold blood? Must our race become
extinct? Will no friends and allies arise and revenge our wrongs? Will
our foes be allowed to prey upon us another year? We have heard of the
power of the "Press;" pray give these few lines a place in your columns,
and let them go forth to the world. I am known among our enemies as
the "Bow-head," but I belong to the Old Greenland family.

> Yours till death,
> POLAR WHALE[170]

And so an appeal to the whalemen, who undoubtedly received it doubled
over in laughter, guffawing; hoarse, hopeless laughter from the rolling forecastles of the greasy whaleships, the men trapped in there as surely as the
whales trapped below. One can only hope that "Polar Whale" sent his
friendly message through someone who really cared. The readership of *The
Friend* cared but little, and knew less, about the parable of whaling and its
diminishing numbers. They only knew that north of the Bering Strait were
whales, ready for the bravest of men to come and harpoon them. Polar Whale
and his right-whale kin on the Northwest Coast did perish in just the way he
feared, so that today the killing of a dozen bowheads by the Inuit peoples
raises a protest around the world. Since the American whalemen killed all the
whales, thirty dead bowheads, one dead right whale—each constitutes a
threat to the survival of its species.

Of the right whales, Scammon wrote: "The whalemen of the Northwestern Coast made such havoc among these colossal animals . . . as to
have nearly extirpated them, or driven them to some unknown feeding
ground."[171] But there were no more secret feeding grounds, only the deep expanse of that same North Pacific Ocean across which sailed the hundreds of
barks and ships from the several whaling countries. Everywhere commercial
exploitation, everywhere overhunting; after the 1850s the Northwest Coast
was depleted. Whaling masters reported plenty of finback and humpback
whales, themselves stolidly awaiting the development of technology to facilitate their capture, but no right whales. Those few that remained could not be
approached, so wary were they of the slapping of oars upon the water.

David Barnard, then master of the whaleship *Abigail*, wrote the *finis* as early as 1846. He did so unwittingly, of course; his last chapter is observable only in retrospect because whalemen continued to visit the Northwest Coast until the turn of the twentieth century. Barnard found that the summer of 1846 failed to equal his providential early visit in the *Magnolia* in 1841. By 26 July he had determined to leave the whaling ground. His decision was more than a month premature, since whaling masters had grown accustomed to remaining on Northwest, "cruising over the ground" until the September gales blew in the fog and rain for good. But Barnard was an experienced master. He had been there before, and he knew enough. On that day in late July 1846 he wrote: "During these 24 hours had fresh breezes at SE, attended with fog   head by the wind to the E   saw fin backs   I now bid adieu to the N.W. feeling satisfied as we have been 23 days and seen no whale."[172]

3

OMINOUS VISITATIONS

*There are two American whalers at the North end
of Vancouvers Island, they have been alone there two
months, and intend to winter there. . . . and what they
get now . . . will induce others to visit the place.*

JOHN MCLOUGHLIN
TO THE HUDSON'S BAY COMPANY (1843)[1]

Where there had once been only fur-trading vessels and the occasional military reconnaissance, there were now whaleships, and there yearly threatened
to be more. On shore, managers of the British and Russian government-
grant trade companies worried about the potential impact of that traffic on
their respective monopolies in the fur trade. What the East India Company
had feared when it sought to restrict British whaleships in the Pacific now
troubled the governors of the Hudson's Bay Company and, farther to the
north, the Russian American Company. Here was yet another, unexpected
branch of mercantile commerce, an unfamiliar group of seamen, probably ignorant of the protocols. They might have a disquieting effect on the natives,
who were accustomed to the bargaining arrangements they had been offered
by representatives of the companies.

Conflicting claims of sovereignty north of the Columbia River had been
obfuscated by long years of argument, negotiation, and discussion. In 1840
territorial prerogatives were not yet permanently adjudicated. From the
Columbia northward, the Oregon Territory was managed bilaterally by the
United States and Great Britain, in accordance with the terms of a
renewable—and revokable—treaty. This co-operation extended to the latitude of 54°40'; the shore beyond belonged to Russia by virtue of treaties
signed in 1824 and 1825 with the United States and Great Britain, respect-

ively. In both the Oregon Territory and in Russian America, any trade monopoly was a façade, sustained by a few scattered trading outposts and armed forts. Neither England nor Russia could afford the troops and machinery to hold the long and complex coastline against any actual threat, be it soldiers of some foreign power or a commercial invasion by the whalemen.

British interests in the territory were vested with the Hudson's Bay Company, which operated a number of outstations for the collection of furs and other valuable export commodities, such as timber. Several factors of the company expressed concern about the whalemen, though others considered them the solution to a problem, rather than the source of one. Whalemen would arrive on the Northwest Coast after perhaps two months at sea, already weather-worn and in need of supplies and perhaps some repairs: surely a lively market could be established in clothing, fresh food, and the well-known Hudson's Bay blankets? Some of the best shipbuilding timber in the world was readily accessible. So it seemed logical for the company to provision sailors traversing the lonely North Pacific, though it was all too clear that some plan would be needed to prevent the whalemen from using these very articles to compete in trade among the native peoples.

The opposition contended that foreign whalemen represented a direct intrusion upon the Hudson's Bay Company charter, which gave to Prince Rupert and his associates "the sole trade and commerce of all those seas, straits, bays, rivers, lakes, creeks, and sounds, in whatsoever latitude they shall be . . . with the fishing of all sorts of fish, whales, sturgeons, and all other royal fishes."[2] Although the company had pushed far beyond the geographical boundaries of the original grant within "Hudson's Straits," the terms of the charter had not been altered. It granted exclusive trade privileges, and even specified the company's right to the whale fishery itself.

The first intercourse between the Hudson's Bay Company and the whalemen took place in the Sandwich Islands. George Pelly, first cousin to the company's governor, arrived there in 1833 with instructions to establish a depot to store and sell lumber and salmon to be shipped from the principal outpost at Fort Vancouver on the Columbia River. He was also charged with developing export cargoes for the company's homeward-bound vessels.[3]

George Pelly has come down as a man of irascible character, impulsive and undoubtedly difficult, but also shrewd, effective, and capable. He continually upbraided the Home Office in London for failing to keep his business supplied, and he was chided in return because of his reluctance to surrender autonomy in his business dealings. His relationships with whalemen seem to have taken on something of his unpredictable, contentious nature. Pelly was most astute, however, in calculating the rise and advance of trade at the Sandwich Islands, and he continually beseeched London to send larger and broader shipments of goods for resale to both whalemen and naval officers. Pelly saw how whaleship masters might free themselves from carrying salt-beef and other staples for a full voyage if the company could guarantee the availability of those items and fresh fruit and vegetables, too, at their far-flung outposts. Salted meat, Pelly told London, would "meet a ready sale here to the shipping (chiefly whalers)."[4]

But he could never ensure that guarantee, complaining regularly to the Home Office that cargoes were too late in arriving. In the spring of 1844, when the company's ship *Brothers* met with damage at sea and was delayed, Pelly fretted because several American merchants had bested him in the busy spring reprovisioning of the whaling fleet. He did admit, however, that the naval stores "are always a commanding article and will go off in the Fall season."[5] Two years later, the spring offerings were once again delayed, and Pelly had to write: "The ready made clothing has been in considerable demand, the finer descriptions, the coarse arrived too late for the whaling fleet of this spring for the N.W. Coast."[6]

Whaling masters who patronized the Hudson's Bay Company often did an indirect business, purchasing from ship chandleries and outfitting houses whose owners bought their inventories at least in part from Pelly.[7] In some cases, however, the whaling masters conducted business directly; this seems to have been a particularly common arrangement among the French captains, including Letellier of the *Ajax* and Backeling of the *Duc d'Orléans*.[8] Whalemen customarily paid by a bill of exchange, a legal paper which entitled the outfitter to receive payment when presented to the shipowner's bank or finance house. Such bills were difficult to handle, since they almost always involved a conversion of currency from the buyer's dollars, marks, or francs into the English pound sterling.

These "whalers bills" would have to be sent to New York or Boston or to some other location specified by the shipowners for collection. The attendant paperwork was troublesome, and much interest on the money was lost through delay. At first, the Hudson's Bay Company had no bank in the United States, and Pelly continually petitioned the Home Office to open an account in New England in order to simplify the financial process. In one such letter, written in January 1845, he passed comment on the oncoming political crisis over the Oregon Territory, suggesting that, "as the American shipping both men of war and whaling is so greatly predominant at present in this Ocean, we think, it would be well for your Honors to recommend us to a respectable house in New York and another in Boston as it will be necessary to invest our funds considerably in American bills."[9]

Sir George Simpson followed developments in the North Pacific with avid interest. The former overseas governor, and co-partner with Pelly in the Pacific, Simpson left the formal employ of the Hudson's Bay Company in 1841 and hired out as a consultant to survey the company's Pacific operations. One

of his goals was to improve trade in the North Pacific.[10] He once again visited
San Francisco, the Sandwich Islands, and Sitka (New Archangel), and at each
stop he examined the company's factories. On his arrival at Honolulu in February 1842 Sir George carefully noted the growth of the whaling fleet; he saw
ten American whaleships at anchor at Lahaina and described that roadstead as
"a great resort."[11]

The enlargement of the whaling fleet captured Simpson's interest. He had
already envisioned a new factory north of 49° to replace Fort Vancouver, and
it was his immediate opinion that the whale fishery, though not previously
considered, might be carried on by the company.[12] In San Francisco, Simpson
learned from Cornelius Hoyer that the Northwest Coast was the "best fishing
ground in the North Pacific," and the news that a French whaleship had
made a good voyage there in 1841 confirmed Simpson's suspicions that a
profit might be made. He wrote the Home Office describing "sperm and
Black oil in small quantities" which he saw at the northern establishments
during the autumn of 1841. The Indians, he reported, "represented those
animals as being excessively numerous."[13]

"For many years," he wrote, "it has been known that the whales were very
numerous about the Straits of de Fuca, and in the Gulf of Georgia, and that
the Indians of Cape Flattery and the Straits of de Fuca were expert, even with
their bone lances, grass lines, and other rude implements of their own manufacture in killing them." Later on, he learned from a Russian sea captain that
whales were large and equally numerous in the Kamchatka Sea and near the
Aleutian Islands, and that the natives hunted them "by means of spears and
arrows shod with stone."[14]

Simpson went so far as to devise the type of fishery that might prove successful. "Vessels employed in the [whale] fishery," he explained, "might run
in and out from month to month, as circumstances might render desirable,
deliver their oil, receive refreshments and other supplies, and thus remain on
their stations."[15] He proposed a series of reprovisioning places which would
allow the Company's ships to follow the whales during their yearly migrations. Unfortunately, Sir George lacked a clear sense of the specialized circumstances of whaling. When he encouraged the company to close its northern forts and instead operate the new paddlewheel steamer *Beaver* as a mobile
trading post and police station, he blithely suggested that this already overworked vessel could be equipped for *ad hoc* duty as a whaleship.[16]

> it appeared to me that a vessel would be specially required for [whaling]
> if entered into, but as I learn there are a great many Right Whales in the
> Straits of De Fuca, in the Gulf of Georgia & the Straits & Canals in the
> neighbourhood of Ft. McLoughlin, Ft. Simpson & Tako, where sailing
> vessels ought never to be sent, every useful object would be gained by

having a few active experienced Harpooners and other Whale Fishers, to form part of the crew of the Steamer, [which] ought to be provided, with two fast Whale boats & the necessary lines & other implements, this branch of business to be combined with the transport & other Coasting duties, merely directing their attention to the fishery as they fall in with Whales in passing up & down the Coast.[17]

Simpson called for a "decked Lighter in tow," to accommodate the bulky products of the whale. The oil would be taken to the proposed new depot at the southern end of Vancouver Island and there prepared for shipment to England.

The new depot, Fort Victoria, was begun in 1843 largely on the strength of Simpson's recommendations. He told Charles Ross—soon to be chief trader at the new site—that if water could only be found in sufficient quantity, Fort Victoria would likely become "a place of much resort to strangers, especially so if American whalers continue to frequent the North Pacific." And, he added, "a profitable business may be made by the sale of provisions and supplies to these vessels."[18]

Simpson's observations underscore the dilemma. On the one hand, the growing number of whaleships suggested the likelihood of a remunerative trade. On the other, any encouragement to the whalemen might create a threat to the very tenuous monopoly enjoyed by the company, particularly in light of the growing pressure of American settlement in the Oregon Territory. In London, far away from the points of pressure, the commercial value of the right-whaling industry was apparent. The 1843 shipment of "whalebone" from the Fort Vancouver depot on the Columbia River commanded £15 per hundredweight.[19] So the administrators of the Hudson's Bay Company had to choose: ignore the whaleships, encourage them, or perhaps pursue the quarry themselves.

They had no sooner dug their shovels into the soil of Vancouver Island when it became apparent that the protective course might prove the most expeditious. A fearsome rumour went around that "300 sail" had been seen off the Northwest Coast,[20] and though there can be little doubt that the number was exaggerated, the northern whaling fleet of 1843 had much increased. Sir George was in a position to substantiate the rumour personally. The previous summer, on a passage from Sitka to Okhotsk, his vessel had encountered an American whaleship whose captain provided interesting and perhaps frightening statistics:

we perceived, in the course of the afternoon, a large ship looming through the fog within a few hundred yards of us. On our nearer approach we distinguished the stars and stripes, while her stock of boats told her business as plainly as her flag indicated her nation; and, on passing close under her stern, we read her name, "Parachute of New Bedford." On our firing a gun, both vessels backed their maintopsails for a parley. A boat was lowered by the American, and [Captain Wilcox] scrambled up among us. According to his account, the Parachute had been out nineteen months, and had got 2200 barrels of oil, 1500 of them the produce of thirteen right whales, taken last summer [1841] between lats. 49° and 56° and longs. 140° and 152° ... she had again, within these few days, reached her old ground, described by our informant as the best at present known, expecting to have about two hundred competitors this year instead of the fifty that she had last.[21]

Fifty now seems an exorbitant number of whaleships up north in 1841. Perhaps Wilcox included the Japan ground sperm-whalemen in the number, or else he simply exaggerated the figure to impress the Englishman. In any case, Simpson was startled to hear such numbers, surprised even to find the *Parachute* in those northern waters. He was also amazed that the Russians made no effort to evict this Yankee entrepreneur from their territory. And he clearly considered the effect such whalemen might have on the business of the Hudson's Bay Company.

Those who feared trade competition from the whalemen had only a short time to speculate upon its effects.[3] On 1 October 1843 Captain Duncan and the crew of the *Beaver* discovered two American whaleships lying in Shushartie Bay, near the village of Nahwitti.[4] This place at the north end of Vancouver Island had long been a refuge from the stormy waters of Queen Charlotte Sound, and it had become an established rendezvous in the business of the fur trade. In that safe haven Captain Duncan found the whaleship *Canada* of New Bedford and its supposed tender, the *Maine* of Fairhaven.

The *Canada* had sailed from Massachusetts on New Year's Day 1843, doing the bidding of a group of New Bedford and Dartmouth owners that included some of the most influential names in whaling.[24] It is not clear whether this vessel was specifically chosen for a Northwest Coast voyage, but it would have made a likely choice. It was forty feet longer than Barzillai Folger's *Ganges*, and at 545 tons, the largest whaleship by registry in the New Bedford fleet. Thirty-five men sailed in it that New Year's Day, the most to leave New Bedford in a single whaleship during 1842–43.[25] At least one unusual crewman was aboard: the *Canada* carried a physician, or at least someone familiar with basic medical practice, and this was singular, since American whaling crews rarely included a doctor.[26] The *Canada* sailed directly for the Northwest Coast; in August 1843 Captain William H. Topham sent a letter home from there to say he had already gathered 500 barrels of oil, intended to cruise two more months, and then "go into the Bays."[27]

The newly built Fort Victoria was shown to the readers of the *Illustrated London News* in August 1848. The Hudson's Bay Company steamer *Beaver* is anchored in the foreground.

The *Maine* was an older and more conventional whaling vessel, 294 tons, built at Kennebunk, Maine in 1819[28] and just ninety-five feet in length. Its participation in the bay-whaling plan might have been the result of circumstance. Shortly after clearing Lahaina, Captain James Makee — often recorded as Magee — was attacked by the steward, who struck him twice about the head with a hatchet. The steward then fled from the cabin, firing a pistol at the second officer and disappearing, presumably overboard. Makee was so badly hurt that the crew were obliged to put back to Honolulu, where medical attention was proffered by the surgeon of H.M.S. *Carysfort*, then in harbour. So severe were his wounds that he could not return to sea immediately,[29] so he promoted the mate, William M. Smith, to acting captain.[30]

Smith sailed from Honolulu on 29 April 1843 and proceeded to the Northwest Coast. At the end of August, the "season having expired," he anchored the *Maine* "in a Bay at Queen Charlotts Sound, for the purpose of taking Whale during the Winter Season."[31] Apparently such a plan was neither known to, nor sanctioned by, Captain Makee. When the *Maine* did not return to Honolulu as expected, Makee prepared a charge of barratry against Smith. Makee almost surely had news of his ship's whereabouts; even in New Bedford the readers of the *Whalemen's Shipping List* knew, because the paper had published a letter from Captain Topham of the *Canada*, reporting the two ships at "Minetta Bay, N.W. Coast."[32]

The announcement by Topham and Smith to the *Beaver*'s crew that they intended to winter at Nahwitti stirred strong emotions among the administrators of the Hudson's Bay Company. The news was quickly relayed to London by James Douglas, the administrator at Fort Victoria:

Two whale ships the "Canada" of 600 tons, and the "Maine" of 300 tons, both American vessels, were found by the Steamer lying at Newitté (say Shushady north end of Vancouver's Island) on the 1st Ins' (Octob'). They had been there upwards of a month, and intended to winter. . . . They have been cruizing on the Coast all summer and made only about 500 barrels of oil apiece. One of these vessels carried an assortment of trading goods, but fortunately so much inferior in quality to ours, that the Indians would not buy them. The American told Duncan that he had picked up about 60 skins, and judging from the "Beavers" trade, I presume his statement is not far from the truth.[33]

"This is an ominous visitation," Douglas wrote, "and I am not without a feeling of alarm about the effects it may have upon our business. If whales are found in the Sounds and Bays of the Coast, the circumstance, when known, will attract so many vessels to these shores as to endanger the security of our trade."[34]

John Kennedy echoed Douglas's concern. From his post at more northerly Fort Simpson, he wrote Chief Factor John McLoughlin in February 1844 to say that the whaleships' crews had exchanged trade goods for any kind of furs. "Captain Duncan informed us that [the whalemen] had procured above 200 Beaver & Land Otters, 20 Sea Otters and a considerable number of Bears, Martins and Minks," Kennedy wrote, "and had distributed among the Indians there, from September to January, above three tons of fine negro head Tobacco. . . . The poor steamer only obtained the gleanings of the Trade there in both her up and down trips."[35]

Kennedy also observed that a whaleship—he did not name it—had made an extensive reconnaissance of the inside waterways in Russian territory during the fall of 1843.

[The vessel] found her way up to within a few miles of Stikein, and though she did not trade any furs (for want of goods); yet she annoyed Mr. Manson very much: by the Captain of her; telling the Indians that she would be back again next season, with a supply of Goods, suitable for trade, on the strength of this, many of the Scamps of Indians are holding up their skins. If any reliance can be placed in Indian report, she paid most liberally, for provisions, for instance; 12 yards White Cotton for a deer.[36]

The *Canada* remained at the north end of Vancouver Island until March 1844 and then proceeded northward along the coast. Captain Topham had only taken two whales over the winter, the animals "having just begun to come into the Bay."[37] The *Maine* remained longer; on 8 March one of Smith's young seamen, Edward Rice from Jersey City, New Jersey, collapsed while pulling for whales. The crew carried his body atop a rise near the eastern shore of Shushartie Bay and laid him to rest close by the grave of John Thompson, a young man from St. Helena who had served as cabin boy in the *Canada* until his death early that January.[38] On 21 April Acting Captain Smith finally sailed from Queen Charlotte Sound "for the off shore ground,"[39] and made his way northward toward Russian America.

This excursion of American whaleships into the protected harbours of the Northwest Coast is documented in at least one surviving contemporary manuscript, a journal kept on board the ship *Milo* of New Bedford. The *Milo* spoke the *Canada* on 6 July 1844, at which time Captain Topham reported he had been "Bay whaling."[40] This same "bay whaling" on the Northwest Coast was described as a failure the following year in a newspaper account published in the *Polynesian* at Honolulu. The author, identified as "M. E. Bowles of the ship *Jane*," wrote:

> Attempts have been made to prosecute the fishery during the winter months, in the bays upon the coast, but none other than the *Scrag Whale* [gray whale] have ever visited these bays, and it is now generally supposed that the cow whales repair to the deep bight toward Behring's Straits, or inside the chain of islands forming the Southern boundary of the sea of Kamschatka.[41]

"There is not," he concluded, "nor is it likely there ever will be, any 'bay whaling' in this fishery."

Bowles probably served in the *Jane* of Warren, Rhode Island, since it was lying in Honolulu Harbor at the time his article appeared.[42] It is not clear whether the *Jane* also entered the inside passage; perhaps this was the whaleship that "annoyed" Manson at Stikein. The identity of Bowles's ship is significant to Pacific Northwest whaling history only because another whaleship named the *Jane*, a bark, sailed out of Sydney, Australia during this same period and because a book published in London in 1849 purports to describe Northwest Coast bay whaling aboard a 300-ton vessel named the *Jane*, otherwise unidentified, which the anonymous author joined in Sydney.[43]

This latter *Jane* allegedly sailed into the Strait of Juan de Fuca, called at Vancouver Island, and laid to in and out of the harbours for nearly six weeks. According to the author, the captain seemed interested above all other things in "turn[ing] a penny on his own account, which was not always in unison with the interests of the owners of the vessel, nor of the crew." The writer sketched the natives as a "fine race," even though "tattooed according to the savage rites of their respective tribes," but made much of their fondness for alcohol, "for which they will exchange their very souls, if possible." That they would trade for beads, pins, or "anything in the shape of metals in a manufactured state" did not mitigate, in his opinion, the fact that they were prone to steal from the ship.[44]

The veracity of this account is suspect. It may have been fabricated from whole cloth, and there is no available corroboration. But its author evidences a clear familiarity with Fort Victoria and carefully describes the purchase of dried salmon from fish-curing depots, which may have been run by the Hudson's Bay Company. A man named Ward has subsequently received credit for authorship, but the entire voyage and its account remains a mystery. The *Jane* of Warren, to which Bowles probably belonged, departed from Oahu on 16 October 1845 en route home, but never arrived. By New Year's Day 1847 both ship and crew had been written off as lost.[45]

Determining which whaleships went "bay whaling" is further complicated: the former Boston-St. Petersburg trader *Kutusoff* of New Bedford hunted whales in 52−56° North during the summer of 1843, and on the

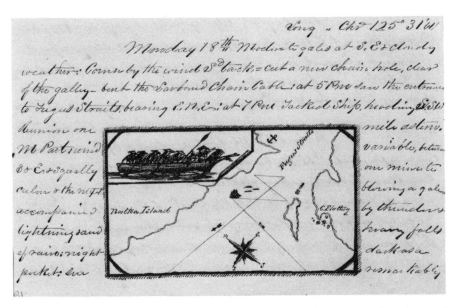

The French whaleship *Réunion* and the *Caroline* of New Bedford traversed the Strait of Juan de Fuca amid thunderstorms and gales, on a night as "dark as a pocket." The course of the *Caroline*, illustrated in a journal of the voyage, reflects Captain McKenzie's decision not to enter the native harbour of Neah Bay.
Kendall Whaling Museum, Sharon, Massachusetts

southward journey came to anchor for one day — 5 August — at Nootka Sound. [46] Perhaps this was the vessel that approached Stikein? The *Emerald* of New Bedford may have also done some trading. The inward manifest, filed on return in May 1843, notes "69 lbs. Beaver Skins" packed among the barrels of oil. [47]

The newly built whaleship *Caroline* of New Bedford also made an extensive survey of the coastline, after a fruitless search for whales offshore in the latitude of 56°. A surviving journal from the summer of 1843 includes a simple sketch of the north cape of the Queen Charlotte Islands, an ink drawing of two natives of "Nootka Island," and a chart of the track of the whaleship entering the Strait of Juan de Fuca. Daniel McKenzie had brought the *Caroline* from Kodiak in company with Captain Aderial Smith[48] in the French whaleship *Réunion*. The French crew was short of food and water and hoped to find a recruitment place along the coast. The two captains agreed to enter the strait together but McKenzie was apparently convinced of the malevolence of the Makah, who met them in large ocean-going canoes and openly welcomed them into their harbour. He made all sail for San Francisco, and Captain Smith, though urgently in need of food and water, decided not to enter the native village alone. The two ships proceeded in company, the malnourished crew of the *Réunion* sustained by Captain McKenzie, who made an admittedly satisfactory trade of provisions for a fine French pocket watch. [49]

Russian America also began to receive whalemen as visitors. One early arrival — an embarrassed one — was William Smith in the *Maine*. After leaving his wintering grounds at Shushartie Bay, Smith sailed northward in search of whales. In the vicinity of Kodiak Island in mid-August 1844 six men fell sick with scurvy, and he put in toward the Russian settlement at New Archangel, making Cape Edgecombe on the twenty-eighth and proceeding toward Sitka Bay.

> [At] 3 P.M. stood in for the Fort, no Pilots coming off, a Boat was lowered and soundings made, in what was supposed to be the channel, & three fathoms found through the passage,   stood in, and at 6 P.M. ship being under Topsail & Topgallant sail going about 1 1/2 Knots per hour, wind directly aft, she struck on a sunken rock,   all Sail was immediately hove aback, and the Kedge got out with 150 fathoms line. Hauled off, and soon after a pilot come on board and took the ship through the same passage into New Archangel Harbour. [50]

Smith remained at Sitka for almost a month, until his men were well enough to sail. On 22 September he departed, the *Maine* leaking a hundred strokes an hour, and came through the channel without incident. He made his way toward the Sandwich Islands, where the vessel's real captain, Makee, was now well on the way to recovery from the injuries he had sustained more than a year before. Acting Captain Smith, having apparently disobeyed Makee's instructions to return to Honolulu the previous autumn, was the subject of a search instituted by the American consul at the Sandwich Islands on Makee's behalf.

As the *Maine* sailed southward, the consul sent a letter to Monterey, California, hoping to learn that Captain Smith had touched there:

> The Am. Whale Ship Maine, Smith acting Captain, of Fairhaven was despatched from these islands in March 1842 [1843], by Capt. Magee, with orders to return to Oahu in the ensuing fall. Instead of obeying Capt. Magee's order, Smith put in to one of the Bays on the North West Coast and remained throughout the Winter and up to this date has not returned to [the Sandwich] Islands. As Capt Magee has some reason to fear that Smith already has, or intends to commit barraty, he requests me to say to you that in the event of the Ship's touching in at Montery, you will take such steps as may appear advisable to insure her safe arrival at Oahu. I am not clear that you would be authorized to take Smith out of the vessel, but I think another man could be put on board to navigate, or see that the vessel comes direct here.[51]

Consul Thomas O. Larkin replied that the whaleship had not been seen. Furthermore, he did not know how to proceed should it arrive there, and he hoped he might have "knowledge before business."[52] In fact, the ship had reached Honolulu on 26 October, well ahead of Larkin's response. Whether its acting captain was finally exonerated is not known, but when the ship was hauled-out for examination, the owners learned that the false keel was badly damaged, as well as eight feet of the true keel. Both had to be replaced.[53]

The *Maine* was not the first whaleship to call at Kodiak Island. The previous season, the aforementioned *Caroline* had approached. Its Captain McKenzie possessed private information about the Northwest Coast whaling grounds—in the form of a journal kept during the 1842 cruise of the New Bedford ship *Roman*—but the information proved useless. McKenzie made his way from 51° to 56° without a sight of a living right whale, only the carcasses of two already killed. These could not be used because of the overpowering smell. The ship ran into interminable fog and rain, so that one of the novices, George W. R. Bailey, complained of the "wet foggy weather" and the "watches lounging about deck [with] nothing to do."[54] When whales were finally sighted, the hunt progressed in typical fashion: harpoons drew from the blubber; a whale stove a boat. All the usual catastrophes barred their success.

The American whaleman. James S. McKenzie, son of Captain Daniel McKenzie,
stood for this daguerreotype in 1856 at the age of eighteen. He had just returned
from a North Pacific whaling voyage in the *Reindeer*. Later he became first mate of
the packet ship *Simoda*, from which he was washed overboard in a gale eight days out
of New York, on 25 January 1862.
The Whaling Museum, New Bedford, Massachusetts

On 12 August 1843 the lookouts raised Kodiak Island, and on the next
day they sighted Cape Grenville. McKenzie sent the whaleboats into a large
bay on the island's south side—probably Ugak Bay—in 57°38' North,
152°30' West, and the men reported that the bay was full of small whales,

which they presumed to be the calves of right whales and humpbacks. Perhaps more importantly, they had gone ashore and found a "deserted" hunt which contained what they took to be wooden traps, "fashioned like a Butte, hollowed out."[55]

> saw some fine "salmon" in a strem of excellent fresh water: one of the B[oat]steerers picked a few flowers, with which the ground was covered & brought them off to the ship, found them to be plants, common at home. They were the "Campanula Erinoides" or Blue Bell, the "Aconitum Uncinatum" or Monks Hood & a species of "Carex."[56]

Later, near Sitka, McKenzie shortened sail to speak a passing English bark in hopes of learning more about the local bays. His mate went aboard and returned with news that the vessel was the *Columbia*, bound for the Columbia River. Her officers warned that the whaleship *William Hamilton* had been driven out of the harbour at Sitka by Russian gun boats. The authorities were not allowing any whaling vessels to enter their ports or bays, "on account of their Fur Trade."[57]

The causes of Russian hostility toward foreign vessels had developed much earlier in the century, during the height of the fur-trade. Colonial managers had begun to complain to St. Petersburg, probably with some accuracy, that the foreign traders were outbidding the Russians by exchanging gunpowder and liquor for the highly prized pelts. Such trade items were expressly forbidden, even during the course of the company's own negotiations.[58] St. Petersburg complained in turn to the United States, accusing Yankee mariners of interfering with Russian commerce.

Receiving neither apology nor restitution, Tsar Alexander issued an imperial ukase in 1821 which proclaimed the Russian boundaries from the Bering Strait to the forty-first degree of Northern latitude in Asia, from the Aleutian Islands to the east coast of Siberia, and southward on the American continent to 54°40', and further granted exclusively to Russian subjects the rights to whaling, fishing, and all other industry. Penalties for infractions were specified, both monetary fines as well as confiscation of ships and cargoes. The ukase also prohibited American and other foreign shipping from approaching within one hundred miles of the shoreline. The Russians seemed ready to support their claim with force; the news of the edict was carried home to the whaling men of New Bedford by Captain Gardner, master of the whaleship *Balaena*, who reported that a Russian frigate was enforcing the prohibition against foreign vessels coming near the Russian-American coast.[59] It is not known if Gardner himself faced eviction; probably he heard the news from a fur-trade vessel, either at the Sandwich Islands or at sea.

The United States, England, and Spain lodged protests against this arbitrary closure of the Russian Pacific, and a long series of diplomatic negotiations eventually resulted in a pair of virtually identical treaties—one signed between Russia and the United States and the other between Russia and Great Britain—which reinstated certain international rights. The convention between Russia and America, concluded in April 1824 and ratified by the U.S. government the following January, agreed that American citizens would not be "disturbed nor restrained" in navigation, fishing, or "resorting to the coasts upon points which may not have already been occupied" for the purpose of trade with the native peoples.

New Archangel (Sitka), on Baranof Island, was the headquarters for the Russian American Company in the Gulf of Alaska. Whaleships visited Sitka Bay in hopes of taking whales there, but were frequently warned away by the officers of the Company. This view, which accompanied Duflot de Mofras's *Exploration*, shows the fortified village as it appeared in the early 1840s.
Kendall Whaling Museum, Sharon, Massachusetts

In exchange, the U.S. government agreed that its citizens would not resort to any location where a Russian settlement existed without first obtaining local permission. Further, the latitude of 54°40' North was accepted as the boundary line, north of which the Americans agreed not to establish any permanent stations, although both nations would share the gulfs, harbours, interior seas and creeks for fishing and trade. The Anglo-Russian convention, signed in 1825, reiterated the same points with respect to marine navigation and commerce, and specifically opened the port of New Archangel to British vessels.

The documents were valid for ten years and renewable. But when the decade expired, the Russian government refused to extend the provisions, instead giving their local governors responsibility for regulating foreign traffic. Whaling masters who arrived in the 1840s found these local officials adamant about refusing permission to land or engage in trade with the villagers. The matter was aggravated by the presence of a new manager, Etholin, who believed that the lucrative business of whaling belonged in the hands of Russians.

Etholin replaced Kuprianov as Chief Manager in 1842, at about the same time whaleships made their appearance in the Gulf of Alaska. He let the whaling masters know, in the strong and often incautious words of a newly placed administrator, that they were no longer welcome in Russian waters. The old treaty terms were no longer in effect. But the whalemen refused to answer his questions or obey his orders to leave. Etholin reported to St. Petersburg and included the comments of Captain Kadnikoff, master of the company ship *Naslednik Aleksandr*, who had been told by an American whaling captain that thirty whaleships were hunting on both sides of the Aleutian Islands. Even more startling, Kadnikoff was told that the whaling had been conducted the previous year with fifty ships; the whaling master's own crew had taken thirteen whales.[60]

Here was a new concern, a potentially disruptive trade and perhaps even worse, for it was said that the whalemen were robbing the local people, who ran away from their huts in fright and left unattended their caches of dried fish and other food. The whalemen were also vilified for trying-out oil on shore. The attendant smoke, the Russians claimed, drove away the valuable sea otters.[61]

Etholin engaged in his *tête-à-tête* with various American whaling masters until the matter came to something of a head in September 1843. Several whaling captains who called at Sitka seeking fresh water and food for their crews were denied permission to land. At least one of the masters, Lewis Bennett in the *Henry Lee* of Sag Harbor, asked permission to hunt whales in the bays but he was rebuffed. Etholin's pronouncement was afterward published for the benefit of other whalemen in the 14 May 1844 edition of the *Whalemen's Shipping List and Merchants' Transcript*:

New Archangel, the 30th Sept. 1843

Sir—In reference to your letter to me of the 29th inst. I inform you that I cannot allow you and neither to give any privilege in the whale fishing, in any bays or gulfs belonging to the Russian territory, under mine administration, viz: the North West Coast of America from 54 deg. 40 min. northern latitude, to Bhering's Straits as also all Aleoutean and Kurilei's Islands . . . because the pursuit of all commerce, whaling and fishing, on all the lands, ports and gulfs within the Russian territories is, by his Imperial Majesty, exclusively granted to Russian Subjects.

In making use of this opportunity I inform you, for your official relation, and request you also to communicate of the same to ship "Ann Mary Ann" and to all commanders of whaling ships you may occasionally meet, that no one has any right to carry on an unlawful whale fishery in the Russian territories and to a certain limit of shore.[62]

The chief manager stood ready to back up his demands with force, or so it seemed, but Etholin had little firepower. St. Petersburg maintained a stubborn reluctance to do more than sign treaties at the bureaucratic level. They were not disposed to spend money on warships to protect their colonial interests at Sitka and Kodiak. The "Russian gun boats" that drove out the *William Hamilton* from the harbour at Sitka were nothing more than a scantily armed merchantman or two—probably flying the Russian naval flag rather than the mercantile flag of the Russian American Company—whose captains were instructed to patrol the areas "where the foreigners were especially annoying."[63] The whalemen largely disregarded these protestations, replying, said one Russian observer, "with threats or contemptuous language." When made to leave Sitka, he noted, they quietly continued their traffic in the bay and "disregarded all protests."[64]

It is clear that some whaling crews were coming ashore, with or without local permission. A few surviving accounts suggest the story: the landing of the men from the *Caroline*, previously described; Erastus Bill's uncorroborated published description of a whaleboat excursion up an Alaskan river in 1845, to explore and "converse with the people";[65] the 1846 visit of the ship *Stonington* of New London, whose crew caught fish at the south end of Kodiak Island and later removed eight boat-loads of wood from Chirikof Island.[66] And then there is the matter of the French whaleship *Narwal*.

The abuses attributed to the American whalemen were not solely their doing. French whaleship crews were equally willing to take advantage of a comfortable shore that offered fresh water and fish for the taking. Captain Antoine Gustave Radou of the *Narwal* apparently behaved little differently from the others when he went ashore on Kodiak Island in July 1846. There he discovered some native huts in what seemed to be the usual state of abandonment; inside, his men discovered a quantity of dried fish which had been left untended. The captain broke into the huts, said one of the crew, and "took the entire provision of fish and different objects belonging to the poor inhabitants."[67]

In place of the fish, Radou left seven or eight pounds of biscuits, but this did not satisfy the anonymous mariner who accompanied him, who admitted in a personal journal that "[the captain's] way of proceeding does not please me."[68] Near shore two days later Radou gammed with an American captain who admitted that the Russians had read him the order forbidding whaling in the bays.

After this meeting Radou continued south of the Russian boundary and landed at Nootka Sound, whose residents had seen many a three-masted sailing ship since the days of Captain Cook. At Nootka, Radou found his ship the object of much attention; he was visited regularly by canoes, some of whose occupants sang and offered biscuits. The singing was described as an "unvaried monotone," and the journal-keeper wrote: "They voice a few syllables and hold them, on the same note nearly until they run out of breath, the sounds are gutteral and, remarkably, they never close their mouths nor do they bring their teeth together in articulating their words. They clap their hands to accompany their singing."[69]

On the first day in Nootka Sound, Captain Radou brought two men and a woman on board and took them to his cabin, which led immediately to some trouble:

> while he talked with [the woman] . . . one of the men put his hand into [Radou's] cigar box and the other reached for his combs; Radou lacking patience this time, kicked one in the buttocks and boxed the other one twice; then escaping . . . they climbed back up on deck and easily reached their canoe. . . . But in spite of all his vigilance our dear captain could not prevent the disappearance during this visit of a Greek skullcap he used to wear to bed.[70]

Two days later, Radou and several others went ashore. They were carried across the beach by the local men and brought to the chief's house. The chief was "nonchalantly stretched out on a straw mat"[71] and gave them a cool reception, which apparently displeased Radou's sense of etiquette. The captain beat a hasty retreat and set sail for California.

Such overbearing attitudes as Radou's disturbed relations between the whalemen and the colonial officials in both Russian and British America. From 1843 to 1847, Hudson's Bay officers at Fort Victoria grew familiar with insistent requests for supplies and assistance from whaling masters whose vessels had come to anchor at Neah Bay or sometimes near the fort itself. Perhaps the most annoying visitor, one whose call touched off high-level diplomatic correspondence with the Home Office, was Louis Jean Baptiste Morin, master of the French whaleship *Général Teste* of Havre.

The daily record at Fort Victoria notes Morin's arrival on 30 August 1847; he came ashore and informed company factors that he intended to winter nearby and begin whaling as soon as possible the following spring. He made a request for fresh meat and vegetables and purchased twelve shirts for cash. The following day he came again, and on this visit bought 481 pounds of beef and nine bushels of potatoes, the bill to be settled by an order on Vidar & Danthuster at the Sandwich Islands.

On the third day the crew was permitted ashore, and the local reaction to their liberty was decidedly negative, if not forthrightly hostile — the daybook notes that the Frenchmen were "lounging about all day."[72] Fortunately for all concerned, Morin decided to cruise in the Strait of Georgia. He sailed on 5 September, after buying all the items he could obtain that were of use to his ship. By that time a Russian brig had arrived, and its master, too, wanted supplies and a cargo of wheat.[73]

Captain Morin is likely to have been one of those whalemen who, through long and difficult experience, had learned not to accept any bluff from any-one.[74] He raised a few hackles at Fort Victoria and even more at Honolulu, where he was detained and fined $500 for failing to apply for the proper port clearances. The King's commissioner of customs there reduced this fine to $25 after hearing a heated protest, but even this did not satisfy the French, who later demanded the return of the fee as well as some adjustment for time lost during the detention.[75]

James Douglas was most upset by the arrival of this unrelenting French-man, and wrote urgently to London for instructions:

> The General Teste, a French whaleship from Havre, arrived at [Fort Vic-toria] in the early part of September, and is now lying off the Island of Feveda [Texada] in the Gulf of Georgia, where Captain Morin intends to remain until the beginning of April 1848. By last accounts he had killed only one [whale] which sunk and was lost, but he is in hopes of meeting with greater success in course of the winter and spring months. . . .
>
> We will have some doubts as to the right of foreign Vessels to Cruize or pursue their occupations on the parts of the North West Coast, north of 49 degrees a practice which will eventually lead to the interference with the fur trade, and affect the interests of the Company. Uncertain as we are about the question of right, we have not attempted to meddle with Captain Morin, but if you think proper, we will publicly assert that right and issue notifications warning Vessels not belonging to the United States of America that they are not [at] liberty to Anchor in the

ports of British Oregon for the purpose of fishing or trade except with the Hudson's Bay Company.[76]

The governor and committee advised that the "guardianship of national rights" most properly belonged to the government, and considered interference in the Morin matter unadvisable.[77] So the captain was not disturbed, and in January 1848 the *Général Teste* reappeared off Trial Island at Victoria and again anchored in the offing. The tour of the gulf had been unsuccessful; Morin had taken no whales. On the following day he bartered 180 gallons of sperm oil at eighty cents per gallon in exchange for 53 1/2 barrels of potatoes and 1,054 pounds of fresh meat. "He left a mem^r with us," the daily journal notes, "of supplies req^d by him on his return in the Autumn together with a list of the principal articles generally wanted by Whalers on the NW Coast."[78] His whaleship remained at anchor at Point Ogden for the remainder of the month, but on 4 February 1848 he traded 10 additional gallons of sperm oil for 200 feet of "pine in boards" and sailed that very afternoon for Cape Flattery.[79]

By the time Captain Morin arrived in the Strait of Juan de Fuca, the partitioning of the Oregon Territory had been accomplished without bloodshed. Oregon diplomacy had reached a crisis in 1845 – 46, as argument raged over the location and nature of the proposed boundary — if there was to be one — between British and American lands in the Pacific Northwest. In some ways the question was more important from a maritime vantage than from landward. Both nations sought suitable North Pacific harbours, in order to prosecute trade with the Orient, and to repair and reprovision naval and merchant ships. A major advantage to the United States would be the acquisition of a friendly port for the repair and recruitment of its whaling vessels, which constituted the largest fleet of its type in history.[80] "Americans are exploiting . . . the whale fishery of the Pacific Ocean, on a most extensive scale," one writer noted, "and thus [are] interested, but in the most positive way, that the Oregon Territory does not pass definitely into England's hands."[81]

Some observers in London intimated that the American whaling fleet might also serve as an effective auxiliary navy in the event of a conflict, to British disadvantage. Eleven thousand American whalemen practiced their nautical trade under the "Stars and Stripes," one correspondent wrote in the London *Times*, and these were experienced sailors "inured to every danger and to the extremes of hardship and toil. These men think lightly of lowering boats after whales on the North-west coast of America, the ship being at the time unable to carry a single reefed topsail."[82] Some London merchants threatened to send "Mexican commissions" to their armed clippers in the

China seas, which would "make prizes of all American whalers they can lay hold upon."[83] But in the event, this was not necessary. The treaty of 15 June 1846 adopted the boundary through the Strait of Juan de Fuca and the crisis of battle passed to Mexico, against whom President James Polk had just declared war.[84]

At Fort Victoria, any animosity between British and American interests was likely to be commercial in nature. The officers of the Hudson's Bay Company fervently wished the whalemen away though they were powerless to stop them, particularly when the Americans called at Neah Bay, sixty miles down the strait. At least by 1844 whaleships began visiting the native village there; two whaling vessels were said to have "looked in at Cape Flattery" during the summer of 1843 and "melted down two whales caught in the offing."[85] And though the Makah told Roderick Finlayson that they had not traded with the sailors, the new chief factor distrusted their word, "as some American blankets were seen in their possession."[86]

Finlayson made every effort to secure promises from the natives that they would not trade with white men other than those at the fort. But these promises, if accepted, appear to have been honoured most in the breach. Again writing to McLoughlin, in June 1845, Finlayson said that the "Cape Flattery Indians" had visited the fort

> and did not appear to be much encumbered with furs, having disposed of them previously to an American Whaler, which lay for about two months in Neah Bay & most likely others will call there in Course of the Autumn about the time the Indians return from their annual trip to Nootka. So that we have but little chance of getting all their furs — they however promised to come direct here on their return from Nootka whither they were bound on leaving.[87]

Fort Victoria maintained a steady trade with the Makah at Cape Flattery, who seem to have been the freight transporters and middlemen in the company's trade with other Vancouver Island natives as far north as Nootka Sound. Oil was purchased from them, in some instances on their visits to Victoria; the constancy of the traffic in whale oil is observable in Finlayson's correspondence and in the daybook kept at Fort Victoria:

> [6 August 1845] the Cape Flattery chief arrived with 4 Canoes direct from Nootka... and... just finished trading 6 large Sea otters 1 small & 10 beaver & Otter.... They have about 70 Gns Oil... not as yet traded.[88]

> [22 May 1846] 2 Small Sea Otters with 3 Beaver & Otters & other small furs & whale oil ... were to day traded from Cape Flattery Indians.[1]

[2 June 1846] 109 Gns. Oil with other trifles were traded... from Cape Flatterey & Sanitch.[90]

[7 December 1846] Cape Flattery Jack made his appearance here this afternoon.... Oil is the principal article of traffic now brought by him.[91]

[18 August 1847] [traded] a little Whale Oil from Kowitchan.[92]

This oil, both whale oil and that obtained from dogfish, was packed with any baleen that had been obtained for transport to England; the large bark *Vancouver* and several other merchantmen were employed for this purpose.[93]

Into that complexity, in 1845, sailed three whaleships which had come across the Pacific from Kamchatka, and whose masters sought provisions from the fort. Their contact with Fort Victoria and the native village at Neah Bay was perhaps more complete than any other, for they not only bought supplies but also left deserters behind who attempted to make their way among the natives. All three had come out from New London, Connecticut, for the Pacific whaling grounds. They had not originally sailed together, but during the summer on the Kamchatka ground their captains had found such poor whaling that they mutually agreed to try their luck on the American coast. They finally reached the southern end of Vancouver Island after a voyage which took them from the Asian shore across the top curve of the North Pacific Ocean, past the Alaskan panhandle and then southward.

The three ships were the *Louvre*, *Morrison*, and *Montezuma*. Two of the masters shared a surname; James M. Green aboard the *Louvre* and Sam Green, Jr., in the *Morrison*. Along with Captain William M. Baker in the *Montezuma* they experienced the kind of voyage that sometimes made wild dogs out of civilized men. The *Morrison* was particularly vexed from the start, and, in fact, was in none too good a condition. Even before reaching the coast of South America, the crew had been forced to make repairs to masts and other wood which had rotted.[94] And though not on his first voyage to the Pacific — he had taken the *Neptune* there during 1839 — Sam Green was completely unfamiliar with the North Pacific and he had never been to the Sandwich Islands. He was nevertheless a youthful and determined man, on his third voyage as master,[95] a man who regularly practised his fine penmanship by writing both his own name and that of Napoleon Bonaparte into his personal sea-journals.

He made every effort to learn about his intended destination and, perhaps fortuitously, the *Morrison* encountered the Fairhaven whaleship *George*[96] before rounding Cape Horn. Its captain proved to be a friendly and accommodating man:

Captⁿ Swift informed us that he had on board his vessel a large volume containing the voyages of several of the most distinguished navigators that have explored the North Pacific. It also contained maps or charts of all the principal bays together with accounts of the natives along the coasts their habits customs &c. The charts of the principal whaling bays he very politely offered to copy himself & present to Capt G if he would come on board [the *George*] tomorrow afternoon.[97]

Sam Green went aboard the *George* and obtained the charts from Captain Swift, doubtless very much relieved to have at least some information about

Flying the house flag of Havens and Smith, whaling agents, the ship *Morrison* sails the Thames River near New London, Connecticut, in 1844. Built as a merchant vessel in Philadelphia twelve years earlier, the *Morrison* was sailed across the North Pacific from Kamchatka to the Northwest Coast during its single voyage as a whaleship. Oil on canvas, by John Ewen, Jr.
Photograph by Claire White-Peterson, courtesy New London County Historical Society

this far-distant cruising ground. But as the voyage progressed, Captain Green received other portents that his trip would not be an easy one. On 18 November he spoke the whaleship *Candace*, homeward-bound from the North Pacific with but two whales, owing to the dissatisfaction of the crew, who allegedly refused to perform the whale-hunt with alacrity. Whether Green knew then or learned afterwards is not known, but he had among his own crew one of the supposed ringleaders of the near-mutiny aboard the *Candace*. He had returned to the States ahead of his former ship and signed on for another voyage, probably to avoid prosecution.

When the *Morrison* reached Tahiti, this man set fire to the ship in an effort to desert from it, but Green successfully staunched the flames and navigated to the Sandwich Islands where he rid himself of the troublemaker.[98] He then sailed to the Kamchatka coast, but the once fabulous whaling there seemed to have fallen into great decline. As in the Pacific Northwest, just a few years of intense hunting had decimated the right whale population. The remaining animals were nervous and unwilling to allow boats to approach them.

The shortage of whales also proved vexatious to James M. Green in the *Louvre*. From the coast of Kamchatka he wrote to his shipowners:

> The cargoes which went home last season and the account they brought must have raised high hopes. I will give you a small sketch   in May I saw Enough whales to have kept boiling steady, in Lat. 46 to 47. Long. 165 to 169 East but it was one continuous round of bad weather. I got two in May sunk two in June saw a few whales in same place, got my last whales 8th day of July   since then I have seen but two whales . . . I have been from berings Island to Cape Lopatka off to 170 and back again and have been up and down the coast off and on for 40 days. You can see from 20 to 15 ships daily   I have not seen one start tack or sheet for 50 days.[3]

"I can not make whals," he concluded.

> I consider myself justified in leaving this coast — if you had been here you would have told me to go [a] month ago. I have been trying to find Sam Green   I have found him and now we are going around the Fox Islands . . . and if whales hang at those Islands late we shall go in, and anchor: after that I shall get to the coast [of North America] if possible   I do not want to go to the Sandwich Islands if I can fetch the coast.

"I have got to stay out another season," he lamented; "that sticks. I can hardly swallow it, it makes me ugly. I give hard looks. I never did go on any [whaling] ground until it was all cut up, but you may be sure when you do see the Louvre to see a voyage in her."[100]

In their determination to "make a voyage," James Green and Sam Green decided to cross the North Pacific to the whaling grounds around the Fox — Aleutian — Islands and Kodiak Island, and from there eastward and southward along the North American coast. They must have also convinced Captain Baker in the *Montezuma*; all three ships were spoken at sea in late Au-

gust, "bound for the Bays."[101] Whereas the two whaling areas had formerly
been discrete—some vessels spending summer on the Kamchatka ground
and others on the Northwest Coast of America—the sudden collapse of the
Kamchatka fishery in 1845 drove a number of whaleships across, thus forming a new diamond-shaped route from the Sandwich Islands to Kamchatka,
across the Aleutian archipelago to Kodiak, and then southward along the
coast to San Francisco or perhaps back to the Sandwich Islands. This
crossing-over apparently continued the following year; in August 1846 the
Luminary of Warren spoke the French whaleship *Jonas* of Nantes, 3,600 barrels and "bound off" the Northwest Coast ground. "She is from Kamchatka,"
the men learned, "and reports the fleet on that side doing but little."[102] The
Narwal, under Captain Radou, also crossed from Kamchatka, pausing on the
newly-opened whaling ground in the Sea of Okhotsk before proceeding
eastward past the "Isle of Ischerikows" (Chirikof Island) and eventually
bringing its master to the native storehouse on Kodiak.

The triumvirate of Green, Green, and Baker did not tarry long at the
Aleutian Islands. By 23 September they arrived in the Strait of Juan de Fuca
and came to anchor among the Makah whalemen in the harbour of Neah
Bay.[103] They may have been surprised to find themselves welcomed to a place
much frequented by whaleships, and by a native chief who spoke English.
This man, known to the Hudson's Bay Company as "Cape Flattery Jack," is
probably the one who conversed with Captain McKenzie of the whaleship
Caroline in 1843 and the same who met Charles Wilkes' U.S. Exploring Expedition earlier in 1845. He is said to have spoken many English words
"quite distinctly."[104]

It is not known if the captains had previous knowledge of the existence of
Fort Victoria, then barely two years old, but in any event they came up the
strait on 1 October "to enquire if any whales were found about this Neighbourhood, in winter." James Douglas met them and happily admitted that
the only whales to be found in the strait were "Fin Backs" and "Black fish"—
probably meaning humpback whales—"which whalers do not consider worth
the trouble of Killing."[105] While at the fort the captains availed themselves of
the island's fresh produce; Sam Green bought two-and-a-half bushels of
potatoes for the equivalent of eleven American dollars.[106]

On 16 October the Hudson's Bay Company bark *Cowlitz* entered the
strait. The whalemen sent a boat's crew from their anchorage near Neah Bay
to request a barrel of powder "and a few other Stores, which was declined."[107]
The reluctant master of the *Cowlitz* must have been chagrined when, three
days later, the wind died and the whaling crews assisted in towing the bark
up to the fort.[108] There the whalemen gave cash for 120 bushels of potatoes,
40 bushels per ship, "all they could carry & we could afford to furnish
them."[109] Each officer paid for his own supply; an undated receipt for the

Louvre's purchase survives: "Bought of R. Finlayson Fort Victoria Vancouvers Island NW Coast of America 40 Bus Potatoes 1/5 £2.16.8"[110] A record of Sam Green's purchase also survives, in an account book kept by him, which notes the 20 October transaction at "Fort Victory": 40 bushels at £2.16.8. He also bought "Flower" and some powder, but not so much, since the bill amounted to only fifty cents.[111] The whalemen requested blankets for the ships' crews, but these the Hudson's Bay people would not supply, for fear the whalemen would undercut their business with the Makah at Cape Flattery.[112]

They told Douglas and Finlayson that many more whaleships would likely call at the fort in the coming year, if only beef, flour, and potatoes could be made available to them. This news caused Douglas to reconsider his previous discouragement of the whalemen:

> In future ... I should be disposed ... to act very differently. I would supply them with winter clothing for their crews at prices sufficiently high to secure our business, and to put it out of their power to trade furs, with any prospect of advantage to themselves, reserving moreover the prime articles of Indian trade, such as the Beaver Blanket, which the Natives chiefly value, in our own hands, and selling them only the "inferior" Blanket, which will not purchase furs in this quarter. By such means, we might greatly improve the business of this Post, which however convenient as a Depot, is not productive in furs.[113]

Forty bushels of potatoes were too many for his whaleboat, so Sam Green hired a native canoe to assist in bringing them from "Fort Victory" to the *Morrison*. For such assistance he paid dearly: one red flannel shirt, ten pounds of tobacco and a musket. He gave a second flannel shirt to the chief at Cape Flattery for recovering a watch that had been stolen from the ship.[114]

The three whaleships remained at Cape Flattery nearly five weeks.[115] During that time, no fewer than fifteen men deserted. Six of them stole a whaleboat from the *Louvre* in hopes of escaping to some American settlement. The other nine concealed themselves in the native village until their ships sailed. Afterwards they were rounded up by Cape Flattery Jack and brought to Fort Victoria. James Douglas took them into protective custody, and, "as it would be imprudent to leave them in a situation, where they might soon become dangerous to us," he determined to send them to the Sandwich Islands in the *Cowlitz*, or else to the Columbia, "as the sooner they are removed from this place the better."[116]

The nine who hid among the Makah fared better than the six who commandeered the whaleboat. In attempting a passage to the Columbia River, they apparently ran short of water and made several unsuccessful attempts to

land. While crossing the difficult and treacherous breakers at the mouth of
Gray's Harbor, the whaleboat capsized; Robert H. Church, Richard C.
Kirby, and Dudley Royce, all from the desertion-plagued *Morrison*, were
drowned. The survivors were assisted by natives to reach Fort George, and
from there the news of the tragedy was conveyed to Fort Victoria by Captain
Scarborough of the schooner *Cadboro*. [117] Their own whaleships meanwhile
sailed past Gray's Harbor on a course for San Francisco Bay. When they came
to the anchorage at Yerba Buena in late November, the captains went ashore
to file consular certificates against "men runaway at San Juan de fuga."[118]

The Hudson's Bay Company arranged for the passage of at least three of
the deserters who had returned with Cape Flattery Jack, and sent them to the
Sandwich Islands on the next sailing of the *Cowlitz*. [119] Among them only one
is known by name, Thomas P. Davis, and it must have been a sight to see the
faces of his former shipmates in the *Louvre* when he returned aboard. He delivered, or had delivered for him, a message to James Green from the U.S.
Consulate, which read:

> Sir,
>   Herewith I send you a seaman, named Thomas P. Davis, who, it
> seems, deserted from the ship "Louvre" some time last season on the
> North West Coast, and who, being found by some of the servants of the
> Hudson's Bay Company in great distress and destitution, was sent down
> here about a month since in their Barque "Cowlitz."
>   Davis, in my opinion, deserves punishment, not only for his deser-
> tion, which from his own confession appears to have been without cause,
> but for one or two petty thefts here at the hospital boarding house where
> he has been staying since his arrival. [120]

It is not known whether James Green exacted any punishment for Davis's
misdeeds. The errant sailor completed his voyage in the *Louvre*, and on return
to New London he received his lay of 515 gallons of whale oil, 23 gallons of
sperm oil, and 160 pounds of the "bone." The whale oil was sold at 20 1/2
cents per gallon, the sperm oil at 92 cents per gallon, and total of his oil and
baleen came to $226, which he received in cash. [121] Davis may not have been
flogged for desertion, but he was surely and swiftly punished at the counting
house: the ship's agents docked him $21.50 on account of his three months'
absence. [122]

Of the rest, virtually nothing is certain. But there remains a mystery.
During the summer of 1846 the residents of Fort Victoria became aware of an
"American deserter" who was occasionally seen in company with the Cape
Flattery Makah. In the forenoon of 17 July, a large trading party of thirteen
canoes arrived with the usual cargo of sea-otter pelts and oil. The white man

was with them and took the opportunity to seek asylum. The daily journal of the fort records his coming:

> The American sailor, deserter, accompanied them; he appears to have Conformed to the habits of the Indian since his sojourn among them, being much interest[d] about their property & securing their Canoes, acting in short as their hired Servant. He has got a *fair one* as partner, & is said to be now in the family way, Consequently we cannot Keep him here at present.[123]

It seems plausible that this man was in fact a slave or serf, and may not have been treated well. Perhaps he acted solicitously toward the Makah until such time as he could beg for rescue. In any event, he hid until the Makah departed for Neah Bay and then made his presence known. He said his name was Bill Edwards, and he received sanction to stay at the fort. There he remained until February 1847, when, "having seemingly got tired of our Company, This morning solicited & obtained leave to joine his countrymen in the Brig Henry."[124]

The *Henry*, of Newburyport, Massachusetts, had reached the Strait of Juan de Fuca in distress on 14 February after running short of provisions on a passage from the Sandwich Islands to the Oregon Territory.[125] Captain William E. Kilbourne remained only a short time, and then sailed—presumably with Edwards aboard—for his intended destination on the Columbia River. On arrival, Captain Kilbourne used the ship's cargo to open a general store in Oregon City.[126] Nothing more is known of the fate of Bill Edwards.

The visit of the *Henry* to Fort Victoria confirmed many unsubstantiated rumours that the Oregon boundary had been settled along the forty-ninth degree of latitude. The news was received with great interest since its meaning was profound: the Hudson's Bay Company was now evicted from the land between the Columbia River and the Strait of Juan de Fuca, and the headland visible from Fort Victoria across the strait belonged to the United States. American permanence in the Pacific Northwest could no longer be doubted or prevented.

It is a pertinent sidelight that Charles Enderby, scion of the great London whaling family, sought and received support from the Hudson's Bay Company in his effort to reinvigorate the moribund British South Seas whaling industry in the late 1840s. Enderby hoped to send his whaling fleet to Vancouver Island once the company received authorization to rule that place as a colonial protectorate. Governor John Pelly informed Enderby that the Company would look favourably on the whaling proposal,[127] but the plan never materialized. For reasons other than the detrimental chill and fog of the Northwest, Enderby elected to establish his new fishery at the Auckland Islands near Australia; the complete failure of that effort ruined any chances of reviving British southern whaling in the nineteenth century.[128]

Before long, the growing Hudson's Bay outpost at Victoria would become the centre of a Crown Colony,[129] host to government and a large naval base at nearby Esquimalt and headquarters for a new kind of whaling which seemed to promise heady profits for anyone — British, American, or Canadian — sufficiently clever and ambitious to develop its untapped potential.

4

GLANCING BLOWS

The problem [of] finwhale hunting led in fact to
the creation of not merely a larger and stronger instru-
ment, but also of one which was new and essentially dif-
ferent from anything previously existent.

ARNE ODD JOHNSEN
"GRANATHARPUNEN" (1940)[1]

Thomas Roys' discovery of bowhead whales beyond the Bering Strait and the simultaneous announcement of California gold forever changed the rhythm of the North Pacific whale hunt. Any shipmaster who called at San Francisco in 1849 quickly found his voyage at an end. Crews deserted to a man, and no captain could hope to find anyone to replace them. Most whaling masters abandoned the Northwest Coast in favour of a quick passage to the bowhead grounds in the Bering Sea. Beginning with the summer season of 1849, western Arctic whaling dominated the industry. Ships pushed farther and farther northward and then eastward, over the top of Alaska and the North-west Territories, searching for these great whales which yielded more oil and longer slabs of baleen than their right-whale relatives on the Kodiak Ground.

Once the gold fever subsided, San Francisco became an ever more important terminus for these Arctic whalemen, offering full-service shipyards, skilled shipwrights, and quantities of supplies hauled overland or brought by sea from the east coast. The development of the telegraph and the completion in 1869 of the transcontinental railroad confirmed the pre-eminence of the city. Afterwards the Sandwich Islands were less often visited by whaleships, although Hawaiian residents maintained their own small whaling fleet. In the last quarter of the nineteenth century whatever remained of American whaling reprovisioned and repaired in San Francisco Bay. Oil, baleen, and men were sent home overland on the railroad rather than returning a ship and crew by the long Cape Horn way.

Entrepreneurs in the eastern states increasingly rejected opportunities to invest in chancy whaling voyages. There were sources of much greater profit in industrialization, railroads, and land speculation. The discovery of ground oil near Titusville, Pennsylvania, in 1859 soon made petroleum and other mineral-based oils widely available. Vegetable oils also undermined the stability of whale oil as a commodity, particularly since the market for whale oil as an illuminant had always been restricted by its price to perhaps 15 percent of American families. The development of natural gas further constricted the available market for oil at a time when right and sperm whales were growing increasingly scarce and the expense of harvesting them rose correspondingly.

When whale oil prices collapsed in the years following the American Civil War of 1861–65, the northern hunt was continued primarily on behalf of fashion, since baleen helped to support hoop skirts, tight-waisted corsets, and Victorian shirt-collars. By 1890 the right whale was commercially extinct in the eastern North Pacific, yet the demand for "whalebone" as a stiffener in garments and accessories elevated the market price as high as seven dollars per pound—each adult right whale might yield a thousand pounds or more—and ensured that the capture of even a few whales would result in a profitable voyage. "Ordinary economic forces," a naturalist later wrote, "which would have made it unprofitable to go to sea for such a small catch, were in this case ineffective."[2]

Nevertheless, whaling companies stopped building ships to replace hulls lost or condemned. Those that remained in the Pacific worked until they were too old to sail any more or until they did not come back from sea. Some of their masters continued to visit the Northwest Coast. Not that the whaling there was anything like it had been during the palmy days of the 1840s; now it was one whale or two when once they had caught twenty or thirty. The fat, slow right whales had become shy. But it was still possible to find one, so captains gambled on a visit to the Kodiak Ground while awaiting the breakup of pack ice in the Bering Sea. A few such men, now very much in the minority, remained an entire season in the Gulf of Alaska, hunting as far south as the Queen Charlotte Islands, cruising back and forth between Cape St. James and Trinity Island seeking that last strike. Such persistence sometimes paid off handsomely. A giant right whale, said to have made 274 barrels of oil, fell to the lances of the *General Pike* on the Kodiak Ground in 1861.[3]

Near the Queen Charlottes in 1867 the ship *Emily Morgan* encountered seven whaleships, among them the *Norman*, whose crew was completing one of the last two French whaling voyages. Later, sailing between Mount Edgecombe and Kodiak, the *Emily Morgan* spoke several more whaleships, including the *Fanny*, the *General Scott*, the *William Gifford*, the *Florida*, and the other of the final pair of French whaleships, the *Winslow*. The crew of the *Emily Morgan* cruised the Kodiak Ground from late April until early summer that year,[4] from the Queen Charlottes well into the Gulf of Alaska. During that time they augmented their salt-beef diet with halibut obtained from "Sitka Indians" in the vicinity of Hazy Island in 55°50' North. Whalemen also fished for themselves; by the 1880s logbooks of Kodiak voyages frequently mention fishing for cod, halibut, and other species, but trade with the natives for fresh provisions had become more or less common along the southern Alaskan panhandle. A narrative of whaling on the Kodiak Ground in the 1860s, published forty years after the fact, describes the arrival of a canoe bearing three men and two women who accepted a pound of tobacco in exchange for a huge halibut "which must have weighed four hundred pounds." One of the native men furnished a twenty year-old letter, written by a shipmaster, which identified the bearer as an "Asset"—probably Masset—chief. It is not stated exactly where this transaction took place, but the presence of a Masset man suggests that the ship was near shore in the Queen Charlotte Islands.[5]

There may have been another marketplace too, one not reflected in the logbooks and journals of whaling voyages to the Kodiak Ground. A report issued by the United States Secretary of War in 1871, four years after the American purchase of Alaska, notes that the women of Sitka, "almost without exception, are prostitutes, and even girls of twelve or thirteen years must be included in the number." The 1870 Sitka census identified 36 of 391 inhabitants as "prostitutes."[6]

If local fishing grounds were dependable, the whaling ground was not. Right whales were much more scarce than ever before, but no tamer. In 1891 a large specimen towed the *Emma F. Herriman*'s whaleboats from 6:45 PM. until 7 o'clock the following morning. The chase was curtailed only when a rising fog forced the men to cut the line.[7] Humpback and finback whales were common, since few attempts were made to catch them, but it was a rare day, indeed, when right whales were seen, and rarer still if a sperm whale happened to come into view. The excitement on such a day could bring a normally patient whaleman near apoplexy:

> At half-past five, we discovered whales spouting to leeward and the captain ordered two boats of us to lower to see what manner of whale they might be. The fourth mate and I were scarcely in the water and our sails set, when the old man shouted:
> "They'r sparm, they'r sparm! D'y hear? They'r sparm, I say! Clear away the other boats! Oh, but they'r big ones! Only think, sparm whales up here in fifty-five. Have a care now, ev'rybody. Look out what ye do.

Finbacks, humpbacks, sulphur-bottoms, right whales and sparm, all in one day. What won't come next? Thar blo-o-ows, o-o-ows. 'Way ye go."[8]

The paucity of right whales and a growing familiarity with the rocky coastline led whalemen almost to the surf in search of oil and "bone." They came very close indeed; the ship *Java*, for example, was sailed to within ten miles of Cape Edgecombe in 1855. In 1889 the bark *Coral*'s crew came close enough to report a heavy breaker, the appearance of a reef, and plenty of driftwood.[9] In 1892 Captain McInnes took the bark *Josephine* toward the coast near Cape Clear, steering so close to Middleton Island—a popular whaling venue—that the captain's wife was able to enter a detailed survey in her journal. "[It] looks green," she wrote, "but through my long glass I find the grass is not very thick, a few bushes, no trees. There are two white men, three indian men and an indian woman living on the island, the report is that they are going to raise foxes."[10]

If Mrs. McInnes could gauge the growth of the grass on Middleton Island, her husband must have been sailing very near shore, indeed. But he found few whales. Nor did other whaleships encountered by the McInneses on their voyage. Among the several cruising within sight of the Queen Charlotte Islands was one whose master complained that he had burnt 120 tons of coal "steaming all around the ground" and had not seen a right whale for a month.[11]

The lofty, snowbound peaks of Edgecombe, Fairweather, and St. Elias and the landfall of Kodiak Island had become beacons for whalemen, who found the largest and fattest right whales in the nearshore waters of the Gulf of Alaska. But even improved charts occasionally failed to alert captains and crews to danger. In 1893 the sealing schooner *Annie E. Paint* crossed over some wreckage in the vicinity of Kayak Island, in 59°50' North, and among the flotsam found a whaleboat with a corpse lying under the thwarts. This news was passed along, and it was soon learned that the whaling bark *Sea Ranger* had broken up on the reef at the tip of the island with 200 barrels and 2,000 pounds of baleen on board. The *Sea Ranger* had sailed from San Francisco in June 1892 under command of Charles H. Foley. After the ship wrecked on the morning of 26 May 1893, Foley took the boats and thirty-seven men and made way for Little Kayak Island, some fourteen miles distant. There they remained until 2 June, not exactly marooned but nevertheless committed to enjoy the enforced hospitality of George Barrett, resident agent of the North American Commercial Company. Eventually the company's steamer *Crescent City* made its call and took the stranded men to Sitka and thence to the United States.[12]

In November 1894 the bark *James Allen*, its Captain Huntley, the first and second mates, and the majority of the crew were given up on a reef extending from Amlia Island in the Aleutian archipelago. Eight survivors made their way to shore and collected roots to live on at sea. Then they continued on to Seguam Island, where native hunters fed them sea-lion meat and conducted them to the Alaska Commercial Company post on Atka Island.[13]

Increasingly, shipowners decided that the Kodiak Ground was a waste of time. As early as 1858, the master of the *George Howland* suffered a mild reprimand from the ship's agent for cruising there instead of in the Sea of Okhotsk.

> *We were much* disappointed in your changing your course last year or season and going to Kodiac instead of the Ochotsk. . . . It was unfortunate for us, that all the ships we had North this year, which was *four*, should go to the Kodiac and the Arctic—*not one* to the Ochotsk—from present appearances all of them did not take more than [1,000] or 1200 bbls of Oil, which is rather discouraging.[14]

On Vancouver Island, the occasional business between the Hudson's Bay Company and the whalemen had not increased. By the 1860s, however, merchants in Victoria began to express hope that their growing port might become the reprovisioning place for ships belonging to the northern whaling fleet. Inflated prices at the Sandwich Islands and the long distance to San Francisco seemed to bode in their favour; the Strait of Juan de Fuca was just a few days' sail from the whaling grounds near the Queen Charlotte Islands.

> by making Victoria the rendezvous [masters] can fit out as cheaply as if they were at home; they can also get any repairs or alterations made at a reasonable price. They are always sure of a supply of vegetables for any length of voyage at one-third the cost of the same articles at Honolulu. Our Douglas pine is found to answer for oil casks equally well with oak, hence their casks . . . can be furnished to them here for less than half the price of those obtained in San Francisco. They will bring their cargoes here and in a few hours communicate with any part of the United States by telegraph and assure the requisite tonnage [for trans-shipment] where it can be had at the lowest rate. . . . The advantages of Victoria over the old rendezvous at Honolulu, are unquestionable; but we contend they are also superior for the purposes required, to San Francisco.[15]

Deepwater merchant captains likewise praised the ports of Puget Sound; the captain of the ship *Chelsea* testified in 1868 that his expenses at nearby Vancouver did not amount to a third of London charges, and, he added, "the crews here are free from the temptations of the grog shop and no idlers are allowed about the place."[16]

"What we have to do," a Victorian merchant wrote, "is to sell a good article at as low a rate as possible, and we need have no fear of finding customers."[17] But the American whaleships did not call at Victoria. A commercial rivalry had led to lively animosity, first over possession of the Oregon Territory; then on account of immigration restrictions during the Fraser River gold rush of 1858–59; and finally because of the disputed international boundary through the San Juan Islands, a diplomatic quagmire not resolved until the 1870s. The gold rush perhaps fanned the hottest fire; *ad hoc* regulations inhibited Americans from travelling freely into the foothills and canyons of the Fraser. The situation caused so much frustration as "to induce every American . . . who has the least spark of national pride, to avoid Victoria, and everything British."[18]

The American Civil War further polarized Anglo-American sentiment in the Pacific Northwest. Anglophiles both tacitly and openly condoned the rebellion of the southern states. Armed cruisers built in British shipyards served the Confederate States Navy, and three of these, *Alabama*, *Florida*, and *Shenandoah* among them captured and destroyed more than fifty whaleships operated by merchants in Union states. Certain British Columbians even offered a 400-ton steamer to the Confederate secretary of state, provided only that he give them a letter of marque allowing them to "harass and injure our enemies."[19] Such shenanigans could not have had a beneficial effect on British-American relationships on Vancouver Island; not when the St. Nicholas Hotel in Victoria's Government Street was a known meeting place for southern sympathizers, nor when Governor James Douglas argued in favour of taking control of Puget Sound in order to "hold the only navigable outlets of the country—command its trade, and soon compel it to submit to Her Majesty's Rule."[20]

But if the Victorians were not going to be given the business of the American whaling fleet, why couldn't they have a fleet of their own? In truth, whales were frequently seen sporting within sight of Victoria. As early as 1859, editorialists in the local newspapers wondered why no one wanted to open a whaling station. "The investment would be profitable," they argued, "as whales are plenty and easily taken in soundings."[21] Six years later, when Matthew Macfie published his widely read prospectus on Vancouver Island and British Columbia, he too commented on "*whales* innumerable sporting in the Gulf of Georgia," and advocated a pelagic whale fishery. Since Victoria was "conveniently located" near the northern right-whale grounds as well as the sperm-whale grounds in the western Pacific, "Colony whalers would have a safe and easy run, with the favouring influence of the trade winds and an open sea."[22]

Americans also looked to the northwestern coast as a prospective location for shore-whaling stations. One Bostonian, James Gilchrist Swan, arrived at Port Townsend, Washington Territory in 1859 to examine nearby harbours with this end in view.[23] And though Swan never carried out his whaling scheme, he nevertheless made every effort to learn about whales from the neighboring Makah, and continued to agitate for a wagon road and later a railroad to move men and products. "For a whaling station," he wrote, "the harbors and bays of the Straits of Fuca present remarkable advantages for ships, while for vessels of smaller size Shoal-Water Bay can not be surpassed."[24]

Charles Melville Scammon, a former whaleman with an intense scientific curiosity about marine mammals, likewise conducted reconnaissances along the coasts of Washington and Vancouver Island. In the Pacific Northwest in 1864 and 1865 Scammon saw innumerable finback and humpback whales—no news, given the observations of whalemen who had preceded him there——and he ambitiously speculated that a humpback-whale fishery might be successful, particularly in the shallow waters of the Alaskan coast. He also described the rarely seen "sharp-headed finner whale" (minke whale) which came temporarily to bear his name—*Balaenoptera davidsoni* Scammon—furiously jotting remarks and measurements as his twenty-seven-foot specimen was flensed by the local Makah, who declared that the meat was excellent food.[25]

Virtually every surviving whaling logbook and journal from the Gulf of Alaska comments on the large numbers of the rorqual whales, particularly the finbacks and humpbacks. Frederick Crapser, on the Northwest in 1843, even composed a bit of doggerel to illuminate the scene:

> Saw Hump Backs and Fin Backs
> And Fish with Broad Tails
> Saw Grampusses and Porpoises
> But saw no wright whales.[26]

Right-whalemen had many times attempted to take humpback and finback whales, with frustratingly little success. Inexperienced lookouts sometimes mistook these for right whales; harpooneers who managed to make fast to one soon discovered that it ran off immediately at great speed. Frederick Crapser set aside poetry to remark: "Lowered the boats twice for Hump Backs but it was all in vain Chain lightning couldn't hold a candle to them." Another sailor wrote, "Early this morning raised whales . . . made them out to be Fin Backs & Therefore of no use to us till we could catch them & that a imposebileity."[27] If a boat's crew did manage to lance one, the great carcass would almost inevitably sink.

Still, crews occasionally expended the effort. They were even known to go after the largest animals on earth, the blue whales, which they called "sulphur-bottoms" on account of the mineral yellow colouration on the ventral side. The *Europa*'s crew did so on the Northwest Coast in 1843, and the *Julian*'s in 1848.[28] But there is no evidence that any of these attempts were successful. About the only way to catch a giant, fast-swimming blue whale was to capitalize on nature's own peculiar bounty, one opportunist robbing from another: "at 11 saw Killers [*Orcinus orca*] Lowered and pulled up and found them eating up a sulphur bottom alive   took him away from the rascals and took him to the ship."[29] Their reward was meager—only sixteen barrels of oil—but the crew must have counted the experience as one of the unique events of a long whaling voyage.

So the singular problem facing whaling men at mid-century was to devise a powerful weapon capable of slaying these large, strong whales, preferably from a distance. The whale fishery was almost entirely dependent on the iron harpoon, hurled by main strength from the forward thwart of a whaleboat which had been rowed almost to a collision with the whale. The blue, finback, and humpback whales remained invincible for want of methods to equal their speed, kill them quickly, and keep their bodies afloat.

At mid-century at least one prescient observer foresaw the future of the whaling industry and the fate of the finback species: "It is not a thing beyond the reach of probability," he concluded, "that this hitherto unmolested sea-rover may yet be brought within the all-powerful grasp of predatory man by swivels or air-guns, that shall fire harpoons into him, or poisoned arrows from a distance."[30] Indeed, the notion of a large swivel-gun or harpoon-cannon was hardly new even in 1850; the concept had been formulated in Europe as early as 1784, when a blunderbuss was designed to shoot a harpoon from a distance of ten fathoms, using "four common tobacco-pipes full of glazed powder" for the charge. It is said that eleven whales were killed with this weapon up until 1787, after which the instrument came "more into use."[31] Between 1850 and 1870 several types of shoulder-guns were developed, including Pierce's, Cunningham's, Eggers's, Brand's, and the British Greener gun, each of which accommodated an explosive "bomb-lance."[32] The bomb-lance consisted of a sharp-nosed brass cylinder, feathered like an arrow for stability and containing a powder charge and a time fuse which detonated upon impact.

The most successful of these instruments was the so-called "darting-gun," an ingenious harpoon fitted with a breech and barrel. A long iron rod extended almost to the harpoon's toggle-point; when forced backward on impact with the whale, the rod tripped a trigger in the breech. After a few seconds, the bomb-lance entered the whale at point-blank range. The concussion of the ensuing explosion and its shrapnel was intended to stun the whale sufficiently to prevent it from diving or running away. Hopefully for whale and whaleman, the shot killed the animal instantly.

With such weapons it became possible to kill at least the smaller rorqual

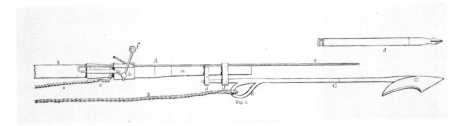

The darting gun efficiently combined the hand-thrown harpoon with a gun which fired an explosive missile on impact with the whale.

whales, particularly the humpback. Some whalemen took this new armament to the Kodiak Ground to try their luck at "humpbacking," but they discovered that whereas they now had weapons to kill, they still lacked the necessary equipment for saving the carcasses. In 1871, for example, the ship *Navy* hunted humpbacks on the Kodiak ground while awaiting the breakup of pack ice to the north. Between 6 June and 1 July the men struck thirteen humpback whales but brought only four alongside for cutting-in. In some cases the harpoon had drawn from the blubber, but three of the thirteen had sunk. A cask was lashed to at least one dead humpback in an effort to keep it afloat, but the next day both cask and whale had drifted out of sight.[33]

Nor did the explosive devices remove all risk from the operation; one wild humpback stove one of the *Navy*'s whaleboats and the men "lost both lines and the Bumgun and a Pearsses [Pierce's] darting gun and lost the Whale."[34] After two particularly trying days, during which two animals escaped and two others sunk, Captain George Bauldry determined to give up humpbacking: "they will only make about 15 bbls," he wrote, "have got to use up 4 or 5 bums & 2 irons so I think it is reather expensive whaling so good by to hump backing here."[35]

Whalemen who tried to harpoon finback whales with the new weapons also experienced disappointment. The logbook of the bark *Sea Breeze*, in the North Pacific on 22 June 1854, reports that the crew "shot one of Brands Boom lances into a fin back to no affect."[36] In 1867 the crew of the *Emily Morgan* shot and killed a finback from the ship—within sight of Mount Fairweather—but the body sank before a boat could be brought alongside. The same thing happened on the following day.[37]

Whalemen who would collect oil and baleen from rorqual whales found themselves in need of some device to keep the whales afloat or bring them quickly to the surface. It was well known that sunken carcasses sometimes returned to the surface as decompositional gasses increased their buoyancy, but this eventuality was no more predictable than the old Koniag way of making fast with a poisoned harpoon and then waiting three days for the animal to "go to sleep." Some whaling vessels carried "humpback irons," which were nothing more than enlarged harpoons which held fast in the blubber while the crew manually encouraged a carcass to the surface with the ship's windlass. But mechanical whale-raising devices were found to be useless at depths exceeding sixty or seventy feet.

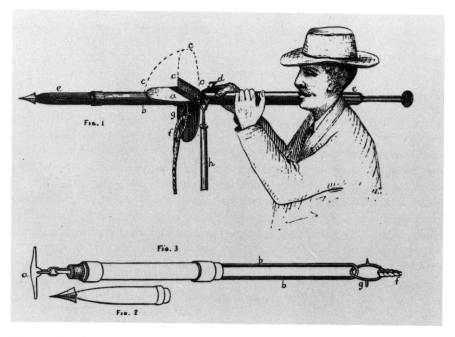

Thomas Roys's whaling rocket attempted to improve upon the technology of whaling and reduce the risk of injury. It was not wholly successful, although rockets remained available for a quarter-century. More than a decade after Roys's failure in British Columbia, the whaling rocket was illustrated as shown here in George Brown Goode's *Fisheries and Fish Industries of the United States*.

After Thomas Roys returned from his discovery of bowhead whales in the Bering Strait in 1848, he developed both a new killing weapon and a mechanical tool for recovering the carcasses. His "whaling rocket" was considerably different from the darting-gun and the various shoulder-guns. Similar in appearance to a modern bazooka, it rested on the gunner's shoulder and fired a bomb-lance. An eye-shield popped up in front of the shooter's face to protect him from powder burns. Roys worked on the whaling rocket in the Bay of Biscay and elsewhere,[38] hoping to solve the problem of ricochet which affected the performance of all whaling bombs. The pointed projectiles would glance the surface of the water or rebound off the slack blubber of a whale.

His solution for raising whales was useful in shallow water. Roys's "whale-raiser" was a large harpoon, ten feet long, weighing perhaps two hundred pounds, fitted with large barbs opening outward and a ring at the rear which accepted first a light codfishing line and then a huge line which fed up to the windlass on the ship's deck. When the whale began to sink, this harpoon was implanted in the carcass and the whaling line from the original harpoon was rove through clasps on its shanks. The whale-raiser sank with the carcass; afterwards the line would be drawn taut on the ship's windlass, and the whale raised. Or so it was hoped. Roys acknowledged that a successful raising apparatus could save the industry $1.2 million in lost whales, and he wrote: "the very day you have proof that Whales can be prevented from sinking, will also prove the Whaling business in its infancy, although existing 250 years."[39]

Technological and financial failures plagued Roys. He established a promising finwhale fishery in Iceland but had to give it up in 1867 after just a few seasons of operation. Then he came to British Columbia. He no doubt remembered the vast numbers of finback and humpback whales on the Northwest Coast; so many, in fact, that one whaleman wrote that he had seen "forty thousand" finbacks and humpbacks there.[40] Roys may have also heard tell of humpbacks sporting in the waters of the Strait of Georgia, for when he began his fishery in British Columbia it was to hunt these animals in their protected domain.

Two investors named Meeker and Arnold underwrote Roys's British Columbia venture; Arnold was probably Edward B. Arnold of Troy, New York, who purchased a portion of the rocket patents when Roys returned from Iceland.[41] From a Victorian ironworker named Joseph Spratt[42] Roys arranged to charter an eighty-three-foot, twenty-five-ton steamer, the *Emma*, a new vessel but one which had already proven itself on a voyage to Sitka under Captain Peter Holmes.[43] Roys equipped it with the whaling rocket and, presumably, the "whale-raiser" and set out. His charter seems to have caused some initial confusion in the local press, which reported in the summer of 1868 that a steamer called the *Ross* had left Victoria under Captain Holmes for "a whaling cruise in the Gulf of Georgia and surrounding waters."[44] Since nothing more is heard of this competitor, one can surmise that the writer confused the vessel's name with that of the charterer.

Roys's real competition came from a group of whalemen organized by James Dawson, a Victorian emigrant from Clackmannanshire, Scotland. In 1866 Dawson and a man named Warren had attempted to prosecute whale hunting in Saanich Inlet; at least three whales were struck, but all were lost in heavy weather.[45] Dawson then joined in partnership with a twenty-seven year-old "Down Maine" sea captain, Abel Douglass,[46] who was recruited in San Francisco along with two other Californians called Bruce and Woodward. The quartet began whaling on 26 August 1868 aboard a small schooner of forty-seven tons, the *Kate*, which Dawson had procured at a Victoria auction just a few weeks earlier. The entire operation was new and untried; the three San Franciscans had only just arrived from the Golden Gate six days before whaling commenced.[47] The tiny *Kate*, not quite fifty feet from stem to stern, made its first whaling cruise to Saanich Inlet, north of Victoria. The crew armed itself with conventional bomb-lances and managed to kill eight whales despite thick fog.

Dawson and Douglass's early success at Saanich Inlet supported an increase in the company's assets. They retrofitted the scow-schooner *Industry* with four 250-gallon tanks, and by the time the first season ended, on 29 October 1868, the *Industry* was ready to bring down nineteen casks of humpback oil —about one hundred barrels—to Dickson, Campbell & Company's wharf at Victoria.[48] This oil was later sold by Lowe Brothers to the Hudson's Bay Company, whose employees readied it for shipment to England.[49] That success occasioned a rousing outburst of praise from the local newspapers: "Mr. Dawson of Saanich," one exclaimed, "in the face of obstacles, which would have deterred most men, and with a perseverance worthy of all praise, has succeeded in placing beyond a doubt the fact that hump-back whale can be killed not only in sufficient quantities to pay, but, to pay well."[50]

Roys's first expedition was less successful. His men suffered a number of mechanical failures and ricochets which spoiled the hunt. During a second cruise, from early September until 6 October 1868, four whales were allegedly struck, and three killed, but all were lost in the dense fogs which continually hampered the expedition. Roys explained that lines had parted and harpoons had broken. By 13 October the *Emma* had returned to the San Juan Island mail-run.[51]

Roys received the sympathy of the press but not its praise. He nevertheless gathered reinforcements and set out to try again. At a meeting in Marvin's Store in Victoria on 22 October he formed the Victoria Whaling Adventurers Company, and quickly sold eighty-two of the one hundred available shares at $100 each. A managing group was named—the store owner among its members—which empowered Roys to go to the Sandwich Islands to attract additional capital and recruit competent whaling men from the North Pacific fleet.[52] Once at the Islands, Roys proselytized for Victoria as the most advantageous port in the North Pacific for obtaining supplies and outfitting vessels.[53] His advice spread outward from Honolulu and Lahaina, and promptly reached even far-away Bremen, relayed there by one Captain Hegemann when he returned from Honolulu in February 1869.[54]

Roys returned to Victoria in January, and the company again sent the *Emma* "outside" with the tools and supplies needed to erect buildings, furnaces, tryworks, and a 150-foot wharf in Barkley Sound. This done, Roys began whaling, but the weather again proved more than a match. Many whales were seen but could not be chased. When at length they made fast to one, the harpoon broke and the animal escaped.[55] In May, after two months of frustration, Roys moved his equipment back into the Strait of Georgia and reinstalled himself at the old haunts in Deep Bay,[56] adjacent to Bowen and Gambier Islands in Howe Sound. There he sighted only two whales, and was chagrined to hear that a large school had been seen at Knight's Inlet, far up the mainland coast above Johnstone Strait. He hurried aboard H.M.S. *Sparrowhawk*, then at Esquimalt, to obtain a tracing of the chart of the inlet.[57] But by the time the *Emma* arrived, the whales had gone.

For the second consecutive year, the Dawson and Douglass Whaling Company outstripped Roys's efforts. Dawson and Douglass removed their shore works from the 1868 site at Saanich Inlet to Cortes Island;[58] and by July 1869 their crew had taken five whales averaging eighty barrels each. It was happily noted in Victoria that the 13,000 gallons thus collected—if sold in New York City—would earn $1.20 per gallon in greenbacks, or eighty-one cents in gold.[59] The late summer of 1869 proved less fortunate. The *Kate* ran on a reef near Mary's Island during a gale and was damaged, though repairs were speedily completed. Despite the mishap, fourteen whales were taken by mid-September, making 20,000 gallons in all.[60]

Three whales harpooned by Dawson and Douglass and then lost were subsequently picked up by "outside parties"—meaning Indians. The *Colonist* newspaper was prompt in reporting, and perhaps also in creating, such outrages; an earlier story accused the Indians of appropriating "to their own use" a whale killed by Roys's men.[61] The native populace regularly suffered bad press at the hands of colonial news editors. Racial tensions were taxed by the rapid increase of population, and even the distant Haida in the Queen Charlotte Islands were vilified, for alleged acts of plunder against white trading parties. The presence of *Sparrowhawk* was occasioned in part by an investigation into the alleged murder by natives of shipwreck survivors from the bark *John Bright*, which had gone ashore three miles south of Nootka Sound in February 1869.[62]

In contrast to press reports, relationships between the natives and Dawson's whalemen seem to have been quite cordial. Native men were occasionally hired, probably for towing carcasses to shore and assisting in the flensing, and some of Dawson's crew lived with, and perhaps were married to native women. An Indian girl who witnessed the whaling party at work reported in her old age that two of the workers, Peter Smith and Harry Trim, were both married to natives.[63]

Roys's Whaling Adventurers Company did not survive the spring. On 4 June the *Emma* lost its propeller off Trial Island while en route to Victoria with passengers who had come from Sitka to Nanaimo in the *Constantine*. The crew had to accept a tow from the large side-wheeler *Caribou and Fly*.

Four days later the company disbanded. Captain Holmes made prompt repairs to the *Emma*; on 30 June he arrived at Sitka with his diminutive steamer in ballast and stayed seven days before departing once more for Victoria.[64] Thereafter the *Emma* did no whaling but remained in the general freighting business under a succession of owners until its shipwreck in 1891.[65] Thomas Roys departed once again for the Sandwich Islands, while the local press lamented his poor turn of luck.

> WHAT strange fatality envelopes nearly every enterprise set on foot in this colony and paralyzes the efforts of our most public-spirited men? Look at the Roys' Whaling Expedition. The party start for the West Coast and kill a few whales — the oil from any one of which would have defrayed two months' expenses of the expedition — but no sooner do they make fast to the monsters than fierce storms arise and the animals, cut adrift, float ashore and are taken in hand and claimed by the savages. Next, despairing of a lull, the party repair to Deep Bay . . . but not a whale enters the bay during the stay of the party. Next, they strike off to Knight's Canal, where shoals of whales are reported to be; but not a fish appears after their arrival. Then they return to Nanaimo, coal and come on to Victoria abandoning the expedition — when, the very day succeeding the one on which they leave the locality of the black diamonds [coal], into the harbor dart a dozen great whales, spouting, sporting and fighting like mad all day long; and, to make the case all the more provoking, if possible, one of the big fish impudently runs aground at low water and actually lies on the beach over one tide — as much as to say, come and catch me, if you can! — before he floats off. Too bad. Too bad![66]

While Roys again canvassed the Sandwich Islands, a third whaling venture, begun by a man named Lipsett, joined forces with the Dawson-Douglass partnership to form the Union Whaling Company. This combined firm cruised three weeks in the Strait of Georgia during the early winter of 1869 and produced 5,800 gallons of whale oil. A second cruise, begun 18 January 1870, brought in a fifty-five foot humpback whale, flukes nineteen feet across, which yielded 100 barrels of oil. But even this whale was insufficient; only four were taken during the two cruises, and as of 3 February 1870 the Union Whaling Company operations were suspended. Lipsett reorganized his own men as the Howe Sound Company, and a party of seventeen set out for that location in early summer.[67] Dawson meanwhile found four new partners in the Victorian business community and registered his new Dawson & Douglass Whaling Company on 27 June 1870 under the provisions of the Joint Stock Companies Act, with capitalization of $20,000 and whaling grounds "universally admitted to be excellent."[68]

Roys remained in the Sandwich Islands, chartering the coasting schooner *Anne* and hunting whales in Kalepolepo Bay for a tryworks on shore at Olowalu, just south of Lahaina. He also demonstrated the whaling rocket in Honolulu[69] before returning to British Columbia at the end of May 1870. The effects of his defunct Victoria Whaling Adventurers Company were meanwhile sold at auction: one boiler frame, a smokestack, fire furnace bars, one whaleboat, five oars, two sails, two boat-hooks, nine paddles, one stove, one blubber mince machine, one hydraulic pump, a powder crusher and pestle, four bomb-guns (probably whaling rockets), one Brand-brand bomb-gun, thirty-three rocket cases, twenty-five bombs and harpoons, twenty-seven bomb-harpoons, six hand irons, five lances, about five hundred dogfish hooks, two rifles, and sundry other items.[70]

In June 1870 an unidentified group of whalemen armed with "the Roys Rocket" chartered the schooner *Surprise* and went off to hunt whales at Barkley Sound. Roys is not mentioned by name in connection with this expedition, but it seems likely that he was the guiding force behind it. At Barkley Sound they commenced building "houses" for the use of the whalemen.[71] Their take was meagre. At the conclusion of the summer whaling season the three firms — the one probably managed by Roys, Dawson-Douglass's, and Lipsett's — had taken only thirty-two whales, which yielded 25,800 gallons of oil worth about fifty cents per gallon at Victoria.[72] Only one of the firms used a vessel equipped with a whaleboat, the others apparently sent boats from their shore camps. Only forty-nine men had been employed, all told.

Late in 1870 the unflappable Roys sailed a third time for the Sandwich Islands, and when he returned, on 10 May 1871, he had new financing and a proper square-rigged vessel for a whaleship. This was the brig *Byzantium*. Built in Southtown, County Suffolk, England in 1844, this eighty-five foot, 179-ton brig had come out to Vancouver Island in 1867 with a cargo of naval stores on what was said to have been an "unlucky" passage.[73] Afterwards it came into the hands of Captain Rufus Calhoun of Port Townsend, Washington Territory, who sailed it all over the Pacific.[74] Roys had become friendly with Calhoun during his transpacific sojourns, and the whaleman's enthusiasm convinced the shipowner of the prospects for success. Calhoun decided to invest $8,000 in the whaling venture and offer his *Byzantium* for conversion into a whaleship.

In Victoria, Roys unveiled his new plan. A whaling station would be constructed in Cumshewa Inlet in the Queen Charlotte Islands and the *Byzantium* refitted in proper style, complete with an on-deck tryworks. Captain Calhoun was praised in the press as "a man of great energy and enterprise, [with] unbounded faith in the whaling resources of these waters and pluck enough to invest a large sum of money in proving that the faith is not misplaced."[75] By 31 May the tryworks had been installed, and shortly thereafter the whaleship was seen by H. L. Langevin, provincial minister of public works, who described it in a report published in Ottawa the following year:

> I saw one of the whalers, the *Byzantium*, in Deep Bay. She was an English brig, commanded by Captain Calhoun, and on board of her was Captain Roys, the inventor of an explosive ball, which is used in the whale fishery, and which, on penetrating the marine monster, explodes, and throws out a harpoon. The first whale, against which this projectile was used, was killed in 1868. In 1869 and 1870 the company made use of a small steam vessel *Emma*; and their success last year [1870] induced them to devote to the trade a brig of 179 tons, manned with twenty hands.[76]

"I was assured," Langevin continued, "that if the expedition proved a success, there is room in our Pacific waters for at least fifty undertakings of a similar character."

The minister also noted the return to Victoria of the schooner *Industry* with a cargo of 300 barrels of oil. The *Industry* seems to have become a pawn in some sort of "divorce" settlement, for by 1871 Captain Douglass had separated from James Dawson to form his own whaling company, in partnership with Victorian vintner and publican James Strachan. Strachan was something of a shipmaster himself; he had visited Sitka during 1870 in a tiny five-ton schooner called the *Major*. In 1871 he assumed ownership of the *Industry*, which had formerly been held by another publican—probably his brother—named Alexander Dalgarus Strachan.[77] The dynamics of the 1871 whaling season were thus significantly altered. Roys teamed with Calhoun and set out for Cumshewa Inlet in the *Byzantium*; Douglass and his new partner Strachan sailed in the *Industry*; while Dawson rejoined with Lipsett to form the British Columbia Whaling Company and departed aboard the *Kate*.[78]

The *Kate* had been enlarged at Meldrum & Small's shipyard in Victoria. Lengthened from 49 to 64.6 feet and increased from 46.93 to 58.11 registered tons, the hull was apparently rechristened *Dominion* in honour of the recent entry of British Columbia into the Canadian Confederation. Old salts consider a midstream name change a bad omen, and the new appellation proved them right; before long the *Dominion*—nee *Kate*—went on the rocks and was damaged. *Dominion* was very soon crossed off the legal registry of hull number 64132 in favor of *Kate*.[79]

Dawson and Lipsett's British Columbia Whaling Company made about 20,000 gallons of oil during 1871, the Douglass-Strachan venture, about 15,000.[80] Whale oil then sold in Victoria for less than fifty cents per gallon, so the companies fared poorly. The British Columbia Whaling Company had invested about $15,000 preparing for the season, and Douglass and Strachan about \$20,000;[81] thus, they spent about a dollar for every fifty-cent gallon they gathered. The grounding of the *Dominion-Kate* confounded matters, and shortly afterward the British Columbia Whaling Company was liquidated. In March 1872 the schooner went on the auction block, together with a 100-acre pre-emption on Hornby Island[82] and the company sheds, wharf, and out-buidings. Vessel and equipment passed to Robert Wallace, who attempted to continue whaling with another former company partner, James Hutcheson, but they had very little success. The last news account of their activities, in July 1873, noted that the schooner had been hunting near Lasqueti Island, but that the whale run was very light.[83] By the end of 1873 the *Kate* was in the hands of Thomas Pennant, a pilot at the port of Victoria.[84]

Roys fared worse. The *Byzantium* reached "Gumshar" in early August and the crew began work on the whaling station. By 8 August the brig was on a cruise in favourable weather, and the Haida near Cumshewa were said to be "docile as lambs." Before the end of September, however, the weather worsened and Roys decided to return to Deep Bay.[85] Near midnight on 18 October the *Byzantium* struck on the rocks in Weynton Passage, Johnstone Strait.[86]

> There came a gale at southeast, very quick and stormy, and right at a little before midnight she struck upon the rocks, stoving two holes in her, and we escaped in the boats, knowing well when the tide turned she would swing off and go down as there were 150 tons ballast on board. This she did, for a few hours afterward she had disappeared altogether with everything on board.[87]

In the whaleboat they made their way to the shore, carrying only the ship's chronometer and some personal articles, and spent a frigid night huddled on the beach. In the morning they returned to the brig despite the continuing storm, and found water within three feet of the beams.[88] Any hopes of refloating the *Byzantium* on the high tide were abandoned, and the men turned the whaleboat in the direction of Fort Rupert—near modern-day Port Hardy—where they afterwards arrived safely.[89]

On 22 October they were put aboard the steamer *Otter* for Victoria. When the *Otter* passed the wreck site, all hands strained for a glimpse of the *Byzantium*, but nothing could be seen. Captain Calhoun, not aboard at the time of the accident, was notified at his home in Port Townsend; an insurance policy for \$5,000 covered part of his investment, but the total loss was expected to reach at least thrice that amount.[90]

A month after the fact, the *Colonist* openly accused the natives of salvaging—*plundering* was their suggestion—everything of value from the brig and then hiding the hull. "It appears," they wrote, "that the Byzantium, after having been abandoned by her crew, did not go down as was supposed. She floated on the tide, when the Indians slipped her cable, took her away out of sight of passing vessels, stripped her of everything valuable, left her, and she has not been seen since."[91] A rival newspaper repeated the *Colonist*'s innuendos and added an inflammatory fillip of its own: the hull, it claimed, had been observed floating out to sea by the "northern passage."[92] The recovery of goods by white salvagers would have occasioned joy and exclamations of heroism, but the native effort — if anything of the sort actually happened — only fueled prevalent racial antagonism. It seems highly probable that the ribs of the *Byzantium* lie there yet, in the dark sixty fathoms of Weynton Passage.

The loss of Calhoun's beloved brig stranded Roys on the beach, worn out from years of trying to rebuild the whaling industry. He took passage to San Francisco in the lumber coaster *Wildwood* in November 1871 and there conducted some business on behalf of his whaling rocket. But he never again operated a whaling company. He next travelled to San Diego to recover his health, and continued on to Mazatlán, Mexico. In January 1877 an American doctor found him "dazed and destitute, and wandering aimlessly in the streets"; he died a week later in the doctor's house.[93] His arch-rival, Dawson, also quit whaling after the 1871 season and died a few years later, at age fifty-six, of "general disability."[94] Captain Douglass's success increased, though not in whaling; he built and sailed schooners and eventually entered the seal-hunting business, in which he remained active as late as 1893.[95]

Roys's whaling-rocket concept lingered in San Francisco for a few more years. He had placed an order with Hawkins and Cantrell's Machine Works for a new rocket press, mould, and rammers when he arrived there in the *Wildwood*, but he was heavily in debt and found it necessary to sell at least some of his patent rights to a Scots machinist named Hugh Lamont. After Roys died, Lamont worked in partnership with Robert L. Suits and Suits's brother-in-law to produce the rocket. From one of Roys's former partners, pyrotechnist Gustavus Lilliendahl, they bought the remaining patent rights and marketed a whale-hunting weapon called the "California Whaling Rocket."[96]

The commercial history of this weapon digresses from the history of Northwest Coast whaling, except that several of the Arctic whaleships may have fired a few on the Kodiak Ground. Only in two respects is its later history curiously pertinent. First, one of its advocates was an ambitious San Francisco shipbreaker and maritime entrepreneur named Thomas P. H. Whitelaw. Whitelaw built a steamer of forty-four tons, the *Daisy Whitelaw*, and successfully employed the California Whaling Rocket against finback whales near the Farallon Islands. Some years later, Whitelaw would fail in his bid to introduce modern whaling techniques to British Columbia.

The second note concerning the whaling rocket was proffered in 1883 by whaling expert James Templeman Brown. Brown wrote that the rocket was in use by the "Northwest Whaling Company" at Sitka, Alaska. His informant, one Mr. Wilson of that town, said that the whalemen fired it from the deck of a small steamer. This company, most likely the Northwest Trading Company of Portland, Oregon, appears to have engaged in shore whaling from its post on Killisnoo Island in Chatham Strait about 1880. The operation enjoyed very little success; a bomb-gun explosion—perhaps it was a California Whaling Rocket—killed two men, and after a few years the company abandoned whaling in favour of the herring fishery.[97]

Following the complete collapse of the infant coastal whaling business in British Columbia after 1871, the provincial Fisheries Commission concluded that the failure "must be attributed more to want of proper appliances than to the scarcity of whales, which are numerous as ever."[98] Before the end of the century several attempts were made to renew the business, but none reached fruition. In 1886, for example, two Victoria merchants—Gutmann and Frank—announced plans to buy a complete "whaling outfit" in San Francisco. They expected to build one or two land stations and employ a 250-ton schooner, which would be fitted out also for the Arctic whale fishery. Frank travelled to San Francisco to obtain a suitable hull, but nothing seems to have come of his efforts.[99]

Harry Trim, who formerly worked with Abel Douglass, may have done some whaling in the late 1880s. Sixty years later, the legendary Vancouver city archivist Major James Matthews was told that Trim had run a shore station with a substantial two-masted schooner and a Columbia River-type fishing boat for a whaleboat.[100] His operation may have been sited on Pasley Island. Afterwards he is said to have had a whaling operation at Jericho Beach in Vancouver, "where the golf links are now,"[101] but the details, unconfirmed at the time of Trim's death on Westham Island in 1922, are unclear.

The most promising proposal came from the aforementioned San Francisco entrepreneur T. P. H. Whitelaw. In 1890 Whitelaw applied for permission to hunt whales in provincial waters with the converted salvage vessel *Whitelaw*. His application bounced from the minister of Customs to the minister of Marine and Fisheries, who turned it down on the grounds that Whitelaw proposed to use an American-registered vessel rather than employ a British hull.[102] He may have been concerned that the product and proceeds would return directly to San Francisco rather than remain in Canada.

Whitelaw planned to refit his steamer with steam-launches and the "latest improved patent guns," presumably the Pierce-and-Eggers-type shoulder gun and perhaps the California Whaling Rocket. He also proposed to outfit the old *Alexander* as a "floating refinery" to serve in lieu of a shore station. Whitelaw thus anticipated the twentieth-century pelagic "factory ships" by more than a decade.

When this became known, a Canadian named Manson is said to have begun construction of a two-masted steam-auxiliary whaling vessel at the Albion Iron Works. To be called the *Thistle*, it was intended to be "the ship in view in the [protest?] entered against Captain Whitelaw's embarking in [the whaling] industry."[103] Another steam-schooner seems to have been built—or begun—in the lower mainland of British Columbia by one Captain Cooper.[104] Perhaps Manson and Cooper need not have bothered; Whitelaw's application was rejected, and there is no evidence that any other attempts reached the sea before the turn of the century.

Except for the few whaling vessels from San Francisco that still looked in along the coast of the Queen Charlotte Islands, the only whaling in the Pacific Northwest seems to have been among the Makah and perhaps a few West Coast people on Vancouver Island. The Makah continued to hunt in cedar canoes, but even they had come into possession of modern tools and iron, which had long ago supplanted the dried mussel-shell harpoon points. The advent of the steam engine added some new wrinkles to the traditional whale hunt, as the crew of the towboat *Lorne* discovered one day in 1905. The chase began in traditional fashion as "60 indians in 6 large canoes" stalked a whale near Cape Flattery. "We managed to get in close," one of the *Lorne* men wrote, "and occasionally a canoe load of indians would dash up alongside the whale and the harpooner would stand up in the bow . . . and, with a shout, drive another shaft into the monster. . . . When the whale was killed he looked like a giant pincushion with all those balloon bladders sticking in him."[105] But then something most untraditional occurred. Another towboat, the *Wyadda A. Prescott*, took the dead whale in tow and pulled it hastily back to the waiting Makah at Neah Bay. Students of Makah society later corroborated the use of powered towboats, which were known to have brought in as many as four whales in a single day.[106]

Despite modern conveniences, the Makah hunt was much reduced. Government sources reported that only three whales were taken in 1892, for example, with a commercial value of but $600.[107] Many young men who would ordinarily have followed in the footsteps of their whaling relatives now found higher-paying employment as hunters in the service of the fur-seal fishery.

Until the turn of the twentieth century, the whaling industry in the Pacific Northwest lay fallow, awaiting those "proper appliances" proposed by the British Columbia Fisheries Commission. The wait was short; even as Roys and Dawson abandoned whaling for want of them, the very tools they required were well along in development halfway across the world, in the waters off Norway.

Norway was a nation caught in the transition from a barter-agrarian economy to one built upon finance, trade, and industrialization. There were few outlets for Norwegian production, yet a rapid population growth glutted the conventional trades. Most labourers struggled under the oppressive circum-

Makah whalemen were quick to include new technology in their hunt. These Cape Flattery whalemen have hitched a ride from a steam tug, which is hauling a sailing ship to sea. Photograph by H. H. Morrison.
James G. McCurdy Collection, Museum of History and Industry, Seattle, Washington

stances of an employer's market, working their lives at minimum wage. Fishing and shipbuilding, natural components of Norway's maritime orientation, enjoyed a period of expansion. And Norwegians went to sea readily, for a sailor might earn half again or even twice the wage paid for comparable work on shore.

One Norwegian, Svend Foyn, pioneered the new technology that would revolutionize whaling and give Norwegians a worldwide control of the industry. Foyn was already an accomplished mariner and sealer when, in the 1860s, he turned his attention to the whale fishery. He was well acquainted with the finback and blue whales which ranged freely off the coasts of northern Norway and Iceland. He also knew that efforts to catch these large rorquals came to nothing on account of their speed and strength. Fortunately, steam had proven itself useful in marine applications; British whalemen had already equipped a few of their Davis Strait fleet with auxiliary steam engines to help push through sea ice.[108] Foyn channelled his resources into the construction of a powered schooner with a minimal sail-plan, one which would employ steam as the primary driving force and utilize a variety of the latest swivel-guns and other whaling weaponry.

He contracted this vessel to Nylands Verksted near Christiania, who produced a racy steam-schooner with two short masts and a prominently large funnel. Foyn named the new craft *Spes et Fides* — "Hope and Faith" — but it soon afterward acquired the loving nickname "Spissa," by which it became widely known. Eighty-one feet long and eighty-six tons,[109] "Spissa" was somewhat shorter and altogether unlike the classic sailing whaleship; it was also fast enough to keep up with finwhales.

At first, Foyn mounted a variety of harpoon-guns, including swivels mounted at the bows and other types set up along the deck.[110] He began whaling in 1864 near the mouth of the Varangerfjord, supplying his processing station opposite the town of Vadsø.[111] Whales were numerous, but for three years he was a complete failure owing to the uncertainty of the weapons. In 1866 he travelled to Iceland to examine Thomas Roys's operation; there Foyn carefully noted both the whaling rocket and a peculiar device called a "compensator" — Foyn dubbed it the "rubber rope"[112] — which employed rubber shock cords, mounted on the mast, which shifted the strain of the whaling line onto the vessel's hull. After a day and a night watching Roys's operation he concluded that the rocket was inferior to equipment he had already tried and was, in any case, too expensive to be practicable.[113] The compensator did attract his interest; later, after Roys's Norwegian patents on the "rubber ropes" had expired, he revamped the compensator idea and patented it. Marcus Bull soon developed an improved spring-accumulator which improved on Roys's early ideas, and before long an arrangement of tension springs mounted deep in the hull became standard equipment in whaling steamers. Such compensators were fastened to the whaling line and absorbed much of the strain.

Despairing of success with the small weapons, Foyn and several colleagues developed the prototype harpoon-cannons. In 1868, whaling over a longer season with an early version of such a gun, he successfully captured thirty whales. In 1870 Foyn took thirty-six whales,[114] and patented a jointed harpoon with four moveable barbs which spread open after impact. Other patents covered a variety of projectile-heads; the exploding *granatharpun*, or "shell-harpoon" quickly proved the most effective design and was adopted as the standard, probably by 1873.

By the mid-1880s the design of the gun and gun platform had been standardized, so a description of the weapon at that time serves also to illustrate the harpoon-cannons that came into general use in North American waters after 1900:

> The harpoon-cannon works on a pivot fixed on the forward end of the forecastle. There is no bowsprit or forestay: a kind of platform projects over the bows on either side, giving room to the harpooner to stand and turn the gun well round to either side. Over the stem projects a square sheet of iron, lying at a slight angle forwards; on this about twenty fathoms of the whale-line ... is very carefully coiled down, and lashed in place with spun-yarn, which breaks directly the line gets the

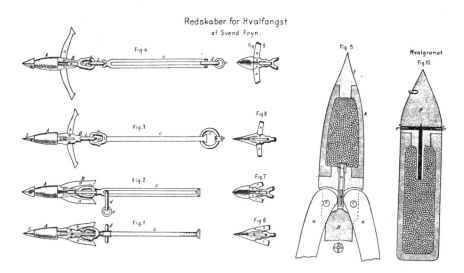

Redskaber for Hvalfangst
af Svend Foyn.

Svend Foyn's trials proved that the exploding grenade was effective in killing the largest whales. The *granatharpunen* shown at Fig. 2 in this early patent drawing proved the most popular type, and its exploding head (Fig. 5) soon became standard equipment.

least strain on it. The sheet of iron is hinged, and when steaming in rough weather through a head-sea, the harpoon is removed from the gun and the iron turned up, so as to protect the gun from the seas.

The cannon has naturally to be very strong . . . in this instance 4 1/4 inches thick at the muzzle. The charge of powder is 15 "lod" and is kept ready measured off in round canvas balls about the size of a cricket-ball. The recoil is taken off by pads of gutta-percha several inches thick at the rear of the trunnions. A pistol-stock shaped handle is fixed to the breech to aim the gun.[115]

Foyn had no sooner made a success with his harpoon-cannons than the first cries were raised about the danger of exterminating the whales and destroying the fishery. In a statement replete with foresight, the editors of a book entitled *The Great Fisheries of the World* fretted: "There seems reason to believe that the Norwegian whale-fishery may extend beyond Vadsø. But it is to be hoped that a prescient legislation will enact a 'close time' for the ocean-monsters, and prescribe certain limits to the whalers in their usual cruises. It would be a serious disaster if this important fishery were destroyed by an imprudent greed of gain."[116]

Svend Foyn's operation killed 506 whales from 1871 to 1880, many of them the huge blue whales that may be counted the largest animals on

Svend Foyn built the first modern whaling station in the Varangerfjord opposite
Vadsø, Norway. It served as a prototype for others that would later be constructed
around the world.
Kendall Whaling Museum, Sharon, Massachusetts

earth.[117] At the shore station opposite Vadsø Foyn employed a steam-winch
for hauling the carcasses onto the shore-station slipway. By the early 1880s
he had helped to develop a system of making "guano" — fertilizer — from the
ground bones and grist left over from the flensing processes, thus adding to
his profits.

Foyn obtained patents on most of his major inventions and set a precedent
for Norwegian inventors. Between 1842 and 1940 a total of 533 whaling-
related patents were issued in Norway.[118] Many of the earliest were exclusively
his for a period of years; when his rights began to expire, beginning in 1882,
other Norwegians quickly entered the whaling business. That year, six whaling stations were in operation around the Varangerfjord alone; by 1886,
nineteen Norwegian companies operated thirty-five steam chaser boats.[119]
Most companies organized themselves as share-companies, the stock being
owned among a very limited number of investors who knew each other.
Company by-laws often restricted further stock sales to the group of shareholders who had initially purchased into the company; the inner circle thus
remained "small and collected in a local sense."[120]

When Svend Foyn died, in 1894, he was confident that Norwegian steam
whaling would take its place among the maritime industries of the world. In

fact, he was already revered in his homeland as the man who revived the Norwegian maritime economy, a saint whose characteristics included "a devotion to work and a hatred of idleness, amounting to a passion; intense and orthodox views of religion; a will, energy, resourcefulness, administrative quality, and power of command over himself and others, which marked him out as a ruler of men and leader in large undertakings."[121] Many were the stories told about him, until his character was expanded into a folk figure not unlike the American legend Paul Bunyan. Foyn larger than life became a superhuman who succeeded at superhuman tasks for the benefit of all Norway. The strong, humble, and impetuous figure that was Foyn—in life "a stout, short man, whose body, in spite of his age, showed strength and flexibility"[122] —served as a role-model for the heroic Norwegian cannon-gunners who would chase whales into the coming century.

Foyn's inventions had proven capable of slaying and securing the great blue and finback whales. The future of whaling, if not the immediate condition of the market, seemed bright. But almost at once the fishermen from the coasts of Finmark and Iceland complained that the cod, capelin, and other fish no longer swam inshore within easy reach of their nets and lines. To catch them, it had become necessary to travel dangerously far to sea. The fault was placed at the doorstep of the whalemen. They were accused of killing the whales that had once driven the fish close to shore. Proponents of the whale fishery argued that the fish had always stayed offshore in certain years, but the fishermen prevailed.

> It is a very old opinion, as is well known, that the approach to the coast of the capelan and herring is solely caused by the whales, and that nature has appointed these gigantic animals to gather the scattered schools of herring and capelan from the different parts of the ocean and chase them towards certain portions of the coast in order that man might get his share of the wealth of the sea. For this reason a sort of veneration was, in olden times, shown for whales, which were considered the special servants of Providence. Our time is less fanciful in its interpretation of natural phenomena, and we hesitate somewhat to assign to the whale the part of a disinterested benefactor of mankind.[123]

Whalemen paid little heed to these complaints, but soon Iceland prohibited Norway from whaling in its home waters. In Norway, too, fishermen took their protest to the *Folkething* (Parliament), and demanded protective legislation. When none was forthcoming they began a series of violent disturbances which culminated in the famous Menhavn Riots on the nights of 2 to 4 June 1904. Then, fifteen hundred people attempted to destroy the facility of the Tanen Whaling Company in the town of Menhavn. This "enchanted drunkenness"[1] took hold again on 13 July, and a complete and systematic destruction was the result. The military stepped in to restore order and many agitators were arrested.[2]

After the riots, the *Folkething* prohibited whaling in certain waters of the north for a period of ten years. Not long thereafter, the fishermen in the Shetland Islands and the Hebrides voiced similar objections to the presence of the Norwegian steam-whalemen, and whaling controls were instituted with the passage in 1907 of the Whale Fisheries (Scotland) Act.[126] By then Norwegian merchants had taken a good look elsewhere. Norway was not the only country in the world with whales swimming in its seas; stock companies were quickly organized to invest in shore-station whaling in Newfoundland, South America, Africa, and eventually as far from home as Australia and Alaska. Initial success led to a phenomenal expansion of Norwegian enterprise. Even in Japan, where national legislation forbade foreign ownership, Norwegian businessmen found a ready market for the equipment and expertise of the new whaling.

It must be remembered that this steam-powered, cannon-armed fishery had not existed before. The steam engine was new to whaling; Foyn's successful harpoon-cannon and exploding harpoon were new; the steam-winch for raising carcasses was new, as was the compensator which took the strain off the whaling line. These technological innovations allowed whalemen to harvest species of whales which had been untouchable since whaling began.

The Americans, who had dominated sperm-whaling and right-whaling in the nineteenth century, now showed little inclination to maintain that leadership into the twentieth. They were busy extracting profit from the development of the American west and in building the United States into the most powerful industrial nation on earth. There were no venture capitalists in New Bedford or San Francisco who considered whaling a viable speculation. The market for marine mammal oils seemed permanently depressed, and the right, sperm, and bowhead whales seemed to be approaching commercial extinction. Norwegians thus jumped into a wide breach, with the equipment developed by Foyn and the others, and obtained a powerful grip on the new industry.

Samuel Foyn—Svend's nephew—promoted this new whaling when he arrived in British Columbia in 1898. His mission was to interest merchants in the Norwegian plan and to seek "concessions" from the government to begin a Norwegian-dominated regional whale fishery. He pretended—not very well—to keep his activities a commercial secret, but he "revealed" his intentions to the first newspaper reporter who bothered to ask,[127] carefully explaining how $45,000 would purchase a steam chaser boat and the beginnings of an oil factory and guaranteeing that stockholders would realize at least a seventeen-percent dividend on their investments. To prove the richness of local waters, he coyly offered a wager that he could set out from Vancouver and return within six hours with a dead whale in tow. He offered to "banquet the board of trade" if he failed—a wise move regardless—but if successful, he insisted upon being paid market value for the whale, which would then be drawn through the streets of the city for the public to see. The proceeds would be given to the city's hospitals.[128]

But Samuel Foyn did not immediately convince anyone to invest in whaling. Perhaps Victorians remembered the troubles of Roys, Dawson, and Lipsett thirty years before. It was true that Canadians rued the success of the American whalemen in the western Arctic—the *Canadian* Arctic—whose annual revenue exceeded a million dollars. In 1899, the *Colonist* would report forlornly that not a single Canadian whaling vessel had been operated, though the whales were to be found off the Canadian coast.[129] But more likely Victorians were busy, for they had invested time and energy in the fur-seal hunt, and the market for the seals' beautiful pelts showed no sign of abating. As early as 1888, Fisheries Inspector Thomas Mowat had given it as his opinion that whaling would not be attempted until the sealing industry had become overcrowded with competition.[130]

At the turn of the new century, however, international politics threatened the livelihood of sealers in the Pacific Northwest. Some who listened carefully to the message brought by Samuel Foyn realized that the future would be found in a new, and yet unregulated, sea-hunting industry.

5

WARRING ON WHALES

*It is to the men of Sandefjord or Tonsberg that
[investors] confide the management of the enterprise, and
it is from among them that they recruit their personnel.
Finally, when . . . [locally owned] fisheries are established, it is still to the Norwegian shipyards that they
turn for the construction and equipment of the whaling
vessels.*

CHARLES RABOT
"THE WHALE FISHERIES OF THE WORLD" (1912)[1]

Legends are sometimes useful as markers, pointing to the beginnings and
ends of eras and the dynamic path of human development. Valuable stories
told and retold become the checkpoints of history and the milestones of times
when mankind turned, for better or for worse, to face a changed and perhaps
expanded horizon. The efforts of the mythic Svend Foyn became the earliest
legends of modern whaling as advancing industry resolved the herculean
joust between man and whale into a unilateral cannon-bombardment. Soon
the romance would be compromised, by accountants, barristers, engineers,
bankers, and, eventually, government bureaucrats who determined to regulate commerce for the benefit of all. Whaling in the twentieth century became more a matter of financing and record-keeping, and less a matter of
life-threatening hazard at sea.

The final saga of the old, freewheeling sail whaling in the Pacific Northwest is but a minor incident, really. Its players are men one might consider
loathsome: vagrants, dope addicts, petty thieves—some of each the victims
of the shanghaiing trade and all brought together in the forecastles of two
aging sailers that met in the Gulf of Alaska for what had to be the last "gam"
there and, in fact, about the last ever.

By 1910 any sort of man was conscripted for the North Pacific and Arctic
whaling service. As there was little opportunity to catch whales, few will-

ingly chose such an ancient but now dishonoured calling. At various places along the coast steam-powered whaling had already begun, and the good berths for sailors were to be found there or in the new freighters and passenger liners that smoked their coal-black trails across the vastness of the world's oceans. There was little pride and rarely a profit in belonging to a sailing whaleship, so the crew lists of those vessels that remained afloat in the North Pacific necessarily included the riffraff and the derelict, the naive and the lost souls of the port of San Francisco.

The last meeting among such men on the Kodiak Ground must have been between the crews of the bark *Gay Head* and the schooner *Latetia* off Cape St. Elias in June 1912. The schooner had sailed from San Francisco in April, a week after the *Gay Head*, and their meeting was apparently prearranged. It could have been a poignant scene as the worn-down crew of the schooner watched the antique square-rigger come up so the two masters of their venerable industry could hail one another. In fact the "gam" lasted barely half an hour; the schooner's captain was anxious to chase a whale that had been sighted earlier. A boat's crew from the *Latetia* delivered their mate to the bark but were forbidden aboard, so they waited in their whaleboat and "swapped insults with the forward crew of the Gay Head."[2]

It was nearly the end for all of them. The master of the *Latetia* died that voyage, and his schooner never sailed again for whales. The *Gay Head* made another cruise from San Francisco, in 1913, but wrecked at Chignik Bay, Alaska.[3] After that, it does not appear that any more pure sail-whaleships followed into the fog of the Pacific Northwest. The new whaling was already under way. Steam chaser boats busily swept along the coast for whales, and oil-fired steam rendering plants on shore processed them into the products of the marketplace. The important business took place at meetings of boards of directors and in the regulatory halls of government, not at sea with the brave man at the bow, hurling a harpoon to snag himself on Leviathan. The American whaling fleet, in 1846 the largest in history, by 1902 was reduced to thirty-eight hulls.[4]

The new whaling technology made use of each carcass far beyond a simple stripping of the blubber and baleen. Steamboats, harpoon-cannons, and steam-assisted rendering machinery together marked the new order, and efficiency its byword; money could be pressed, boiled, ground, and dried from the giant's body. The shore station not only processed oil but also manufactured fertilizers, bone meal, animal fodder, glue, and eventually edible meat, vitamins, and the stuff of lipsticks, perfumes, soaps, crayons, and even stringing for tennis racquets. Steam winches, pressure cookers, rotary-drum dryers and other mechanized tools reduced the whale to profit in quick time, with more complete utilization of raw material, improved quality control, and better working conditions than could be had at sea.

The new whaling steamers covered a large area of water quickly. The masthead observation barrel improved upon the unprotected crosstrees of the sailing whaleship, while the engine eliminated the lack of responsiveness, windward liability, and the sweated-labour aspect of the wind ship. The harpoon-cannon made prompt work of whale killing with far less danger, since the explosive-laden 125-pound harpoon counterweighted the contest in favour of the whalemen. This was much appreciated by all hands, even though some of the challenge of whaling—and much of its romance—was sacrificed.

The introduction of improved whaling processes and equipment paralleled developments in other industrial arenas which expanded the utility of whale oils. The traditional uses for whale oil, notably for lamplight, machine lubrication, leather tanning, and the production of coarse soaps had been much diminished by the growing availability of less-costly vegetable and mineral oils. After 1869, when margarine was first made from animal fats, chemists were encouraged to find ways of making solid fats from liquid oils, including vegetable and whale oils.

Between 1902 and 1912 a superior process called hydrogenation was developed, which converted unsaturated fatty acids and their glycerides—obtainable in large quantity from whale oils—into solid fat. The oils derived from the large baleen whales were particularly suitable; sperm oil, being more of a wax than a fat, could not be hydrogenated. The process removed the strong smell and taste of whale oil, as well as most of its colour. A small quantity of hydrogenated whale oil could thus supplement animal fats in margarine production, and, after 1929, more sophisticated techniques made it possible to vastly increase the percentage of whale oil in edible fats.

Removing the smell and colour also increased the value of whale oils in soap manufacturing. It was well known that the liquid part of whale fatty bodies was convertible into soap, but its smell and its tendency to discolour had restricted its use to industrial-grade cleansers; hydrogenation made whale oils usable in better cosmetics. Glycerine, a byproduct formerly discarded, was now saved for several uses, not the least of which was in the manufacture of explosives. The increasing demand for whale oils was encouraged by the development of an international marketplace centred in Glasgow.[5] Glasgow gained ascendancy in part because Scottish sail-whaling and its whale-oil market were active concurrently with the beginning of modern whaling in Norway and in part because a Norwegian immigrant to Scotland, Christian Salvesen of Leith, established business relations with Svend Foyn and assisted in marketing the Norwegian oil through Glasgow. Makers of margarine, lard compounds, and soaps demanded more and more whale oil; accordingly, its price began an upward and relatively stable climb after 1900, and supported a renewed international interest in whaling.[6]

Several factors soon conspired to bring a modern Norwegian-style whale fishery to the Pacific Northwest. First, whales were abundant. It was not considered unusual to see humpback whales breaching and spyhopping in the Straits of Juan de Fuca and Georgia. Visitors travelling the coast by passenger steamer regularly sighted large whales, and herring fishermen occasionally snagged them in their seine nets.[7] Second, local merchants at Victoria continued to bristle because the Americans were capturing whales from the Queen Charlotte Islands northward to the Canadian Arctic, and the bulk of this oil and baleen was borne away to San Francisco. This was tantamount to poaching, and they encouraged the dominion government to evict the pirates or at least require licensing of foreign whaling vessels operating in Canadian territorial waters.[8]

If asked, Norwegian whalemen could already demonstrate good results on the Asian Pacific shore. A Russian company founded by Count Keyserling — called the Pacific Whaling Company — had begun operations around the Sakhalin Islands in the mid-1890s using two modern steam chaser boats, the *Nicolai* and the *Georg*, constructed at Christiania, Norway. Keyserling's operation took perhaps 220 whales by 1897 — some of the meat was canned — and a large freighter was made over as a floating factory for trying-out the oil.[9] The Japanese also began a modern whale fishery, in 1899, using the chaser boat *Choshu Maru*. A decade later, twelve companies competed there with twenty-eight boats, most if not all employing Norwegian captains and harpoon-cannon gunners.[10]

With all this in mind, it seems surprising that Victorians did not immediately take up Samuel Foyn on his offer of 1898. But the west coast merchants who were likely to be interested in whaling were caught up in the hunt for fur seals. Since the late 1860s the sealing business in Victoria had grown tremendously. Hunters took pelts in the spring and early summer as the migrating herds made their way to the breeding grounds on the Pribilof Islands in the Bering Sea. Despite increasing protest from both the United States and Russia, vessels from Canadian ports continued to hunt these animals. The Victoria fleet, which in 1876 totalled "at most a half dozen vessels,"[11] by 1891 numbered at least forty; one leading sealing historian has estimated twice that number.[12] Sealing became so lucrative for the Makah and West Coast native hunters that their traditional whaling expeditions virtually ceased. They signed on, instead, as hunters of the fur-seal.

The annual catch was large. In 1894, for example, some hundred thousand skins were taken, despite increased patrols by Russian and American cruisers and the occasional confiscation of ships and cargoes.[13] It appeared as though the seals might be exterminated. The situation became so politically charged that an international tribunal was convened in Paris in 1893 to take testimony on the issue. Stiffer regulations were forthcoming; in 1897 the

U.S. government forbade its citizens to engage in sealing at any time or place, and in 1911 the North Pacific Fur Seal Convention finally outlawed all pelagic seal hunting in the North Pacific except that undertaken by native hunters utilizing traditional weapons.

Many investors were injured on account of seizures of cargoes, imprisonment of crews, or levying of fines. If they were astute, they began to search for some other maritime livelihood, since it appeared certain by the end of the nineteenth century that the golden days of sealing were over. Modern whaling supplied that potential, and there was no statute book to say when, where, or how whales could be harvested. Samuel Foyn's message seems to have inspired the principals of the beleaguered Victoria Sealing Company, and none more surely than its captain, George Washington Sprott Balcom.

Sprott Balcom grew to a seaward orientation following a path worn smooth by family history. The Balcoms claim lineage from the hereditary village of Balcombe, West Sussex, England, but his ancestors resettled in the Massachusetts colony at Salem and moved again at the time of the American Revolutionary War. Many of them went to Halifax, Nova Scotia, where they engaged in the maritime professions, and it was into this branch that G. W. Sprott was born, at Sheet Harbour, Nova Scotia, on 19 September 1850. The American connections were preserved in his christened name. At that time his surname was spelled "Balcam," and Sprott used this spelling for many years, probably until he moved to Victoria.

He married Jessie Blackwood Smith about 1879 and she bore them a son they called Harry. Soon thereafter she died and he returned to the sea full-time, obtaining his master's certificate in Halifax in September 1884. This entitled him to assume command of sailing vessels of 150 tons or less.[14] He entered the seal-hunting trade, a Halifax enterprise already active in sending schooners as far as the Falkland Islands and Tierra del Fuego. These schooners shipped goods southbound and returned with skins, which were often sold at U.S. ports. Sprott's brother Reuben also served as master of sealing schooners, and another brother, Sam, was financially involved in the business together with H. A. Balcam of Halifax and various members of the Donahoe family.[15]

Sprott remarried, and by 1891 he commanded a schooner named the *Dora Sieward*. It was probably in this vessel that he rounded Cape Horn and arrived in British Columbia in 1892.[16] For reasons now unknown, he and other members of his family traded the "am" spelling of their surname for "om," and, as Sprott Balcom, he took the sealer *Marie* to Asiatic waters where he was overhauled by a Russian patrol vessel. Balcom and his crew were seized and, by his own account, imprisoned for nearly six months in a "chicken-coop" jail near Petropavlovsk before they escaped to Japan in the schooner's boat.[17]

Sprott Balcom (right) with his son Harry.
Graeme Balcom, courtesy Vancouver Maritime Museum

His new wife and family came overland by rail to join him in Victoria, but he maintained connections with the Halifax sealing business, particularly through his son Harry, who became a sealing master in his own right. Sprott remained in command of sealing vessels at least through 1899,[18] but in 1900 he returned to Halifax and stayed long enough to become acquainted with the modern whaling industry underway in Newfoundland. It is not clear whom he met in Newfoundland or what he saw of this new kind of hunting, but it is a certainty that he anticipated the end of sealing on the North Pacific. Whaling may have seemed like a potentially fabulous replacement for it; by mid-1904 he was back in Victoria, determined to devote his talents to

the development of modern whaling on the Canadian Pacific coast. He en-
listed the interest of his old colleagues at the Victoria Sealing Company, particularly William Grant, who, like the Balcoms, had owned and commanded
sealing schooners. Both men possessed an encyclopedic knowledge of the Pacific Northwest coast and the habits and haunts of whales. Captain Grant
soon invested in the new venture and gave it a home, the Victoria Sealing
Company docks at Point Ellice in the Inner Harbour at Victoria.[19]

If Balcom and Grant anticipated an unregulated fishery, they were soon
disappointed. Until recently, Ottawa had shown no interest in the affairs of
the whalemen, but now they began to take a hard look. The outcry to establish Canadian sovereignty in the far north had grown louder and more stri-
dent. The Low Expedition to the eastern Arctic in 1903 had been instructed
to note the presence of whaleships, and Joseph Bernier's 1904–05 tour confirmed Canada in those northernmost territories. But it was more than territorial prerogative that led Ottawa to consider whaling legislation. The modern steam-whaling industry which had begun in Newfoundland was at first
regulated only by economic forces. The unfortunate result was overexpan-
sion, cutthroat competition, and consequent economic disaster as the whale
population was extinguished. After 25 June 1898, when the steam-chaser
Cabot of the Cabot Whaling Company took the first whale, the catch rose
from 47 whales in 1898 to 1,275 in 1904.[20] Shore stations were constructed
in close proximity to one another, and the several companies' chaser boats
competed aggressively for nearby whales. Economic pressure was intense.
One writer in the popular press was openly torn, extolling the virtues of the
"wonderful whaling by steam" while simultaneously prophesying the oncoming debacle. "It is impossible to suppose," he wrote, "that the present
enormous catches can be maintained for many years. The gradual extermination of the herds is inevitable, and the killing out of the rorquals as of the
northern [right] whales will surely be the outcome."[21]

The government at St. John's—Newfoundland was then a separate British
colony—hurried to legislate controls. Whaling companies were compelled to
pay $1,500 annually for a licence to operate, and new shore stations were
prohibited within fifty miles of any existing plant. St. John's hoped to weed
out marginal operations, reduce the competition to a self-sustaining level,
and so preserve the industry. This problem had already been encountered in
Norway when Svend Foyn's patent rights expired. There, the competition
became so fierce that an observer remarked: "Everyone concerned is ready to
acknowledge that [the whalemen] are treading seriously on one another's
heels."[22]

Despite the new regulations, the catch continued to fall. The precipitous
drop was first blamed on a shortage of krill, the small shrimplike in-
vertebrates that are a prime source of food for rorqual whales. But the decline

of 1905 was halved again in 1906, from 892 to 439 whales.[23] Losses of $2 million were reported. Several firms were ruined, and others promptly sold their chaser boats and equipment out of the country, some, prophetically, to Japan. By 1910, only five whaling stations would remain active in Newfoundland waters.

Prior to the Newfoundland troubles, Ottawa's disinclination to legislate Canadian whaling had become "policy." Several requests for exclusive whaling rights had been set aside. In 1902 the deputy minister of Marine and Fisheries had informed one applicant—seeking rights to a portion of Hudson's Bay—that the department had "never been in the habit of issuing licenses for seal, walrus, or whale hunting, especially for exclusive privileges."[24] When Toronto businessman John Leckie sought the "sole right" to catch whales and non-edible fish on the British Columbia coast "from Seymours Narrows south to the United States line,"[25] his application was summarily and repeatedly refused by the minister of Marine and Fisheries, Joseph-Raymond Fournier Préfontaine. One of Préfontaine's advisers, Minister of Inland Revenue William Templeman, did not see any purpose in acquiring such a privilege: "As it is not at all likely to become an industry out there, or that there will be any competition, exclusive rights of that kind seem to me objectionable."[26]

But Préfontaine gradually became uncertain. Newfoundland whaling was coming to grief for want of controls, and it seemed as if the problem could be avoided in dominion waters. He let it be known that he was considering an amendment to the Fisheries Act which would regulate whaling. The advice he had previously received was proven wrong. In the spring of 1904, even before Parliament began to consider his amendment, two groups solicited the whaling licences that would eventually be issued for British Columbia. The first group included eight discrete applicants, each of whom requested permission to erect a whaling station at a specified location along the west coast. Filed under the aegis of Aulay Morrison, member of Parliament for New Westminster, B.C., in the House of Commons, the wording of each was identical, save for the applicant's name and his chosen site, which in every case was separated from the other applicants' sites by at least fifty miles. Taken together, the requests constituted a de facto whaling monopoly.[27] Morrison personally conveyed the applications to Préfontaine—his secretary probably typed them, since they were all identical—and the minister rashly promised Morrison, "immediately the proposed Whale Fishery Regulations are passed, to direct that the licenses be issued . . . in accordance with your wishes."[28]

Préfontaine soon had cause to rue his offhand goodwill, when he discovered how many competitors planned to file for the few available sites. In July a letter arrived from one R. E. Finn, barrister and solicitor of Halifax,

Nova Scotia, representing Captain Sprott Balcom and William Grant of British Columbia, as well as H. J. Balcom, Reuben Balcom, and James Donahoe of Nova Scotia. Finn claimed for his clients a "right to prospect" for a suitable whaling station site between 48°30' North and 54°40' North, including the west coast of Vancouver Island, Queen Charlotte Sound, the Queen Charlotte Islands, and the Straits of Georgia and Juan de Fuca.[29] He further asked for eighteen months to allow his clients time to complete such a survey, with an option to shift their locations at will in the event that one or another place proved unsuitable.

Finn went so far as to beg a government "subsidy" or "assistance" to offset the large and uninsured investment that would be necessary to begin this entirely new industry. As an alternative, he suggested that the government might be willing to give the company "one free site" when his principals determined to work the whale fishery on a permanent basis.[30] This free site was no small concession. Finn surely anticipated a very pricey licence, perhaps equivalent to the $1,500 fee levied in Newfoundland.

But the ministry was not prompt in responding one way or the other. Finn cabled impatiently for an answer, and received only a cursory response: "Nothing can be done before law is passed."[31] Préfontaine did give a hint; the new regulations would include the fifty-mile separation rule restricting shore-station sites. Finn promptly submitted a proposal for four specific venues, including "Fitz Hugh Sound" on the mainland; "Rose Harbour" in Esperanza Inlet and "Ucluelet Arm" in Barkley Sound, both on the west coast of Vancouver Island; and a fourth site either just north of Nanaimo on the island's east coast, or else at a point on the opposing mainland facing the Strait of Georgia.[32] His proposal covered only the southern half of the provincial coast; clearly, Sprott Balcom was not yet thinking of whaling in the Queen Charlottes.

Now Finn and Balcom introduced the diplomacy of frustrated preparedness which would stand them in good stead in the bureaucratic campaigns to come. Instead of waiting to learn the government's wishes, Balcom and Grant forged ahead. Finn told Ottawa that his clients had already engaged the services of Charles Smith of St. John's, Newfoundland, an expert in the construction of whaling stations, and that Smith and Balcom were on their way to British Columbia to begin construction of the first station.[33] Balcom intended to establish pre-eminence by proceeding, even though the government had not yet outlined its new whaling policies. Going ahead regardless was a successful ploy he would use again and again in the coming years.

The issue of licences was further complicated by the arrival of an application from one William J. Corbett for a site at Nootka Sound. In fact, Corbett was affiliated with the eight previously acknowledged applicants proffered by Aulay Morrison.[34] Another correspondent, from Brooklyn, New York, anticipated building a whaling station at Safety Harbour on Calvert Island and had written Ottawa for information.[35] The minister's office began to require that all applicants provide specifications for their proposed processing methods and plans for station buildings.

In August 1904 Préfontaine's amendment was added to the Fisheries Act.[36] It required licensing of each shore station by the Department of Marine and Fisheries, with final approval vested in the hands of the minister. No station could be erected within fifty miles of another or close to habitations where there might be "danger or detriment to public health."[37] Unlike in Newfoundland, each operating factory would be limited to the use of one chaser boat, nor could towboats be used for hauling carcasses from the open sea to the station. Whale catching was prohibited within one-half nautical mile of any vessel not at anchor and one nautical mile of anchored vessels, and the use of any device other than a harpoon with a line attached was also prohibited.[38] The law additionally demanded that whales be completely and thoroughly processed within twenty-four hours, without dumping "noxious or deleterious matter" into the waterways.

The licence itself would be issued for a maximum term of nine years, renewable annually on payment of a fee of $1,200. A subsidy was offered, in the form of fee reductions — to $800 and $1,000 respectively — for the first two years of operation, and a proviso allowed the governor-in-council to accept two per cent of the annual gross earnings in lieu of set fees. Violations could result in a forfeiture of the licence without refund, and in some cases a fine of as much as $200.

Sprott Balcom submitted complete plans for the station under construction; they show a main building fifty by forty-five feet and thirty feet upright, housing two boilers each measuring fourteen by five feet, a twelve-horsepower engine, three thirty-ton-capacity cooling tanks, six fifteen-ton-capacity blubber pots, two forty-ton storage tanks for oil, two oil-and-boiler feed pumps, a slicer, an elevator, three steam winches, a canting winch — for turning the carcass over — two "sausage machines," plus assorted tools and "utensils."[39] Together with the plans, Balcom submitted a request for a licence at his chosen site "at the outlet of Crawford Lake, Sechart Channel, Barclay Sound."[40]

Despite his promise to Aulay Morrison, Préfontaine issued Whaling Licence No. 1 to Captain Sprott Balcom, for the Sechart site, in May 1905. Captains William Grant and Reuben Balcom of Victoria and Henry J. Balcom and James Donahoe of Halifax were also named as licencees.[41] How Morrison's men lost the advantage of their early application is not completely clear. Préfontaine may have appreciated a man of action. Or perhaps their inability to retain Morrison's attention had to do with his elevation to Puisne Judge of the Supreme Court of British Columbia, an event which transpired in September 1904—an inopportune time for his nine whaling constituents.

Immediately the first licence was issued, new applications began to arrive. W. H. Riddall applied for Freeman's Pass on Porcher Island in the name of West Coast Whaling Company Limited, and John O. Townsend made application for Plumper Harbour, in Nootka Sound.[42] Townsend's application seemed to carry some weight; his request appeared on the stationery of the Canadian Pacific Railway Company, and he introduced himself as a CPR employee. He was, in fact, master of the company's coastwise steamship *Queen City* and reportedly owned 160 acres of land at Plumper Harbour which would be suitable for a whaling factory.

William Sloan, M.P. for Nanaimo, supported Townsend's application, but the Nootka Sound venue had already become a morass, with three other applications pending for sites within fifty miles of that place. These were Corbett's, Balcom's, and a third submitted by G. F. Pearson of Halifax. Neither Corbett nor Pearson had filed the requisite plans and specifications, and even Sprott Balcom was delinquent in that respect, although a ministry official was quick to point out that the plans already on file in connection with Balcom's Sechart application might be regarded as a compliance with the conditions. The ministry nevertheless notified the others that unless good reason were given to the contrary, John Townsend's application would be approved.[43]

Sprott Balcom responded promptly, submitting plans and specifications on 27 November 1905 for a second factory—duplicates of the Sechart plans—and a cheque for $800 in payment of the first year fee for the Esperanza Inlet site claimed by barrister Finn in the 1904 application. Esperanza Inlet fell within fifty miles of Townsend's property, and Balcom's claim took chronological precedence, so Sloan was told that it would be impossible to acknowledge Townsend's application there, "or at any other point at a less distance than fifty miles from Esperanza Inlet."[44] And in January 1906, Sprott Balcom received his second whaling licence.

Raymond Préfontaine did not live to adjudicate this issue. He died in Paris on Christmas Day 1905 and was succeeded as minister of Marine and Fisheries by Louis-Philippe Brodeur. If Brodeur thought his most pressing whaling problem had been solved by the disposition of the Esperanza Inlet matter, he was much mistaken. On 19 July 1906 he received word from Charles Lugrin, barrister and solicitor from Victoria, that others were now interested in entering the whaling business in British Columbia. Lugrin represented anonymous clients who were disappointed to learn that Balcom's company held the "exclusive privilege of erecting stations along 300 miles of the British Columbia coast."[45] They had discovered, however, that his influence did not extend to the Queen Charlotte Islands, and so Lugrin filed formal application for shore stations at Virago Sound on the north shore of

Graham Island and at Tasco Harbour on the west coast of Moresby Island.[46] The process ensnared Lugrin in what later became known as a "Catch-22." His clients, "certain capitalists of Scotland,"[47] refused to go to the expense of "organizing themselves into a company" unless they could be assured of receiving permission to hunt whales. And the dominion government could not give that assurance without receipt of plans and specifications for their whaling plant.

Almost simultaneously, Brodeur received application from Captain Theodore Magneson of Victoria for a site at Virago Sound,[48] and the matter was further confounded by the arrival of a third application—from Vancouver barrister John Bostock—for a station in his own name at Rose Harbour on Moresby Island.[49] Furthermore, the original applications made in 1904 by Aulay Morrison's group—never formally rescinded—stood ahead of all of these by chronological precedence. Suddenly the British Columbia coast was inundated with would-be whalemen. Brodeur must have wondered why he had not heard from Sprott Balcom as well.

Balcom may have been too busy to respond. About $260,000 had been spent on development of the Sechart station, and operations had begun about 1 September 1905. The new enterprise, called the Pacific Whaling Company, had raised $200,000 of this amount through the issuance of preference and common stock shares; additional moneys were taken out of first months' operating income.[50] Once capitalized, there were myriad logistical problems to be solved, not the least of which was obtaining a suitable steamer for whaling. No Canadian shipbuilder would contract to build such a unique vessel without model or specifications; "in fact," barrister Finn told the minister of finance, "they absolutely knew nothing about the construction of such vessels."[51]

To obtain a proper chaser boat, the company contacted Isak Kobro, an agent in Christiania, Norway,[52] whose business it was to supply ships and men to the world's growing steam-whaling industry. Kobro arranged for the shipbuilding firm of Akers Mekanisk Verksted to complete one of their fine whaling steamers for the Pacific Whaling Company. Akers and a competitor, Nylands—builders of the original *Spes et Fides*—had constructed the majority of first-class chasers in world service.

The salient features of these vessels were their high, trawler-style bows which warded off heavy seas; a round bottom to aid in manoeuvring; low bulwarks, especially aft, where water could easily come aboard but just as easily leave; and a tall forward mast supporting the lookout barrel, which was reached by means of ratlines and a short rope ladder. Scotch marine boilers provided steam, and the whales would be played out and reeled in with a steam-winch located forward of the wheelhouse. The whaling line, some of it six-inch-diameter manila rope, would be fed into a line locker below deck.

Another characteristic of these steamers was their scanty accommodations. A small forecastle forward housed firemen and deckhands; at midships a tiny mess and a galley were provided; aft were two bunks for the second engineer and mate and cramped cabins for the chief engineer and the captain.[53] A final attribute, best known to their crews, was the chaser's peculiar way in the water: "Climbs up and down waves like a mountain goat," one traveller wrote, "the usual forward and back motion . . . a terrific roll plus one or two special movements which defy categorization."[54]

Reuben Balcom travelled to Christiania to inspect and take delivery of Akers' hull no. 237, a 94'1" vessel, 17'4" wide, 109 tons gross,[55] which the whaling company had christened the *Orion*. He signed a local crew of whalemen under Norwegian shipping articles and loaded with coal for Cape Horn. The little steamer cleared Christiania on 10 December 1904 on its long and perilous voyage around South America, a passage which replicated in many respects the century-old itinerary of sailing whaleships outward-bound for the Pacific.

Though powered, the *Orion* was not as large as many of the old ships and barks. The New London whaleship *Louvre*, for example, measured 116'6" by 27'7".[56] Nevertheless, it was around the Horn. There was no other way; the Panama Canal was still a visionary dream, and the Good Hope and Suez Canal routes were much too far to consider. Besides, Reuben Balcom had plenty of seafaring experience near the Horn and a clear idea of what the *Orion*'s crew might find there.

Most of the available space on board was crowded with coal. There was little room for the men and none for the whaling gear, which had to be shipped out as cargo aboard the steamship *Ping Suey* by way of Glasgow.[57] The heavily burdened *Orion* proceeded to Dartmouth, Nova Scotia, where the Balcoms and the Donahoes were no doubt on hand to provide a warm welcome. From there Balcom steamed to the Cape Verde Islands, then to Montevideo for coal and provisions, and thence around "Cape Stiff." The crew worked the chaser boat through the Straits of Magellan and made their first Pacific port at Valparaiso, where many an errant whaleman had gone astray in the previous century. After a stop at Callao—another infamous seafaring town—the *Orion* steamed on to San Diego, California, for more coal, and then came on to Victoria. That passage would be duplicated many times by whaling steamers, but the *Orion*'s was an epochal trip, the first of its kind by a chaser boat bound for Canada's west coast.

It was probably a wild ride. Aside from the acrid smell of coal smoke and the fuming of the ashes from the fires, the design of the chaser's hull made it prone to pitch and roll; steady sea legs were required. "The wind had risen and I was deathly sick," one experienced sailor later wrote of his whaling cruise in the *Orion*. "Even the best sailors lose their 'sea legs' on one of these

The *Orion*, first steam chaser boat in British Columbia. It was photographed southeast of Cape Beale in January 1906, while its crew searched for shipwreck victims from the S.S. *Valencia*. The harpoon-cannon is turned sideways and the foregoer pan has been withdrawn.
Vancouver Maritime Museum

little round-bottom vessels. The *Orion* twisted and writhed about like a wild thing."[58] But an engineer who worked in British Columbia nearly twenty years later remembered the *Orion* kindly; it was so easy to fire, he said, that the stoker "could practically sit in the [lookout] bucket and toss the coals in!"[59]

The arrival of the chaser boat and the whaling gear fueled more bureaucratic trouble. Dominion Customs refused to allow the unusual machinery to enter duty-free, even though it could not be obtained domestically. Barrister Finn argued cogently, though unsuccessfully, that the company had made every attempt to have the vessel built in Canada but failed because no one had the expertise to construct it. Finn even offered to show the *Orion* to Canadian shipbuilders, theorizing that its design would prove to be "of immense benefit to the ship building trade and our mechanics."[60] But the government was unwilling to set a precedent in this new industry of whaling and insisted upon the duty.[61]

There were other problems. The men who came from Norway in the *Orion* were given the opportunity to remain in British Columbia in the service of the Pacific Whaling Company. Several of them elected to stay, particularly since Canadian shipping articles paid much more than the Norwegian. A second engineer, for example, who earned twenty dollars per month in Christiania stood to make thirteen Canadian dollars—about sixty dollars Norwegian—every month.[62] But some British Columbians objected to the signing of foreign crew. They thought the new jobs ought to be given to Canadian nationals. One such objection came from a New Alberni resident, who petitioned the dominion Department of Labour for an adjustment of the situation:

> A Cap Balcome of the west Coast of this Island Has Lately Had Built a Steel Vessel at About 103 Tons Burden At Norway. He Fitted Her Out there with all the Aplyances for Engaging in Waleing And On the Arrival of the Steamer Here He Imediately Tried to Put Her Under the Canadien Flag.
>
> Last, but not Least, He Brought a Crew of Men From Norway to Engage in Hunting and Fishing for Wales to the Detriment of Canadien Fishermen.

"At Least Sutch," he concluded, "Is the Report in the Papers."[63]

That charge of unfair competition made its way by degrees to William Lyon Mackenzie King, then deputy minister of Labour, who passed it along to Marine and Fisheries, whose Deputy Minister Gourdeau explained that the new technology was "wholly experimental" in Canada; Captain Balcom had "found it necessary to procure skilled assistance to enable him to initiate the work."[64] Nothing more seems to have come of the objection, and the *Orion*, together with some of its Norwegian crew, went to the work of whale hunting.

Reuben Balcom did not accompany them. He returned to his former occupation of sealing, assuming the master's cabin in the schooner *Edith R. Balcom* on a voyage from Halifax to the South Shetland Islands. Shortly after his return from that trip he traversed the southern hemisphere aboard the former Balcom sealing schooner *Agnes G. Donahoe*—now owned by E. Donahoe and Son—and returned 4,014 seal pelts, claimed to be the largest cargo brought into Halifax up to that time.[65]

The government's insistence that whale carcasses be processed and disposed of within twenty-four hours proved as problematic as finding chaser boats and experienced men to run them. The only machinery adequate to the task was held under patent by a German-American chemist and engineer, Dr. Ludwig Rissmüller. Rissmüller's reduction methods, and only his, appeared to make the difference between success and failure in modern Canadian whaling, and this was the machinery that Balcom soon obtained.

It is not known when Rissmüller and Balcom met, but it was probably during 1903 or early 1904. Balcom may not have taken the chemist for a partner in the early phase; he seems to have begun on his own, only engaging Rissmüller's services after experiencing some technical problems.[66] In any event, the Rissmüller equipment was probably in place at the Sechart plant before the 1 September 1905 start-up date. The formal arrangement between the two parties is contained in a memorandum of agreement dated 29 November 1905, by which Rissmüller transferred to the Pacific Whaling Company, in perpetuity, all rights to his processes, methods, and machinery as well as his personal services as consulting chemist and engineer. Rissmüller also agreed to train and supervise assistants—here there is more evidence that Balcom may have begun alone—to perform the necessary "alterations" of the "whaling establishments and oil refineries of the . . . Company."[67] In exchange, the company gave Rissmüller all 1,982 issued shares of its common stock. The value of the stock was $97,800,[68] and the arrangement made Rissmüller a major shareholder.

A more enigmatic personality than Ludwig Rissmüller can scarcely be imagined. He has entered the industrial history of Canada as the father of its modern whaling enterprise, yet little documentation exists about the activities that earned him his title. He shrouded his work in secrecy, probably, like Svend Foyn before him, to retain exclusive control of his ideas. "To him," one commentator wrote in 1907, "is owed the utilisation of every part of the whale, including the flesh, the blood, and the liver, and parts of the skin which were only regarded as wastage a few years ago."[69] Another writer called him "the whaleman of the world."[70] The developments which canonized Rissmüller were these: he worked out simple ways of rendering fatty oils from whale flesh, and he devised means of making fertilizer from dried and macerated bones and visceral waste. From his inventions came additional revenues from a more complete and diverse rendering of the giant bodies daily towed to shore by the whaling steamers.

As early as 1898 Rissmüller devised methods for making "guano" from dried and ground fish offal; this was sold as chicken feed and cattle fodder.[71] In 1900 he set up the Colonial Manufacturing Company at Cape Royal, Newfoundland, to manufacture guano from the ground and dried carcasses of whales purchased from nearby whaling stations. The whalemen were pleased to sell them. Formerly discarded back to the sea, now they might be sold for cash at a little advantage.[72] Rissmüller knew that a considerable amount of high-grade oil remained encased in the unprocessed skeletal bones, and he devised means of extracting it.

He also established the Fish Industries Company, which carried similar contracts for the purchase of carcasses—the Cabot Whaling Company initially provided whales for this firm—and soon he was able to build a guano factory at Bonne Bay, on the south coast of Newfoundland. It became the chemist's headquarters and a "school to learn how to make guano."[73]

Rissmüller eventually controlled more than twenty whaling stations in New-foundland, presumably by the exchange of his inventions for company stock. His whaling station in the Saint Lawrence was described as "the most perfect plant for the manufacture of whale products."[74] He obtained for it the fine Nylands-built steam chaser boat *Saint Lawrence*.[75] It may have been Sprott Balcom's exposure to this station which made up his mind to begin a similar industry in the Pacific; some claim that Rissmüller personally convinced the great Nova Scotian sealer to enter the whaling business.[76]

If Rissmüller's as yet unpatented innovations were labelled "the great secret of the whaling industry,"[77] his personal life remained equally secret. He was affiliated with the F. A. Rissmüller Chemische Dünger-Fabrik[78] in Germany and preserved close kinship ties with at least one brother, Julius, who lived there. In America, he was a traveller who provided transient ad-dresses on important legal documents. In August 1904 he gave the Hotel St. George in Clark Street, Brooklyn, New York, on a Canadian patent docu-ment.[79] The preceding month he had listed an address in Quebec on a similar paper.[80] And though he married—his wife Pauline outlived him at Vic-toria—he seems to have been busy at this time organizing his rising empire. At the turn of the century Ludwig Rissmüller was nearly fifty years old.

Rissmüller's greatest innovations are contained in three patents. The first of these, registered in Canada as No. 88351 and granted at Ottawa on 19 July 1904, provided a methodology for separating those oils and fats in animal flesh which could not be extracted during the "trying-out." Chemical tests had shown that certain "albuminous substances" in the meat sealed the pores during heating and prevented release of the free fats. The addition of an alkaline such as caustic soda or potash to the rendering vats prevented this from happening, and the chemical process could be neutralized at any time by adding sulfuric acid or some other mineral acid to counteract and destroy the alkalines and the soaps.

Rissmüller designed special wooden tanks, twelve feet long, six feet wide, and four feet deep, and lined them with lead. Five pounds of caustic were added to every five tons of meat heated in one ton of water. Steam was employed to raise the mix to the boiling point. After three hours of cooking this infernal recipe, the acid was added to destroy the alkalines and emul-sions, leaving the free oil and the fat.[81]

The second patent, Canadian No. 88875, dated 23 August 1904, was per-haps the most significant, since it covered the apparatus used in making fertilizer. By methods covered in the patent, Rissmüller ground and dried meat and other portions of viscera that had formerly been discarded, and produced from it a product which everyone called "guano," presumably bec-ause it resembled the dry, dusty material gathered from islands frequented by birds. A rotary-drum dryer with a built-in crusher dried, pulverized, and

Large Rissmüller-type oil storage vats under construction in May 1911 at the West-port station, Gray's Harbor, Washington.
Graeme Balcom, courtesy Vancouver Maritime Museum

screened the fleshy material; a furnace heated the air used for drying.[82]

The most readable description of this technical apparatus appeared in the popular press in 1911. Evaluating the construction of a new whaling station equipped with Rissmüller's gear, the writer explained the basic chemical concepts employed in guano-production:

> As soon as the bones are taken from the [rendering] vats the meat and of-fal of the whale . . . are led by chutes into a deep pit. From this pit the macerated mess is elevated into a press, which squeezes out most of the water, and is then fed into the drier. The latter is a huge steel cylinder about 30 feet long, which is mounted on cog bearings and set at an angle of approximately 35 degrees. The upper end is connected with a blast furnace and the pressed meat is fed into it and dried by terrific heat as it slowly gravitates to the lower end. When taken from the drier the meat, etc., resembles peat in appearance. It is converted into fertilizer by a crusher, which powders it as fine as salt, and is then carried, by means of an air-current, through big zinc pipes to the guano-house.[83]

Rissmüller gave a Montreal address in 1905 when he signed Patent No. 94083, covering the construction of wooden vessels with a double shell. These wooden vats prevented leakage when boiling the acid-alkaline solutions described in 88351. Though simplistic today, his solution to the leakage problem was ingenious; the two walls of the vat were separated by two inches of space filled with calcium, barium, and strontium — ingredients found in certain commonly available materials such as Portland cement. Should there be a leak of acid solution, the minerals in the cement formed an insoluble salt which plugged the leak automatically. Any leakage of fats or oils would be plugged also, since fatty substances are insoluble when heated with oxides of the three elements; such a leak would result in the formation of an insoluble soap at the point of contact with the cement.[84]

With these basic concepts under his personal control, Rissmüller was free to unveil the great whaling "secrets" in exchange for blocks of whaling stock. His patented systems made it possible to use far more of the whale carcass than had previously been feasible. The blubber was fully cut by a revolving slicer and fed to a bucket elevator which lifted it to the rendering tanks. From there, it was rendered by steam heat and later by sulfuric acid; the liquid resulting from boiling the blubber could either be sterilized with boracic acid and converted into glue or else pumped to the drying machine for conversion into fertilizer. The bones would be rendered in steam heat and water and then ground into a bone-meal fertilizer. Whatever meat was not processed as edible food was converted into fertilizer by the alkaline-acid process followed by crushing and drying. The blood was allowed to coagulate, and was later introduced into the drying machine to make bloodmeal, a high-grade ammoniated fertilizer. Even the entrails were used; the oil was separated out and the rest dried for fertilizer. The intestinal casings were used also, Rissmüller explained, transformed into "a leather-like product of great strength and pliability."[85]

The machinery covered in these patents was soon in use in no fewer than seventeen Newfoundland whaling stations and received the approval of St. John's as "the most efficient... known and used in the whaling industry."[86] One writer anguished that the doctor's inventions were not properly appreciated. "In a society whose one aim and object is the rapid accumulation of money," he wrote, "many things of this world that are of real importance and interest are scarcely noticed... and so people go hurrying on, only to find too late that they had a great man in their midst without their knowledge.

"If Dr. Rismuller had made a fortune rapidly out of his discoveries," he concluded, "people in America, Canada, and Newfoundland would have thought him a wonderfully 'cute' fellow, and would have placed him on the pedestal of fame allotted to successful trust magnates and other human sharks."[87]   .

The writer remarked that Rissmüller was still "comparatively poor," but in truth his apparent poverty may have been the result of his willingness to waive a short-term cash royalty in exchange for a potentially long-term profit. Patent rights gave him great leverage; his "Rissmuller Patented Whale Reduction Process" appears to have been the only means of meeting Ottawa's stringent processing prerequisite. The great entrepreneur must have envisioned a fleet of shining new whaling steamers chasing and capturing an abundance of finwhales, and everyone rich on their bounty. He must have dreamed of the company bonuses and incredible stock dividends which would accrue as a natural economic result of his labour.

An early catch at Sechart, photographed not later than 1907. A rorqual whale is being hauled under the slipway awning, while two humpback carcasses float nearby.
Kendall Whaling Museum, Sharon, Massachusetts

Fantastic dividends *were* paid, at the beginning. At the first general meeting of the Pacific Whaling Company in January 1907, Sprott Balcom announced net earnings of $45,552.11 for the first year of operation, and a tremendous dividend of 23 per cent on preference shares and 16 per cent on common shares.[88] There had been the inevitable delays and problems in starting the new plant, including unfamiliar equipment and some defective barrels which had leaked out their contents while in transit to market in Glasgow. But as of 28 July 1906, a total of 176 whales had been caught.[89]

The results were so promising that Sprott Balcom made every effort to secure his other sites along the coast. But the halls of the Department of Marine and Fisheries were now blocked with stranded whalemen, each hopeful of obtaining the precious licence that would give him the right to take oil from the sea. With the lucrative west coast of Vancouver Island largely sewn up by Balcom's applications, the quest had become focused around sites in the Queen Charlotte Islands and certain unclaimed harbours on the mainland. Captain Magneson's application for Virago Sound, barrister Bostock's for Rose Harbour on Moresby Island, and Lugrin's at Virago Sound and at Tasco Harbour on behalf of his Scottish investors — these were yet outstanding, and before they were resolved, more applications arrived on Minister Brodeur's desk.

One was from Cereno Jones Kelley in Victoria. Writing on behalf of Frank Hall, Alexander McCrimmon, and himself, "all loyal British subjects, residing in Victoria, B.C.," Kelley asked for Milbank Sound, Estevan Sound, Nepean Sound, and the north end of Porcher Island.[90] The minister also heard again from John O. Townsend, once defeated by Sprott Balcom in his bid for a whaling site at Nootka Sound. Now Townsend applied for Quatsino Sound, and to show he had learned a thing or two he sent full plans and specifications for the proposed shore station.[91] William Templeman, minister of Inland Revenue at Victoria, had been disappointed by the rejection of Townsend's previous application, and he protested against what he perceived to be a monopoly on west coast whaling. He had objected to exclusive rights from the very beginning, and now forced Brodeur to admit, indirectly, that the licensing situation had become a quagmire. "I find at the moment," Brodeur stammered, "that it is impossible to give any definite decision regarding the various and conflicting applications, especially in view of the fact that my predecessor had promised a number of sites, applied for through Mr. Aulay Morrison, when a Parliamentary Representative in the Province."[92]

The matter was complicated by the Ministry's decision to extend the fifty-mile separation rule to one hundred miles. It is not completely clear how this concept gained acceptance; perhaps someone had put the word into Brodeur's ear, loudly enough so he thought it wise to hear it. If so, then the hailer must have been Sprott Balcom. In a recapitulation of North Pacific whaling published in Norway in 1917, it was said that Balcom "petitioned" the minister to extend the fifty-mile rule. In this way, the Norwegians complained, there were so few suitable sites that a monopoly could easily be obtained.[93]

Under the proposed hundred-mile separation, three locations proposed by Charles Lugrin were in conflict with one another; two of the original Morrison applicants were also in collision, and Bostock's application for Rose Harbour crossed Lugrin's for Tasco Harbour. Kelley's three requests stood in conflict, and all were compromised because of Andrew Holcross, who had petitioned with the other Morrison applicants in 1904. The Holcross site also conflicted with Sprott Balcom's proposed station on Calvert Island.[94]

Brodeur decided to send a letter to each interested party demanding an assurance within thirty days that his factory would be capable of converting a whale to commercial product within twenty-four hours. Specifications of the plant would again be required, as well as payment of the initial $800 licence fee. He also gave formal notice that the separation rule had been extended, "owing," he said, "to the experience in the industry in Newfoundland waters" and, intriguingly, "in view of strong representations to the Ministry."[95]

These stipulations effectively favoured the Pacific Whaling Company, which had plans already drawn, profits from which to pay any fees, and, best of all, patent rights to the Rissmüller methods, which virtually guaranteed compliance with the processing requirement. The "strong representations" mentioned by Brodeur almost certainly came from Sprott Balcom, and his involvement is further suggested because he seems to have had advance warning of the impending changes in the law. Just one week before Brodeur's ultimatum, Balcom made formal application—and paid—for a station site on Page's Lagoon, "a point about five miles North of Nanaimo on the East Coast of Vancouver Island."[96] Of course, this plant would be built to the specifications previously submitted to the ministry and—once again invoking a *fait accompli*—Balcom gave notice that he already had a second chaser boat en route from Newfoundland to service the new station.[97]

William Templeman responded angrily to Brodeur's letter. He told the minister that the demands were "too drastic," knowing surely that the short notice would deter any but the Pacific Whaling Company from applying and thus give Sprott Balcom the monopoly. He warned that unless the requirements were eased, competitive factories would be built on adjacent American soil, either in Alaska or in Washington State, to catch and process whales "our factories might have captured."[98] He particularly assailed the restriction on dumping "noxious and deleterious" matter: "If the factories have a waste that must be disposed of, what better place is there for it than in the salt water? The sewerage of our cities goes into salt water and no ill effects result."[99]

Templeman may have been voicing the frustrations of Charles Lugrin, who still hoped to obtain a licence for his clients in Scotland. Shortly after the ultimatum from Ottawa, Lugrin solicited Templeman's assistance to obtain an extension of the thirty-day deadline; the short notice seemed equivalent to a refusal, since his clients could not possibly organize an appropriate response in such a short time.

"I am therefore practically out of court," he said sadly. "I have the money ready to embark in the business, but cannot get a reasonable chance to comply with the law." Lugrin particularly blamed the obstructionist tactics of Ludwig Rissmüller, who had allegedly warned the Scots clients that only the Rissmüller method would be accepted by Ottawa. Therefore they would have to "purchase certain rights" before their business might commence. "He has us all in the hollow of his hand," Lugrin complained, "unless we can get time enough to mature our plans irrespective of his patents, or make arrangements with him to use his patents. If Dr. Ritzmuller had [been involved in] the framing of the law, he could not have drawn [an] act more in his own interest than [the law] under which we must work."[100]

William Sloan, the Liberal M.P. who had previously supported John Townsend's Nootka Sound bid, also responded angrily to Brodeur's October missive. "I do not propose to be bound," he stormed, "by any promises made by previous parliamentary representatives, *who had no jurisdiction in this District* and who have no right to *parcel out the whole coast of British Columbia in the interests of their friends.*"[101]

Meanwhile, one of the applicants, Cereno Jones Kelley, quietly abandoned two of his requests but pressed suit for the site at the north end of Porcher Island. He submitted the application fee and indicated, not without emphasis, that Rissmüller's equipment would be employed at his new plant.[102] Sprott Balcom's response was equally cool. In another of his characteristically flattering letters to Ottawa — letters which always assumed a deferential tone, as if whaling was too little a matter to concern the ministry of Fisheries — Balcom reiterated the 1904 request for "Namu, Fitz Hugh Sound" and added a new request for a site "three miles north of Gale Point, Bank's Island."[103] He attached a single-page letter signed by Ludwig Rissmüller which verified the availability of the doctor's reduction equipment for these stations.[104]

Minister Brodeur bent under the pressure. At the end of the thirty days he allowed Charles Lugrin's clients more time to develop specifications for their whaling factory, although he insisted on immediate payment of the first-year fee.[105] Then he called in Sprott Balcom and R. E. Finn and issued them licenses for Page's Lagoon and Fitzhugh Sound, instructing Balcom to remove the latter site as far to the south as possible in order to make room for a station on Campania Island to be held in reserve for the frustrated John O. Townsend. Balcom's request for Banks Island was too near to Campania Island; the ministry's decision would be delayed until the CPR captain made up his mind.

In the meantime, the Pacific Whaling Company had got Sechart nearly to full steam, and construction of the second plant had begun on the northwestern coast of Vancouver Island. Although constructed on the Esperanza Inlet licence, issued in January 1906, there is little to suggest that machinery was ever erected there. Instead, a site was chosen at the mouth of adjacent

Kyuquot Sound. Just as Sechart was situated close to La Pérouse Bank to the southwest and Swiftsure Bank to the southeast, so the Kyuquot site was immediate to even larger shallows, some barely fifteen fathoms deep. The upwelling of currents on these shallows brings nutrients from the sea floor to the surface, and the invertebrate animals and small schooling fish that make up the food chain provide a vigorous supply of food for visiting filter-feeding baleen whales. The circumstance promised years of good whaling. Ludwig Rissmüller, growing ever more aware of the value of aggressive public relations, would soon praise the station as the "world's largest,"[106] with 50 per cent greater capacity than Sechart.

Now the completion of the third station became critical. To be built on Page's Lagoon north of Nanaimo, it promised winter employment for the boat crews and shore workers from both west coast operations. Migrating whales and failing autumn weather forced a halt to the business "outside," but a whaling station in the protected waters of the Strait of Georgia could employ many men who would otherwise drift away to other jobs or spend the winter "on the beach." The great expense of time and money involved in reorganizing the whaling each spring might be avoided.

Construction commenced early in 1907. A strong underground water source was tapped for use in the steam power plant, and the company deflected its first labour dispute—occasioned by an extension of the work day from nine hours to ten—by increasing each workman's pay.[107] As the station neared completion, Balcom foresaw a more serious obstacle: he had arranged for a second chaser boat to service the station under construction at Kyuquot, but when the two west coast stations were idled by winter weather, the one-boat-per-station clause in the Fisheries Act would specifically prevent him from using both steamers at Page's Lagoon.

Already the new steamer was more than halfway to British Columbia; it was Rissmüller's own *St. Lawrence,* which had cleared out of Newfoundland on 22 October 1906 to duplicate the passage of the *Orion* around Cape Horn. Balcom had promised work for the new crew, and he feared they would sue for breach of promise if that offer were not fully met. He sent at least two messages to Ottawa requesting permission to employ the new boat alongside the *Orion* at Sechart until the northern station opened. Otherwise, he promised, the men would quit, since the bulk of their pay came from "bonuses" earned on each whale killed, and without their special expertise the chaser boat could not be employed effectively.[108] Surviving documents do not record the ministry's answer, but the question was fortunately rendered moot; Kyuquot reached operational status, so Balcom was able to send the *St. Lawrence* there.

At 111 tons, the *St. Lawrence* was a bit larger than the *Orion*, but the differences were the kind that only seamen would notice: there was an enclosed wheelhouse in place of the *Orion*'s open bridge,[109] and the *St. Lawrence* fired from the engine room whereas the *Orion* was constructed with a separate stokehold.[110] Whaling was successful at once; the first shipment of the 1907 season, brought south in the Canadian Pacific steamship *Amur*, included 639 barrels of oil and 815 bags of "guano" from the new station at Kyuquot, plus 625 barrels and 655 bags from Sechart.[111]

An unidentified whaleman poses heroically atop the slain leviathan. The photograph was probably taken at Sechart about 1907. Ludwig Rissmüller's chaser boat, the *St. Lawrence*, is seen in the background. Its enclosed wheelhouse has already been replaced by an open bridge.
Kendall Whaling Museum, Sharon, Massachusetts

The Page's Lagoon station also promised rich returns. Many whales had been encountered throughout the inside passages during 1906; herring fishermen snagged them in their nets, and at Stewart, at the head of the Portland Canal, a large whale entangled itself in steel anchor cables securing a floating storehouse built atop planked eighty-foot logs. The confused animal towed logs, cables, planking, and storehouse on a seventeen-mile "Nantucket sleighride" before drowning with exhaustion in his bondage.

"Here was more than magic," a reporter wrote. "For three months the raft had been securely anchored. Yet, without a moment's notice it suddenly started off headed full speed for the open sea."[112] A prospector who watched the raft go by warranted as how he thought "the bad man himself" had got into the boat house. "There was no sign of propelling power," he said, "yet the wharf was making great time."[113]

Work proceeded in expectation of this bountiful harvest, and by late October the station neared completion. The *Nanaimo Free Press* picked up the contagion; since passing vessels had seen a "plentitude of whales," they speculated that the opening would be auspicious.[114] During the first week of November the *Orion* came around from Sechart, followed closely by the *St. Lawrence*, and now the employment situation that concerned Sprott Balcom came to pass: the company possessed two fine chaser boats and two trained crews, but only one operating whaling station. In a clever effort to circumvent the wording of the Fisheries Act, Balcom asked permission to use the *Orion* as a "training vessel" to teach Canadians to shoot whales. He argued — not without truth — that the Norwegian gunners refused to let anyone else learn "the expert work, i.e. the shooting." By using the *Orion* in this way, it would soon be possible to employ Canadian citizens as gunners and make the company independent of foreigners. Their wages and bonuses would be spent in Canada, rather than being sent home to savings accounts in Norway.[115]

In order to train new gunners, of course it would be necessary to shoot whales; how else was a man to learn, except by practice? Processing of these carcasses followed as an appropriate and prudent step. To legitimize the scheme, Balcom proposed that the training operation be placed under the "special supervision" of Edward Taylor, the Fisheries inspector at Nanaimo.[116] Taylor had already aligned himself with the Pacific Whaling Company; perhaps because of his active interest in the ways of whales and other marine mammals he seemed more than willing to support Balcom's ambitious and sometimes audacious ideas. In exchange, he could avail himself of first-hand knowledge that was otherwise virtually unobtainable.

The *Orion* meanwhile began the hunting from Page's Lagoon on Friday, 15 November, and the seafaring duties were apparently shunted from one steamer to the other. By Sunday noon five humpback whales had been brought in,[117] and it looked like Balcom's wish for twelve months of whaling would be granted. The *Free Press* printed glowing reports: "The success which has attended the Pacific Whaling Company has surpassed all records in the annals of the industry," they said, "and that this success will now be continuous through summer and winter for some years to come is established by the first week's success at Page's Lagoon." Ludwig Rissmüller capitalized on the free publicity by entertaining the reporters with engaging stories, which the newsmen quoted *verbatim*. "Ritsmuller, the authority pre-eminent," they wrote, "says that in a decent sized whale, the tongue alone will be as big a load as a fair-sized team of horses would care to draw." And though they do not attribute to Rissmüller, it was probably he who gave

them the topical simile that, "Like the Chicago pig, everything is utilized
with the exception of the squeal, only in the case of the whale it is not the
squeal, but the smell."[118]

This remark became better understood by the residents of Nanaimo about
12 December, when a wind shift brought the pungent aroma of the plant
into town. The *Free Press* dutifully noted the odour but excused it in a mood
of commercial bonanza, with a little help from an anecdote which again
smacks of the inventive mind of Dr. Rissmüller. "There is considerable con-
solation," the paper said, "in the fact that this does not happen very often,
and that whale smell is considered very healthy. Indeed in some parts of the
old country, consumptives are sent to reside near whaleries as the smell is
considered a sure cure for cases not too far gone."[119] The *Free Press* writer did
not say which of the "old countries" was meant. Whalemen had come to understand that the public knew little of whaling. They could say anything,
and few would look amid the slime and the stench to see it with their own
eyes. If it concerned whaling, well, the crazier story was the more believable.

A well-dressed man thought to be Ludwig Rissmüller stands beside a humpback
whale brought to Page's Lagoon near Nanaimo. One fluke tip has been removed at
sea to facilitate towing.
Provincial Archives of British Columbia

Ottawa seems to have ignored Balcom's "training ship" proposal, but he used both chaser boats anyway; one worked near the mouth of the Fraser River while the other steamed off Comox, north of Nanaimo. The two boats brought in thirteen whales during the first week of December, enough to glut the station and force the boat crews to lay over a day. By 11 December, forty whales had been taken.[120] The first shipment of oil from Page's Lagoon —400 barrels—was placed aboard the CPR steamship *Otter* for delivery to Vancouver, where it was trans-shipped to the S.S. *Oanfa* of the Holt Lines for Glasgow.

The local press raved.

> This company, by securing the sole right of whale fishing in the Gulf of Georgia, has solved the whaling problem on this coast. This company is and will remain the only whaling concern which has a double whaling season and can keep their steamers and skilled men employed through the whole year.[121]

Unfortunately, a few vociferous residents objected to this "sole licence" and, in fact, objected to the license altogether. A week after Page's Lagoon opened, the manager of the Terminal Steamship Company of Vancouver petitioned his parliamentary representative, R. G. Macpherson, to enter a protest at Ottawa against the "Sechart Whaling Co." and its hunting in Howe Sound. The petitioner, J. A. Cates, claimed that Howe Sound was a feeding and breeding ground for whales, and his appeal against the whalemen was couched in phrases similar to those heard from conservationists a half-century later. But he had less than conservation on his mind. The whales in Howe Sound had become a popular attraction among tourists visiting Vancouver, and the Terminal Steamship Company provided the "whalewatch" boats.[122]

Macpherson took Cates's protest to the Department of Marine and Fisheries, which answered that the Pacific Whaling Company did not, in fact, enjoy a "charter" to kill whales in the "Straits of Georgia."[123] To this, Macpherson responded: "There is no use in mincing matters, whether they have got a charter or not they have killed in the last four months, over 20 whales in Howe Sound."[124] The wrangling soon involved Inspector Taylor, who was asked to provide scientific corroboration of whales breeding in the sound. Taylor answered that he did not consider the place "a special resort for whales"; neither did he believe that the whaling was seriously detrimental, since it was conducted only a few months of each year.[125] So for a brief time, any humpback whale that happened into the Strait of Georgia proved fair game for the harpoon-cannons of the Pacific Whaling Company. There were two very prosperous weeks of whaling in December 1907 followed by one exceptional week in January 1908. Then, whaling suddenly ceased.

Whether the whales were extirpated is not known, but it is clear that the hunting ended on 25 January 1908 and was not resumed. Afterwards the Page's Lagoon columns in the company's Daily Report are blank.[126] And though the Pacific Whaling Company continued to pay the annual licence fee for the station—leading some to believe that whaling continued there for perhaps three winters—there is nothing to suggest that it was conducted there after January 1908. Whaling in British Columbia would remain a seasonal occupation, requiring that the previous season's employees be called up again each spring. Some would be lost to the industry, particularly those who needed winter work to support their families. But others would live from season to season along the waterfronts of Victoria and Vancouver, perhaps helped by understanding landlords who would carry them on credit between whaling seasons.

It is an old sailor's epithet: "Jack's money comes hard, and goes away easy,"[127] and so it would be among the shore workers and boat crews of the Pacific Whaling Company. Some years later, an older whaleman who had come from Norway to work for the company was asked what became of all the money he made. He spoke for many of the shore-whalemen when he answered: "Oh, I spent some for whiskey, some for women, and the rest, it went for foolishness!"[128] Had the whales been present in larger numbers, perhaps the Pacific Whaling Company could have broken the detrimental cycle of laying-off and later rehiring its best workers.

Meanwhile, the company called its second annual meeting. There was certainly some disappointment over the shortfall in the catch at Page's Lagoon, and the decision to shut down may have been reached by consensus since whaling ceased the following day. But Sechart and Kyuquot had enjoyed great success. Additionally, the company held licences to build additional stations in Fitzhugh Sound and on Banks Island, and though it may not have become public at the meeting, plans were being formulated to further extend the geographical boundaries of the business. President William Grant and Managing Director Balcom cited earnings of $119,657.29 in 1907,[129] more than double the previous year, and announced a fabulous 25 per cent dividend on preference stock and 18 per cent on the ordinary shares (the latter held solely by Vice-President Rissmüller). Little wonder, then, that the stockholders re-elected the entire board of directors.

Despite the failure of the Page's Lagoon station, the year 1907 closed in Sprott Balcom's favour. He had constructed two successful whaling stations and obtained a second chaser boat. He had also circumvented several serious challenges to the de facto monopoly; at the beginning of 1907 Ottawa had juggled several competitive applications, but during the following eighteen months Balcom and Rissmüller would see that each was dismissed or else coopted by the Pacific Whaling Company.

John O. Townsend, who had been offered the conciliatory six-month option on Campania Island, was simply outwitted. When the option expired in
the late spring of 1907 he made a forced play, submitting station plans and
the $800 fee at the last hour. He did not give any precise location for his con-
struction site, since he understood himself to be in geographical conflict with
the Fitzhugh Sound licence already granted to Sprott Balcom.[130] Balcom was
wired at once; could he say exactly where the Pacific Whaling Company
shore station would be built?

Surprised that Townsend had accepted the option, Balcom wired an
urgent and provocative answer. First, he apologized for not providing the information sooner, but he had been fully occupied *erecting the two new whaling
plants*. Then he implored the ministry to grant his newest application for
Banks Island—also in conflict with the Campania Island site—rather than
honour its promise to Townsend. Balcom alleged that Townsend intended to
"sell out . . . to a Norwegian concern."[131] Conveniently forgetting the
Norwegians in the employ of the Pacific Whaling Company, he assaulted
Townsend's plan as a threat to Canadian sovereignty:

> The Norwegians . . . have the right to bring into Canada their vessels
> free of duty, while I had to pay duty, having my Norwegian built
> steamer brought into Canada under English flag, and they can, without
> being liable to prosecution, bring in alien contract labour signed on as
> crew of their vessels, at rates of wages so far below current wages here as
> to demoralize rates of wages in the industry, to the serious injury of all
> workmen engaged in it. This will give foreigners a great advantage over
> our Company, and the ninety-four Canadian shareholders of our Com-
> pany feel indignant about such possibilities, and hope that Town-
> send . . . will not be given a preference over Canadians.[132]

Richard Hall, a marine insurance agent in Victoria, supported the objection.
The Norwegians had fished themselves out of Newfoundland, he said, and
were coming to British Columbia "by reason of the tremendous diminution
of the herd" in the east. He argued, as did Balcom, that "all the profits arising from this enterprise would go to a foreign country."[133]

William Templeman, the Inland Revenue minister who consistently objected to a whaling monopoly, overcame his competitive preferences and also
recommended in Balcom's favour. So the Ministry of Marine and Fisheries
once more rejected Townsend and granted a whaling licence in September
1907 to the Pacific Whaling Company for its proposed station on Banks Island.[134] The good captain was once more outwitted by Sprott Balcom's shrewd
politicking.

Balcom then challenged Cereno Jones Kelley, and outmanoeuvred Charles

Lugrin. The Ministry of Marine and Fisheries had attempted to resolve the several geographical overlaps by requesting various applicants to move their sites. When Cereno Jones Kelley agreed to shift from Porcher Island to a site he called "Stanley Harbour," on the south side of Big Dundas Island, Ottawa issued a licence. At Kelley's request, this licence was filed in the name of Dr. Walter Francis Hall, who had collaborated with Kelley on the original application.[135] Sprott Balcom soon enquired whether the Kelley-Hall operation planned to utilize Rissmüller's patented equipment. "I think it is only fair," he wrote, "that other Companies should be treated just like [the Pacific Whaling Company]. . . . When we found out, that no other process could do [what it promised], and we have tried them all—we had to secure the Rissmuller process, which is the only one, which gives perfect satisfaction from a sanitary standpoint and in every other respect and we had to pay heavily for it. We therefore think, that it is only fair, that other Companies should not be allowed to use inferior processes, by which the waters would be polluted."[136]

The question of whether or not Rissmüller's patents enjoyed the exclusive approval of the Ministry of Marine and Fisheries challenged the integrity of the ministry's decision making. In February 1908 Charles Lugrin complained to Ottawa that his clients in Scotland had been visited by Rissmüller, or one of his representatives, who left them with the clear impression that they would be required to take him into the company or pay him a royalty. They objected to the idea that they could be "held up by one man and compelled to adopt his methods" when theirs were equally as good. "The inference has gone out," Lugrin charged, "and has been spread abroad, either rightly or wrongly, that no other method will be approved or accepted."[137] The Ministry of Marine and Fisheries vigorously denied the allegation,[138] but Lugrin's men nevertheless decided that Canadian waters were too murky for the business and withdrew.

Captain Theodore Magneson did likewise, leaving Kelley and Hall in possession of the only active whaling licence in British Columbia not held by the Pacific Whaling Company. But if these two men considered competing against Sprott Balcom, they held on to their conviction only long enough to form a stockholder corporation and sell the bulk of the stock—to Balcom and Rissmüller. They began this process in the summer of 1908 when they founded the Prince Rupert Whaling Company in Bastion Square, Victoria. Citing the handsome dividends paid to stockholders of the Pacific Whaling Company, investors were advised that the new firm would use the same type of reduction equipment. The station would be built at Stanley Harbour on Big Dundas Island, where, Kelley's stock prospectus noted, whales were "so numerous at times as to seriously interfere with the operation of the fishermen."[139]

In Prince Rupert, the whaling plan was heralded with great fanfare: "During the winter months whales abound in the waters of PRINCE RUPERT Harbour; these, with all other fish industries . . . are capable of the greatest growth and advancement, and will be a great factor . . . in the building up of this city."[140] A share in the venture might be had for five dollars. The issue of 80,000 shares — nearly all common stock — was well subscribed; the major buyers were the principals of the Pacific Whaling Company who, by the summer of 1909, owned $50,000 of it.[141] By this method Sprott Balcom and Ludwig Rissmüller came into control, if not outright ownership, of the Dundas Island whaling licence.

Balcom had yet to apply for a licence to build in the Queen Charlotte Islands. Perhaps he anticipated serious charges of monopoly if he did so. But by utilizing some of his lesser-known colleagues, he soon came into possession of such a licence. In September 1907 — the same month Ottawa selected Balcom over Townsend — the ministry received application from one George A. Huff of Alberni for a whaling station site at the entrance to Skidegate Channel on the west coast of the Queen Charlottes. Huff's application referred to the Rissmüller reduction methods "now in use by the Pacific Whaling Company"[142] and implied that the doctor was affiliated with the new project. Unfortunately, the site was less than one hundred miles from Banks Island; this problem was brought to Huff's attention and before long — miraculously — the Pacific Whaling Company agreed to remove far enough to make up the required separation.[143] In the ensuing months Huff also relocated his proposed station, to Rose Harbour on Moresby Island. This site was granted in March 1908.[144]

In fact, George Huff and Sprott Balcom were partners long before Huff submitted his application. Huff was one of the founders of the Pacific Whaling Company, and the affiliation could not have been secret; in September 1908 the *Colonist* reported that the two men had set off for the Charlottes to select a site for a whaling station, and Huff afterward visited Sechart, where, the *Colonist* gossiped, "he is making some experiments with regard to the treatment of whale meat [for human consumption]."[145] A year later, in December 1909, Huff incorporated the Queen Charlotte Whaling Company Limited and issued 5,500 shares of stock at fifty dollars each.[146] Balcom obtained an unknown number of these shares and belatedly admitted to Ottawa that he was "interested with Capt. Huff in this company."[147]

But a twofold disaster, barbed like the tip of the old whaling harpoon, beset the British Columbia whaling industry in 1908. Fewer whales were caught, and the facts pointed clearly to a decline in the numbers of large blue and finback whales, species which had been sought only since 1906. At the same time, the international oil market came to grief in a general business depression which critically impeded the efforts of the Glasgow agents to sell

Canadian oil in the European market. Factories in that city operated only half-time in 1908, and the price of oil fell to about twenty-five dollars per ton, off more than four dollars per barrel.[148]

In fact, whalemen in British Columbia killed 592 whales in 1908; more than in any previous season. Most of them—462—were humpbacks, which, Sprott Balcom explained, "although belonging to the whale family are comparatively very small animals and hardly pay the expenses of manufacturing."[149] For want of larger specimens and because of the simultaneous business depression, the announcements at the third general meeting in 1909 were less than enthusiastic. Profits were reduced to $71,070.36; dividend payments on preference shares fell off from the 25 per cent paid in January 1908 to a comparatively unimpressive 14 per cent.[150]

The Tyee Company's whaling station at Murder Cove, Admiralty Island. The first modern whaling plant in Alaska, it was used to process whales from 1907 until 1913.
Cobb Photograph, Special Collections Division, University of Washington Libraries

A competitor had also appeared on the scene. In 1907 the Tyee Company of San Francisco opened a shore station on Admiralty Island in the southeastern corner of the Alaskan panhandle, at a place unhappily called Murder —or Murderer's—Cove. Here was the reality of William Templeman's threat of unequal foreign competition, of freewheeling Americans outhunting overlegislated Canadians. Tyee's whalemen worked on a free rein since the United States government placed no restrictions on Alaskan shore-whaling operations. Any number of boats could be used, and the oil could be imported into the forty-eight states without payment of duty, whereas British Columbians could not afford to send their oil to the United States; the duty—eight cents per American gallon, $24.96 per ton of 2,240 pounds— amounted to more than one-fourth the total value of the oil. Most of it had to be shipped to Glasgow. "We are producing about 3300 Tons of whale oil," Balcom explained, "and by entering this quantity into the United States, we would have to pay over $81,000 . . . duty. Thereby the United States, the best oil market in the world, is almost closed against us and we are compelled to ship the bulk of our oil to Europe."[151]

Balcom may have complained about the tariff, but he and Rissmüller nevertheless had their hands in the Americans' pocket. John Barneson, the Norwegian-American founder of the Tyee Company, and his colleague, Captain I. N. Hibbard, began construction of the Admiralty Island whaling station with expert assistance from Rissmüller, who agreed to accept 100,000 shares of Tyee stock, worth one dollar each at par, in exchange for certain "exclusive rights" to his plans and specifications. Tyee's principals gained fifty years of exclusivity in Alaskan and Pacific coast U.S. waters, provided that they erected an operating whaling station *south* of the forty-ninth parallel before 31 December 1910.[1]

At face value, Rissmüller violated his 1905 memorandum of agreement with Sprott Balcom, since the two undertakings would compete for the same whales. Perhaps Balcom and Rissmüller intended to take over the Tyee Company by controlling its stock; more likely Balcom was somehow "interested" in the company as he was with George Huff in the Queen Charlottes. For whatever reasons, the connections between the Pacific Whaling Company and the Tyee Company were contractually incestuous. Rissmüller even agreed in writing to provide plans for appropriate steam vessels—presumably both the *Orion* and the *St. Lawrence*—in order that some American shipbuilders might construct a chaser boat for the Tyee Company. And if Sprott Balcom did object, he could lodge no complaint; a clause in the 1905 agreement permitted "any party representing [Rissmüller] to examine and take drawings of the steamer 'Orion' the property of the [Pacific Whaling Company]."[153]

The Tyee contract was awarded to the prestigious Moran Shipbuilding Company of Seattle, Washington, which was just then constructing steel warships for President Theodore Roosevelt's expanding navy. Moran's shipwrights completed the *Tyee Junior*—*Tyee* was already taken—in 1907; it was the first steam-powered chaser boat built in North America. A long, narrow, seventy-one-ton vessel, 97.9 feet by 17.7 feet by 11.8 feet, it was 8 feet longer and more shallow in draft than either of the Norwegian-built prototypes. An oil-fired engine was substituted for the coal-fired power

The first whaling steamer constructed at an American shipyard, the *Tyee Junior* was considered unusual by Norwegian sailors, but it operated successfully for the Tyee Company and later, as the *Tanginak*, for the American Pacific Whaling Company.
Provincial Archives of British Columbia

plants in the *Orion* and the *St. Lawrence*, and it turned two propellers rather than the customary single shaft.[154]

Upon completion, the *Tyee Junior* steamed to Victoria, where eight whalemen recently arrived from Norway were collected. These men had been working at Sechart prior to completion of the Admiralty Island plant.[155] During its visit, a party of local businessmen enjoyed a ride aboard the new steamer, and among them were William Grant, president of the Pacific Whaling Company, and Sprott Balcom.[156] After their inspection the *Tyee Junior* steamed north to the new station at Murder Cove. One of the first observers to walk its deck there was Roy Chapman Andrews, the young curator from the American Museum of Natural History in New York, who investigated humpback whaling in August 1908.[157] The *Tyee Junior* proved itself to be the fastest chaser boat on the coast, with a top speed of thirteen knots, but the Norwegian seamen called it *eiendomelig*, an expression which conveyed an impression — not entirely flattering — of peculiarity or uniqueness. Nevertheless, they went aboard eagerly; the promise of large bonuses payable on unexploited Alaskan whales far outweighed any injured national pride occasioned by shipping in a non-Norwegian hull.

Disregarding the interrelationship between the Tyee Company and the Pacific Whaling Company, Sprott Balcom used the spectre of competition as a lever to force repeal of the one-chaser-per-station rule in the Fisheries Act. Not only was Tyee operating near the Alaska-British Columbia border, but others, he said, were planning similar operations in Washington State. One of these would be located at Neah Bay and another at Port Angeles, just across the Strait of Juan de Fuca from Victoria. None of these competitors would be legally constrained to a single chaser boat. Inspector Taylor verified Balcom's report by sending the ministry a newspaper clipping which announced the construction of a whaling steamer in Washington for a "Whale Products Company" operating from Port Angeles. Taylor also confirmed that a Seattle-based company was considering the Neah Bay site.[158] These would compete directly with the Sechart plant.

Balcom also alerted Ottawa to the possibility of an invasion by Norwegian-owned floating "factories," large processing ships which, like the vanishing American sailing whaleships, were prepared to render whales at sea. These factories, he warned, were already on the "South Pacific coast" and no doubt would be moving north.[159] In February 1909 he proposed a modification of the Fisheries Act which would limit Canadian whaling firms to one chaser boat for each *paid license*. The effect of attaching the vessel to the licence rather than to the station benefited both the company and the government at the expense of the whales, since Ottawa was likely to receive fees for disused or unbuilt station sites which the company would pay to justify the employment of additional chaser boats. The Pacific Whaling Company held five valid licences but operated only two whaling stations; under the proposed change five chaser boats could be run from Kyuquot and Sechart instead of the two to which the company was presently limited.

Balcom argued convincingly that the legal ratio of one boat per station was highly uneconomical; more steamers could bring in more whales and therefore increase revenue. In order to add more chaser boats, a Canadian whaling company must also invest in shore stations, even when whales could as easily be processed at the existing ones. And even if a new station were built and its chaser boat added, no additional profit would accrue; the annual cost of operating a single station just then amounted to nearly fifty thousand dollars.[160] Balcom wisely sidestepped the intent of the one-boat-per-station rule; namely, to avoid overtaxing the local stocks of whales, as had happened in Newfoundland.

R. N. Venning, then superintendent of Fisheries, recommended the amendment and cited a ministry precedent of aiding struggling whaling companies. The Quebec Steam Whaling Company, he noted, continued to operate despite its inability to pay its annual licensing fees during the three previous years. "It does seem clear," Venning wrote, "that [the Pacific Whaling Company] is in a very serious plight, and it would be embarrassing in the extreme to the industry in British Columbia for this Company to go under."[161]

The ministry initially refused to reshape the law, but in 1910, Subsection 9 of Section 4 of the Fisheries Act, Chapter 45 of the Revised Statutes, was repealed, and new language was substituted which allowed whaling companies to operate one steamer for every fully paid licence.[162] By this change the Pacific Whaling Company was granted permission to use five chaser boats and there would be no further objection if all five operated from the two existing shore stations, or either one alone.

With the new law in hand, Sprott Balcom and his associates set out to acquire more chaser boats and more licences. They began the process of stock manipulation to bring the Pacific Whaling Company, the Prince Rupert Whaling Company, and the Queen Charlotte Whaling Company under the umbrella of a single corporate structure. This expansion required substantial additional funds, and the moneys were soon forthcoming from a most solvent source: the builders of the Canadian Northern Railway. The Canadian Northern had begun construction of a direct transcontinental route from Pacific Canada to the Atlantic with its western terminus located at Prince Rupert—Vancouver already being in the hands of the competing Canadian Pacific. The ambitious builders of this line, William Mackenzie and Donald Mann, set out to develop any business that would create freight for the new road to and from the Pacific Northwest. One of these was whaling, which might use the Canadian Northern to transport oil to the Atlantic seaboard for trans-shipment to Glasgow.

Mackenzie and Mann stood ready to use their considerable power and money to assure the success of the Canadian Northern, and because they were employed as consultants by the railway, they were free to acquire "whatever assets seemed desirable" and organize "a variety of promotional, financial, and developmental functions which the railway company could not perform under its own charter."[163] The whaling industry, particularly if centred in the northern part of the province, seemed likely to be a prime user of the new railroad.

Before 1909, the only rail involvement with whaling in British Columbia had been through the use of the Canadian Pacific coastal steamers *Amur*, *Otter*, *Queen City*, and *Tees*, which brought cargoes of oil and fertilizer from Sechart and Kyuquot to Victoria. Mackenzie and Mann now envisioned a double profit, one from dividends paid on their shares of whaling stock and a second from the business of shipping whale products. They probably approached Sprott Balcom early in 1909, although it is conceivable they had provided advice to Balcom and George Huff the preceding year. In any case, in 1909 Mackenzie and Mann's colleagues introduced a supposedly independent bid for a whaling licence in the Queen Charlotte Islands. This application, filed by John M. Macmillan of Vancouver for Rennell Sound on the west coast of Moresby Island, would eventually provide Sprott Balcom with another licence.

Macmillan's very formal, complete, and handsome application stated that $150,000 would be available to begin construction at the site and that the reduction equipment would be of the "most approved type"—presumably meaning Rissmüller's. For references he named William Murray, manager of the Canadian Bank of Commerce in Vancouver, and Alexander D. McRae.[164] The Canadian Bank of Commerce maintained close business ties with Mackenzie and Mann, while McRae served as a land commissioner in Winnipeg and general agent for the Canadian Northern's land immigration and industrial departments and handled land transfers for the railway both in Canada and the United States.[165]

Not surprisingly, Macmillan's request was courteously and promptly approved by the ministry in November 1909.[166] Six months later he petitioned Ottawa to have his licence transferred to the Queen Charlotte Whaling Company Limited—the firm begun by George Huff.[167] That request, too, was quickly approved, and now the several newly licensed whaling companies in British Columbia stood in line, waiting to be absorbed into the conglomerate that Sprott Balcom and the railroad builders had already designed.

At the fourth annual meeting of the Pacific Whaling Company, in February 1910, Balcom announced 18 per cent dividends on preference stock and 11 per cent on common shares, on a net profit of nearly $70,000. He also pleased the newcomers, and staggered everyone, by announcing a doubling of the stock and the issue of a new $60,000 series, to be sold in proportion to the shareholders' current portfolios.[168] In addition, the company would soon offer extra shares and debenture certificates to pay for new projects— 800,000 five-dollar shares, and nearly $2 million in debenture certificates. The debenture mortgage, established with the British Empire Trust Company Limited, obligated the firm to service the £400,000 principal at 5 per cent annual interest. But if Balcom was worried about supporting the repayment, he appeared unconcerned. He must have anticipated that the expansion of the whaling business would reap an increased harvest to pay the bills. And some of those projects were already underway, including construction of a large whaling station in the Queen Charlotte Islands, purchase of a fleet of whaling steamers to service the company's new licences, and acquisition of a shore whaling station and chaser boats in the United States.

The legal steps leading to the creation of the umbrella whaling corporation were convoluted. In August 1910 the Queen Charlotte Whaling Company— which now held Huff's and Macmillan's licences—was sold to the Pacific Whaling Company; stock transfers made up the bulk of the sale.[169] A similar arrangement was reached with the nascent Prince Rupert Whaling Com-

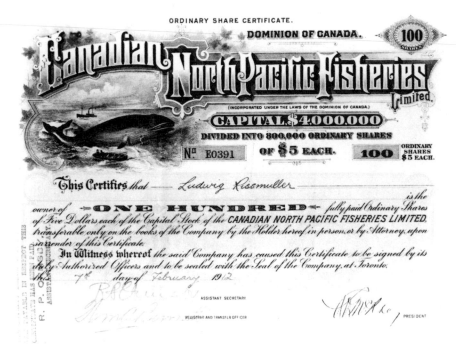

The Canadian North Pacific Fisheries offered several stock issues to finance their expansion plans. This share of common stock was part of Ludwig Rissmüller's portfolio.
Graeme Balcom, courtesy Vancouver Maritime Museum

pany, which had already drawn some fire from Ottawa for not beginning construction of its whaling station. The merger was reported in the local press in September, and the new investors were described as "millionaires from Winnipeg and Vancouver."[170]

The umbrella corporation styled itself the Canadian North Pacific Fisheries Limited and subsumed the operation and stock issue of the several whaling firms. Legal and administrative offices were moved to Toronto, where the railroad interests could keep watch, but operations remained at Victoria. Alexander D. McRae and Andrew D. Davidson signed the papers of incorporation; Sprott Balcom and Ludwig Rissmüller continued in their former management positions. Ottawa expressed concern over the implications of monopoly, but the government's own analyst saw no objection because, as he put it, "the public interest will not be in any way contravened."[171]

It seemed as though the expansion might extend even beyond provincial borders. Balcom and Rissmüller devised a plan to establish two whaling stations in the Canadian Arctic and petitioned Minister of Fisheries Brodeur for

Sailor-artist Charles Robert Patterson (1878–1958) served aboard the *Orion* and completed its portrait sometime before April 1908. Though remembered for his depictions of square-rigged sailing ships, this painting was often reproduced on postcards and in advertisements for the West Coast whaling companies.

Cape Bathurst and Herschel Island, near the old haunts of the San Francisco-based sail-whalemen.[172] The time seemed ripe for Canadian entrepreneurs to appropriate this lucrative fishery for Canada, but, in fact, the time was considerably overripe. By 1909 the fearsome American whalemen were virtually out of the business, even then struggling through their penultimate voyages into the ice for the increasingly uncertain returns of baleen from the mouths of the last bowhead whales. Even so, the press lauded Balcom's proposal as a blow against the "poachers." "Conjecture is rife," one reporter said, "as to what policy the Dominion government will adopt in these northern waters, now that Canadian capital is about to exploit the industry. There seems to be no doubt that the American poachers ... can be warned not to invade Canadian waters if the Ottawa authorities assert their rights."[173]

Balcom earned additional support from within the Department of Marine and Fisheries: "There can be little question," one advocate said, "that the operations carried on from such a base ... could be done more economically than from San Francisco, and the liklihood is therefore that in a comparatively few years the United States Fishing Vessels will be largely all out of the business."[174]

The two licences were promptly granted in September 1910 to the new

Canadian Arctic Whaling Company, which was identified to all as an "offshoot" of the familiar old Pacific Whaling Company.[175] Sprott Balcom's real design seems far more pragmatic than to conduct an exercise of Canadian sovereignty in the far north. The cost of building and supplying an Arctic whaling station — let alone two — would have been astronomical, particularly in light of expansion plans at home. The licences, however, enabled the company to possess two additional chaser boats. Nothing remained in the Fisheries Act to stipulate where such boats must be employed.

By Christmas 1910 the Canadian North Pacific Fisheries Limited held ten permits to operate whaling stations: the former Pacific Whaling Company's licences at Sechart, Kyuquot, Banks Island, Fitzhugh Sound, and the disused station at Page's Lagoon; George Huff's location on Moresby Island and Macmillan's on Rennell Sound, both of which had belonged to the Queen Charlotte Whaling Company; the Prince Rupert Whaling Company's license for Big Dundas Island; plus the two Arctic sites. Taken all together, they gave the CNPF the right to operate no fewer than ten whaling steamers.

Balcom had already acquired a few of the needed chaser boats. A new vessel had been ordered from Akers in Christiania, and in a novel attempt to avoid the wear and tear of a Cape Horn passage, it was arranged to construct hull number 290 in sections and ship these to Victoria aboard the Blue Funnel steamship *Titan*. There they were riveted together at the Victoria Machinery Depot.[176] Christened the *William Grant* in honour of the president and business manager of the Pacific Whaling Company, this chaser boat was similar to the others, 91.2 feet long, 106 tons displacement.[177] Grant's friend William Heater assumed command, confidently expecting to begin at the new station at Rose Harbour in the Queen Charlottes. But the Victoria Machinery Depot outstripped the whaling company's construction schedule, and Heater had to run his service trials from Sechart.

During 1909 Balcom also acquired two second-hand whaling steamers. One of these, the *Sebastian*, had been built by Akers in 1904 and was bought in Newfoundland, where it had served the Micmac Whaling Company. A crew steamed the 103-ton chaser boat through the Strait of Magellan, rounded by Punta Arenas on 2 February 1910, and arrived in Victoria in the spring. Balcom's second purchase, the *Germania*, was already assured a place in whaling history, for it was the first steam-whaling vessel constructed for a German company. Begun in 1902 for Germania Walfang-und-Fischindustrie A/G of Hamburg, it measured 94.3 feet by 16 feet, and 105.95 tons gross.[178] The *Germania* was initially placed in service at the company's station at Faskrudfjord, Iceland, but its Icelandic career did not last long; it was transferred in 1906 to the Sociedad Ballenera y Pescadora, a Chilean-German whaling company which operated a whaling station at Puerto de Corral, near Valdivia, Chile. That firm was reorganized in 1908 with Norwegian financing, and in the early spring of 1910 it sold the *Germania* for service in British Columbia. The vessel arrived at Point Ellice in July.[179]

These three, plus the *Orion* and the *St. Lawrence*, made half the number of chaser boats permitted under the revised Fisheries Act. Apparently knowing no fiscal bounds, Balcom committed a fortune—perhaps as much as $300,000—to the construction of five Akers-built chaser boats. Isak Kobro again brokered the transaction, and Reuben Balcom returned to Norway to supervise the passage of the new steamers to the North Pacific. Ninety-six feet overall, eighteen feet in breadth and drawing just over eleven feet,[180] these 102-ton chasers were each fitted with a powerful 330-IHP triple-expansion steam engine which promised a top speed of about twelve knots. A steam-steering gear would allow the helmsman—working in consort with the engineer—to turn the vessel within, or almost within, its own length.[181]

Their general arrangement was typical of whaling steamers of this period. The crew lived forward in bunks in the forecastle, under the gun platform; between the forecastle and the wheelhouse was the cargo hold and a chain locker to receive the whaling line as it fed in from the double-barrelled drum winch on deck. The wheelhouse was open to the elements, and contained little more than the wheel, an engine room telegraph, binnacle, and two lifebelt storage boxes. Meals were prepared in the galley, directly below the wheelhouse on the port side, and consumed in the mess room to starboard. Fuel bunkers, water ballast, and boilers were placed amidships, the engines just aft, and officers' accommodations in the stern. The latter included two small cabins for the captain and the chief engineer and a cramped saloon with a U-shaped sofa and table. Each chaser carried two boats on davits above the walkways on either side of the engine room house.[182]

During construction the new hulls were known around the Akers yard by their work numbers: 306, 309, 310, 312, and 313. But as launching time approached the workers began to fret; no names had been provided for the christening, and it would be nothing but bad luck to launch a nameless ship. They waited and worked but heard nothing. In Canada, the matter was blocked by a donnybrook between two of the most influential shareholders, each of whom insisted on selecting the names. Ludwig Rissmüller wanted to call them after the rivers of his German homeland, but his opponent,[183] while agreeing on rivers, preferred instead the soft rolling names of streams in his native Scotland.

There could be no consensus. Would the boats be called the *Elbe*, the *Ems*, and so forth? Or would *Tay* and *Tweed* prevail? The matter rested at the highest diplomatic level within the company. No one dared attempt to break the deadlock. And while the two men disagreed, the shipwrights worried the vessels toward completion. Not long before their launching, an anonymous arbitrator stepped in with a resolution: the chaser boats would be given the

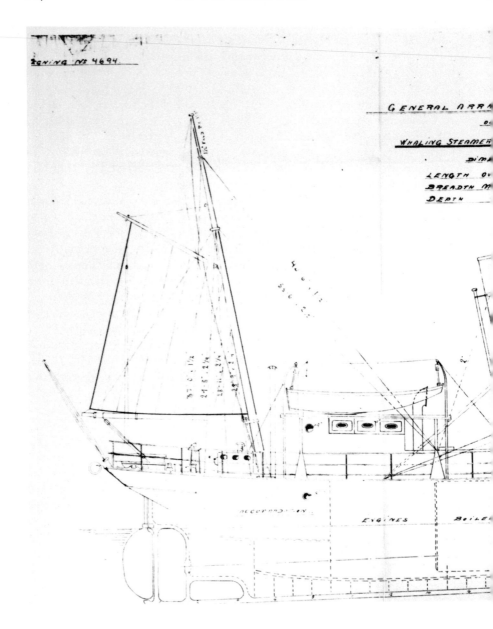

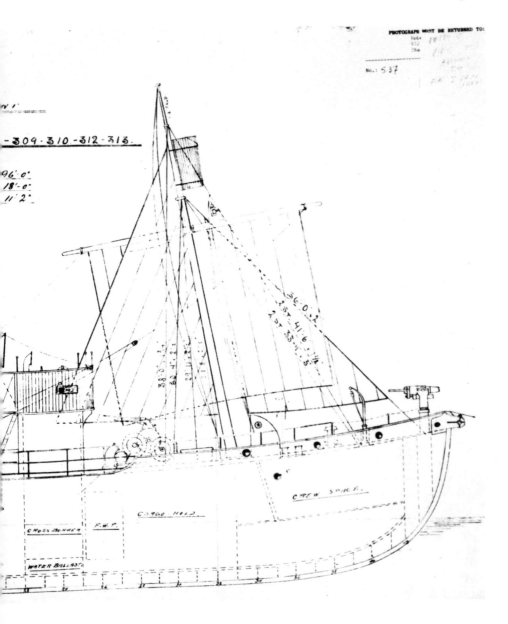

A builder's plan for the "colour" fleet shows the forward mast fitted with a yard and square sail. These were not used for whaling, and may never have been installed. Large circles forward of the wheelhouse represent the steam whaling winch.
Aker Engineering A/S, Oslo, Norway

names of colours. It was a simple idea and had the practical advantage of making five identical vessels distinguishable at sea by means of a coloured flag or a painted stripe on the hull. Balcom family lore holds that Sprott Balcom resolved the dispute, and this seems likely, since only he carried sufficient weight to overrule his two mighty colleagues.[184]

So the "colour" or "rainbow" fleet came into existence. They would remain a fixture in British Columbia waters for the next thirty-five years. Hull 306 went into the water as the *Green*, closely followed by the *Black*, *White*, *Brown*, and *Blue*. Early newspaper accounts suggested six vessels, one to be called the *Rose*, but this steamer never materialized. The reporter may have confused the vessel with the station under construction at Rose Harbour. Another newspaper, the Prince Rupert *Evening Empire*, published the correct count but not the correct colours, identifying the new fleet as the *Black*, *Green*, *White*, *Red*, and *Yellow*.[185] The names chosen by the company may have suggested morbid coolness and sobriety, considering the sort of adventures that were anticipated; still, management may have wished to avoid the connotations of bloodletting and cowardice implicit in the brighter hues of the *Evening Empire*'s spectrum.

Eleven men crewed each of the new steamers, including a captain-gunner, a mate, four seamen, two engineers, two firemen, and a cook. Their Cape Horn passage may have been sail-assisted; surviving builders' plans show two sturdy masts. The mainmast, forward of the wheelhouse, appears with a crossyard designed to carry a square mainsail. The wooden lookout barrel was mounted above the yard toward the truck, its top rim thirty-seven feet over the deck below. Aft of the engine-room house was a second, raked mast and boom carrying a fore-and-aft sail. Both the after mast and the square mainyard — if ever installed — were removed soon after arrival in Victoria.[186]

Controversy surrounded even their passage to the Pacific. Legend insists that at least two of the skippers raced from Panama to San Pedro, California, and that both expended their coal and had to be towed in under the lip of Point Fermin.[187] Nevertheless, they arrived safely in Victoria Harbour in the early spring of 1911. Each was described as a veritable "water witch,"[188] and the fleet was expected to increase substantially the catch statistics of the Canadian North Pacific Fisheries Limited.

Balcom purchased one more ship. In 1907 the Pacific Whaling Company directors had announced the impending acquisition of a cargo freighter to carry men, equipment, and products to and from the far distant whaling stations. But the purchase was deferred, probably because of the economic depression of 1907–8. In December 1910 Balcom obtained from the Northern Steamship Company a 182-foot clinker-built freighter called the *Petriana*.[189] At first its name was retained, but in a spirit of consolidation the ship was renamed the *Gray*, by which color it would become known to every

The Canadian North Pacific Fisheries Limited photographed its entire fleet, including the *Green*, shown here, during the expansion of 1910–11. The *Green* outlived the others, and finally sank at its berth in Victoria Harbour in the late 1960s. *Vancouver Maritime Museum*

coastal resident in British Columbia for the remainder of a very long career.

At the same time, construction was begun on the first whaling station in the Queen Charlotte Islands. Originally sited by Balcom and Huff at the south end of Moresby Island, it was determined that a better location existed across the channel on Kunghit Island, and the plant was built there, despite a strong tidal current in the channel. It cost perhaps $270,000; this was the figure provided to the press, but it may have been exaggerated for effect.[190] Much of the reduction equipment was second-hand. The company finally wrote off the Page's Lagoon station, dismantled it, and sent the machinery north to Rose Harbour. Nanaimo was abandoned for good, and in 1912 the property was sold by O. H. Burt & Company to a Mr. Clelland for $50,000.[191] Clelland had no interest in whaling; he wanted the property as a land speculation.

Rose Harbour was the most ambitious station yet constructed, with large rendering buildings; separate storage sheds for oil, bone meal, and fertilizer; bunkhouses for the crew; and small but comfortable cabins for the station manager and his guests. The storage and rendering buildings were constructed of wood framing and heavy galvanized iron sheathing set on wooden pilings over the water.[192] A two-section inclined slipway was built for the hauling-up and disarticulating of whales, and a long wharf allowed for the coaling of vessels and the loading of supplies and products. Business proved to be all that could be desired. Reuben Balcom, visiting at the Prince Rupert

STATION NUMBER 3

Rose Harbour proved to be the most successful of all modern whaling stations in British Columbia. Buildings and docks in this architectural rendering (circa 1910) are identified in Appendix B.

Graeme Balcom, courtesy Vancouver Maritime Museum

Inn in September 1910, told interested reporters about a large catch of eighty whales in the first six weeks of operation; 150 men were employed at the station, and the coastal steamship *Amur* was chartered to carry the first lots of oil and fertilizer to Vancouver.[193]

Immediately the Rose Harbour station was completed, construction crews moved to the north end of Graham Island, about as far as they could go in the Queen Charlottes, and began a second whaling station at Naden Harbour in Virago Sound.[194] Men and equipment were ferried to the site aboard the chartered steamship *Henriette*, a former sailing vessel belonging to the shipping arm of the Grand Trunk Pacific Railway. Seventy men went north on the first trip, together with preliminary construction needs and a complete electric lighting plant.[195] The steamer's captain was instructed to lie to as a floating hotel until the bunkhouses were completed. On subsequent trips the *Henriette* carried lumber from the Victoria Mills as well as forty or fifty additional workmen and the vats, dryers, and other necessities of modern whaling. Naden Harbour largely duplicated the Rose Harbour station and became operational before the close of the 1911 whaling season; several of the new "colour fleet" were sent to work there.

In less than twenty-four months Sprott Balcom and his railroad-building associates organized the new corporate structure, obtained ten whaling licences, purchased seven whaling steamers and a freighter, and constructed two whaling stations. But these works were not yet the sum of Balcom's unrelenting enterprise during 1909–11. With the assistance of Ludwig Rissmüller and the new investors, he also built a third whaling station on American soil and outfitted it with American-made chaser boats. This was Westport, located within Gray's Harbor, Washington, and known variously as "Westport," "Gray's Harbor" or "Bay City." A firm called Hall and Company had first obtained the site for a whaling-station; probably this was the same Frank Hall who was affiliated with Cereno Jones Kelley on Big Dundas Island. If so, then it is likely that Balcom was part of the plan from the beginning. If not, he very soon became involved because the pattern of its development followed a timetable observable also in the execution and transfer of the Prince Rupert and Queen Charlotte Whaling Companies.

Before any construction began, Hall and Company changed its name — both auspiciously and suspiciously — to the American Pacific Whaling Company, Incorporated. Headquarters were established in Seattle. In an affidavit taken in 1921 Sprott Balcom affirmed that he and Rissmüller had acquired the property, built the station, and arranged for the construction of two steam whaling vessels at a Seattle shipyard, "all of which we afterwards turned over to the American Pacific Whaling Company as payment on account of its stock."[196] In fact, Balcom and Rissmüller signed a memorandum of agreement with Canadian North Pacific Fisheries on 25 August 1911 which provided the two men with shares of American Pacific Whaling Company stock as a "guarantee" against their expenses in constructing the Washington plant and in having the two steamers built. By the same memorandum the CNPF agreed to repay the construction costs from earnings accumulated prior to 31 January 1912, the exchange of cash to be effected when Balcom and Rissmüller countersigned their stock shares and deposited them in the Canadian Bank of Commerce at Victoria for delivery to the CNPF directors.[197]

The Bay City whaling station was potentially invaluable, since it could serve as an entry for Canadian oil into the duty-restricted American oil market. Construction proceeded quickly and was well along by May 1911; surviving photographs dated that month show the large vats in place though not yet functional, and main buildings erected. Fifty men were hired to work at the plant upon its completion, and the inventory of equipment included the now-familiar two-section slipway, haul-up winches, wooden cooker vats, a skimming trough, centrifugal separator, canting winch, and the usual assortment of flensing knives, crosscut saws, and other tools of dismemberment. Steam from an oil-burning plant would power the machinery.[198]

The chaser boats *Moran* and *Paterson* were constructed by Moran Shipyards in Seattle. They carried a crew of nine at 9.5 knots.[199] In 1912 the *Aberdeen* and the *Westport* were completed by the Seattle Construction and Dry Dock Company and added to the Bay City fleet.[200] Together these four boats pursued the whale within a 135-mile radius of Gray's Harbor, from the Canadian border south to Cape Blanco on the Oregon coast.

By the end of the summer 1911, the CNPF's four whaling stations were operational, and enough progress had been made at Bay City to begin processing there as well. Their combined harvest totalled 1,806 whales. The per-station average reached 361, but three of the five plants exceeded it: Sechart (474), Naden Harbour (424), and Kyuquot (416). Only the incompleteness of the Bay City plant—which processed 102—restrained the statistics.[201] Twelve chaser boats brought in the catch, which exceeded by threefold the 1908 total from Sechart and Kyuquot and more than doubled the results from either 1909 or 1910.

The resulting production statistics were equally impressive. It appears that 59,268 barrels of oil were collected—about 9,000 tons—as well as 3,000 tons or more of "guano."[202] Humpback whales made up the larger percentage of the kill; 403 were taken into Sechart in 1911, making almost the entire catch at that location, and Kyuquot's workers disarticulated 293 more. But the gunners shot a remarkable number of the larger rorquals: 208 finbacks and 38 blue whales were taken to Naden Harbour. Kyuquot received 86 of the giant blues.[203]

The value of the cargo apparently justified all the preparation. In 1914 it

The new chaser boat *Moran* on the ways at Moran Shipyards in Seattle.
Puget Sound Maritime Historical Society, Inc.

was calculated that a $140 profit accrued from marketing the products ob-
tained from an average humpback whale of twenty-seven tons. A fifty-ton
finback cleared $338, while a sixty-ton blue returned $572 in profits.[204] The
cost of chasing and processing, previously deducted, turned out to be $206
per animal. Such statistics suggest a handsome return in 1911 because of the
large number of blue and finback whales in the catch, although it is likely
that costs associated with construction and maintenance of the isolated
northern stations severely undercut the actual profit margin.

The aggressiveness of the hunt led one popular writer to call the Queen
Charlotte Islands operations of 1911 a "war on the whale." He questioned
how long such hunting could reasonably be expected to continue. "Seventy-
five per cent of the whales taken are cows," he wrote, "and most of these have
young, either unborn or unweaned. The time required for pre-natal growth
of the whale . . . is generally understood to be not less than a year. It will
thus be seen that the war on the whale has been truly turned into a massacre

Stacked oil barrels await loading aboard the freighter *Tees* at the Kyuquot whaling station, about 1910. The large cargoes of oil and "guano" were carried from the stations by coastal steamships until the company put its own freighter, the *Gray*, into service.
Vancouver Maritime Museum

in the interests of commerce."[205] As in Newfoundland a decade earlier, the negative effect of intensified hunting began to appear at once. The 1912 catch fell from 1,806 to 1,374, despite a substantial increase in production at Bay City from 102 to 265 whales. The 1913 season was even more disastrous; only 916 whales were landed, about half the 1911 total. In just two years the average catch per station fell from 361 whales to 183; in 1914 it sagged further, to 151.[206] Like every whaling operation, improved efficiency actually worked against the company's best interests.

And whereas oil worth $575,000 was marketed after the 1911 season, only $252,000 was obtained just two years later; fertilizer production declined similarly, from $120,000 in 1911–12 to $40,000 two years later. The combined value of all products fell from $760,000 to $303,000.[207] The slackening of Minister Préfontaine's well-considered controls, and a complete absence of limitations based on numbers, sex, or size of whales taken led to an immediate and serious depletion.

Several schemes were proposed to offset the growing debits, particularly the steep interest payments due on the debenture mortgage. One plan called for the charter of the chaser boats to the dominion government for use as fisheries patrol vessels. A trial charter was effected in 1911, when Fisheries officers sailed on board two of the steamers. The CNPF received $100 per vessel per day, from which were drawn the ships' operating expenses. A surviving copy letter, undated but presumably prepared in 1912, proposes to renew and expand the charter at $17 per day during the whaling season, $100 per day, as before, during winter months, and suggests that the captain or pilot of each chaser be appointed a Fisheries officer, empowered to chase and seize any poaching vessels found inside the three mile limit. "This would be the best fisheries protection the Government could get. The cost would be a trifle, compared with the amount of money the Government would have to expend by building special steamers for fisheries protection, and man and operate them." The whaling steamers, the letter explains, could form "a kind of drag net all over the coast of British Columbia."[208]

As much as $40,800 might have been realized on such a charter for the eight summer months, assuming the government took all ten chaser boats in the province. This amount, plus the winter fee at the higher rate, would have more than met the interest on the debentures and would have represented a federal guarantee of those bonds. But Ottawa does not seem to have accepted the offer.

In 1913 Sprott Balcom surrendered his managerial interests and took an extended vacation to his native Nova Scotia. There he was regaled as "the most outstanding figure in the fur seal industry of four oceans" and the guiding light behind the whaling industry in the Canadian west.[209] He seems to have retired. Soon after, when called to testify before a royal commission examining the Bering Sea sealing controversy, he was asked: "What are you doing now?" He answered: "I am doing very little of anything just now."[210]

In fact, he was busy with other investments, turning some of his whaling and sealing money into Victoria real estate and gazing with interest at the charter freight business. Eventually he owned shares in the Lincoln Steamship Company of Vancouver, until that firm liquidated its assets in 1918. At the time of his death, on 21 December 1925 at the age of seventy-five,[211] he owned a fine house on Howe Street and was respected among the Victoria business community. His assets were said to have exceeded $2 million.[212]

From 1910 onward, the annual whaling license fees were brokered from Toronto by Harvey Fitzsimons, a general agent formerly in the employ of the Canadian Northern Railway. Payment was always accompanied by a cover letter on Canadian Northern stationery. Early in 1914 Fitzsimons again set his pen to that stationery to pay the fees, but the whaling company's cash flow had fallen into arrears and the firm was near collapse. An association of shareholders met hastily in London and proposed to buy the assets, but the British Treasury refused the syndicate an application to organize. The matter was soon in the hands of Yorkshire Guarantee and Securities Corporation Limited, receivers.

At home, the 1914 whaling season was dramatically troubled by spells of bad weather which locked the chaser boats in harbour. George LeMarquand, manager at Naden Harbour, reported a heavy sea during the first week of July—prime hunting time—that made it "too rough to do anything."[213] Even the most southerly of the Canadian stations was affected; in late April, manager Alfred Gosney reported that the Sechart whalemen had suffered "a very bad week of weather and . . . had no chance to fish."[214]

Poor weather hindered the whaling, but the larger problem of overhunting was not addressed. Financial mismanagement, perhaps based on a highly optimistic sense of success, had also played a part. Arne Odd Johnsen and J. N. Tønnessen, among the most knowledgable of all modern whaling historians, attributed the ensuing collapse to rapid and unwarranted expansion and to a careless expenditure of the $5 million capital. "What had started so promisingly," they wrote, "and with a little moderation could undoubtedly have developed into a business showing steady profit, appears to have been subject to pure speculation."[215]

Mackenzie and Mann seem to have taken everything that could be had, showing little interest in long-term effects. Everywhere in the history of whaling, the proximal cause of failure in a whale fishery proved to be the annihilation of the prey species, and in British Columbia the mercurial rise and fall of the Canadian North Pacific Fisheries proved no exception to the rule. On 24 June 1915 the assets of the CNPF were sold by the receivers at public auction.[216] Shore stations, chaser boats, even the office equipment went on the block. The total was purchased by one Charles Rogers Brown, the general manager of the defunct company, but Brown's role was that of a front-man; financing for the purchase came from William P. Schupp, an ambitious American insurance agent and business associate of Mackenzie and Mann who would soon become the "kingpin" of commercial whaling in the Pacific Northwest.

There is one sidenote concerning railroad empires and the modern whaling industry in British Columbia. While Mackenzie and Mann engineered the takeover of the Pacific Whaling Company, Charles M. Hayes, president of the competitive Grand Trunk Pacific Railway, met with British fishing magnate Sir George Doughty to investigate the potential of whaling in British Columbia. Doughty publicly advocated a more active Canadian fishing industry to compete against Japanese fishermen, who were almost alone harvesting the resources of the Pacific coast. In Doughty's mind the future of the fisheries was great, and he noted with open hostility and unabashed jingoism that the bulk of the fish were caught by Americans and other non-nationals, most of them Japanese.[217]

Sir George proposed to bring steam-trawlers from Grimsby, England as part of a grand venture to catch all commercial species and process them at a plant to be built in the Queen Charlotte Islands or on the mainland near Prince Rupert. His son Wilfred and a retinue of advisers arrived to scout for a site and begin the process of hiring white fishermen, and these fishermen told the press that Sir George intended to alter three of his trawlers for whaling.[218]

Ottawa was unable to support his interest. While politicians openly applauded his plan, still the law was law and where whaling was concerned, "the whole coast of British Columbia [was] taken up."[219] But Sir George stubbornly persisted. Even after his first application was rebuffed he re-applied, in the late summer of 1912, for a whaling station site at Skidegate or a similar point on the west coast of the Queen Charlottes.[220] He mistakenly believed that the fifty-mile separation rule was still in effect, and when Ottawa once again turned him down it was in a carefully worded letter from Assistant Deputy Minister Stanton, who regretted sincerely that the department was "not in a position to favourably consider the application."[221] So nothing ever came of his effort, and the proposed railway-fishing collaboration was inhibited to the last degree by the death of Hayes during the foundering of the *Titanic*.

It is perhaps only a coincidence, but the emergence of modern shore whaling in the Pacific Northwest coincided with the departure of the last American sailing ships from the Arctic, Anadir, and Okhotsk Seas and the Kodiak Ground. Faced with the reality of the commercial extinction of right and bowhead whales, the crews of the legendary square-rigged whaleships *Charles W. Morgan*, *Alice Knowles*, and *Andrew Hicks* eked out what existence they could, and then returned to New England. On its penultimate Pacific voyage, the *Andrew Hicks* cruised northward from Tahiti to the Gulf of Alaska. Its crew chased whales in view of Kayak Island, Middleton Island, and Cape Clear, and saw many finback and blue whales as well as killer whales and "grampuses." But by 8 May 1906 they had taken no right whales, and the captain's wife was prompted to seek divine intervention: "Pray God," she wrote, "for good weather and something to fill our empty casks."[222] But their take was slim; by the first week of June only one whale had been caught, and that a sperm whale lanced near Kayak Island.

Then they spoke Captain St. Claire in the sealing schooner *Vera*, who reported news of the San Francisco earthquake and "also loss of 150 souls."[223] In November the *Andrew Hicks* returned through the Golden Gate to that shaken city, and just one month later sailed again on a twenty-one month voyage which finished in New Bedford, Massachusetts. Other old sailers were sold out of the business from San Francisco or abandoned on foreign shores. The schooner *Latetia* and the bark *Gay Head* met off Cape St. Elias in 1912 and then parted for the last time, leaving the troubled whaling stations to reap an unsustainable harvest of rorqual whales.

6

SKYTTEREN

If the gunner knows his job, he is the big man on board. His movements during the chase will be watched with interest. If he fires at the right moment and the whale is caught, he will receive the credit; and if he misses the mark, he will receive the blame.

KNUT BIRKELAND
THE WHALERS OF AKUTAN (1926)[1]

The businessmen who inherited Svend Foyn's industry watched Canada and Newfoundland with avid interest. The growth of whaling at St. John's and Victoria had provided work for Norwegian shipyards as well as for the manufacturers of explosives and the special hemp foregoer lines. There was work also for Norwegian harpoon-cannon gunners, engineers, and other specialists. As legislation threatened to bring North Atlantic whaling under strict control, Norwegian wherewithal sought out unexploited locales, in South America, Africa, and even in the North Pacific, where untapped populations of large whales could be hunted with impunity.

The initial success of Sprott Balcom and the Pacific Whaling Company drew out the scouts of the great Norwegian whaling financiers. As early as July 1907 a delegation from Norway visited Sechart, where they learned that 231 humpback whales had been taken since March. This fact surprised the visitors, especially since, in their considered opinion, the chaser boat *Orion* was not a "first-class" vessel. Their amazement was all the more profound when they learned that the plant could not always handle the catch; when the *Orion* brought in too many whales, the hunting would be temporarily suspended. Even more astonishingly, its crew were instructed to give preference to the smaller humpback species, since it was considered too difficult at that time to process the larger carcasses of blues and finbacks.[2]

Norwegian capitalists assumed they could obtain concessions to hunt whales in western Canada, just as in South America, Africa, and elsewhere. But a strong chauvinism decried foreign involvement in the local fisheries, and Ottawa's hundred-mile separation rule finally defeated all attempts. So the whaling merchants of Sandefjord transferred their aspirations to south-eastern Alaska. One of their advance scouts, Larsen Bjornsgaard, travelled as far as Juneau in 1907, all the while continuing his effort to obtain two stations on the British Columbia coast.[3]

Alaskans knew all about their whales and are said to have wondered why no one had shown up any sooner.[4] The Norwegians took a long look and then announced their findings: the seas about the Aleutian Islands were "vastly more prolific in whales" than the waters near Vancouver Island. "Instead of a whale being seen now and then, they are never out of sight, and it frequently happens that a hundred whales can be seen at one time."[5] Scouts also reported that krill, an important food for baleen whales, was available in large quantities in the Gulf of Alaska.

In 1910 the Sandefjord whaling entrepreneur Peder Bogen raised $3 million dollars through stock sales to finance the so-called United States Whal-

The *Star I*, largest of three chaser boats built for the United States Whaling Company, here makes its service trials on Lake Union, in Seattle. After a decade in Alaska, the *Star* boats were refitted for a pioneering Norwegian whaling expedition to the Ross Sea.
Puget Sound Maritime Historical Society, Inc.

ing Company.[6] This enterprise, incorporated under American laws, would operate from a shore station to be built at a place named Port Armstrong — before long it became known as "Smellstrong"[7]—on Baranof Island near Sitka. Bogen fitted out a "floating cookery" to work in partnership with the three steel-hulled chaser boats under construction at the Seattle Construction and Dry Dock Company. One of the chasers would be 117 feet long, the other two 95 feet. Registered in Ketchikan as the *Star I*, the *Star II*, and the *Star III*, they soon proved to be efficient, fast boats, though like their Norwegian predecessors they were very short in human comforts. The larger *Star I* was later praised by sailor-author Alan Villiers, who recalled that its sweeping lines and seaworthy characteristics "excited the liveliest comment."[8] Villiers was quick to point out, however, that the *Star* boats were not for the weak of stomach, since their round-bottomed design frequently provoked a violent *mal de mer*.

The chasers arrived at Port Armstrong in late April 1912 and joined the 6,000-ton factory ship *Sommerstad*, which had arrived on the fourteenth. The first whale was taken on 2 May and in six weeks, 71 whales were caught and 2,000 barrels of oil produced on board the floating factory. An additional 600 barrels of "inferior" oil was made on shore during this period, though the plant was not fully operational.[9] The *Sommerstad* soon removed to the Shumagin Islands, west and south of Kodiak, where two of the *Star* boats took 184 whales. An additional 3,000 barrels of oil were obtained before the U.S. government objected to the presence of a foreign factory ship in American waters. The managers of the United States Whaling Company were forced to concede disappointment with the season total of 314 whales, even though 8,500 barrels of oil were eventually obtained.[10]

Indifferent early results and high operating costs did not prevent the establishment of a second Norwegian whaling company in Alaska, this one under the direction of Lars Christensen, son of the respected whaling financier Christen Christensen. Young Christensen toured American whaling ports in 1907, at the age of twenty-three, and he may have been among the entourage that visited Sechart. He determined to have a station in the Aleutian Islands; although the Aleutians were as far removed from Norway as any place on earth, his scouts had sent glowing reports. His new firm was established in October 1911, with a capital of $315,000. Like most other Norwegian whaling merchants, Christensen carefully adopted a company name which disguised the source of ownership while reinforcing local and regional symbols. Just as Bogen had chosen "United States Whaling Company," he settled on "Alaska Whaling Company."

A site was selected on volcanic Akutan Island, and Moran's was chosen to provide two whaling steamers, the *Kodiak* and the *Unimak*, which were completed in 1912 and registered at the company's American corporate head-

Norwegian whaling entrepreneur Lars Christensen, photographed during a visit to New Bedford, Massachusetts in 1911. Christensen failed to obtain permission to hunt whales in British Columbia, and so financed the construction of the Akutan station in Alaska.
Pardon B. Gifford Collection, Kendall Whaling Museum, Sharon, Massachusetts

quarters in St. Paul, Minnesota. They were 100 feet long and 99 tons each.[11] Christensen also employed a small factory ship, the 1,517-ton *Admiralen*, and purchased the barkentine *Coronado* from Barneson and Hibbard — principals of the Tyee Company — in San Francisco.[12] It is not known how this vessel served the Alaska Whaling Company; probably it ferried supplies and men.

There were numerous problems to be overcome on Akutan. The setting was notoriously stark and disagreeable, with a "barren and gloomy appearance," as one manager later reported, "not a tree nor a bush . . . to be seen, only stones and solid rock, with a little patch of green appearing now and then."[13] The plant itself nestled on a strip of flat ground on a point of land. Bunkhouses and outbuildings had to be situated on the hill that rose behind it. Supply and shipping costs were much higher than anticipated, and there were long delays in obtaining advice from senior managers in Norway. The first catch, in 1912, did not meet the expectations of the advance parties, although blame for this was laid upon the death of much of the krill, a happenstance which was charged to volcanic activity near Kodiak.

A wide-angle view of barren Akutan shows the volcanic desolation that greeted the whalemen every spring. Little level ground was available for building the station, bunkhouses (far left), or oil storage tanks (on the hill, right).
Special Collections Division, University of Washington Libraries

After the initial season, whaling was halted. Norwegian accounts of the Alaska Whaling Company failure usually place blame on the incompetence of a local manager, a Norwegian-American pastor who allegedly "had never seen a whale before."[14] But time, distance, and changeable weather added to management's woes. Christensen lost perhaps $150,000.[15] There was no 1913 season at Akutan; in April of that year the Canadian Pacific Railway purchased the *Coronado*, only to lose it seven months later when it foundered on the lower British Columbia mainland coast near Sechelt.[16]

The United States Whaling Company on Baranof Island experienced similar difficulties. Bogen and his directors decided to resume the hunt during 1913, and production did increase, from 8,500 to 9,333 barrels of oil. Nevertheless, a loss of 85,240 kroner had to be reported.[17] A merger of the two companies was proposed but not enacted, and the Norwegian whaling newspaper *Norsk Hvalfangst-Tidende* predicted the imminent collapse of the entire venture. There were other concerns, too, centred on the possibility of exterminating the whales. Even at this early date, the Norwegian consul at Montreal wisely recommended that regulations be instituted to control catch; otherwise, he said, the number of whales in Alaskan waters would be reduced to "zero within ten years."[18]

Lars Christensen's idled company was reorganized, and a considerable amount of stock was sold to Americans. The operation fell under the auspices of the North Pacific Sea Products Company, an umbrella corporation registered in Minneapolis which was involved in several types of commercial fishing. A major shareholder, and soon to be head of both this company and the remains of the Canadian North Pacific Fisheries Limited, was the insurance broker William Schupp.

The takeover was announced to Alaska Whaling Company shareholders at the general meeting in Minneapolis on 30 January 1914. Investors were asked to approve the change of management by exchanging shares of the old for shares in the new. As a bonus for being agreeable, they were promised additional stock gratis, to the value of 50 per cent of their current holdings.[19] Knut Birkeland, a respected whaling man, was brought from Norway to serve as the station manager, with instructions to cut delays and costs.

Studies of the 1912 and 1913 seasons had demonstrated that the majority of the catch — about 80 per cent of the whales — were taken between June and September, though the stations opened in April or May and remained active into October. At the April 1914 general meeting of Bogen's United States Whaling Company, his stockholders voted to adopt a shorter season, and this same decision was reached by the North Pacific Sea Products Company.[20] By concentrating effort, both firms completed the 1914 season showing a modest profit,[21] and the Akutan operation actually remained open longer than planned to take advantage of late catches.

Success at Akutan hinged on a discovery. During 1912 Christensen's crews had hunted only in the Bering Sea, because the channel between Akutan and Akun Islands was not considered navigable. Early in 1914 a chaser boat successfully passed through on a high tide, and once on the "Pacificside" found the very large blue whales and finbacks which made up the improved statistics.

Far to the south, at Murder Cove on Admiralty Island, Barneson and Hibbard's Tyee Company had fared even more poorly. As early as 1909 the catch had fallen well below expectation, even though the firm seems to have employed a flotilla of small vessels made over as chaser boats. A barge called the *Diamondhead* may have served as a floating factory.[22] But the station was far away from the large whales. At the beginning of May 1910, Ludwig Rissmüller was advised that the company would probably wind up its affairs and sell the station.[23]

One week later, one of the firm's whaling vessels, the *Lizzie S. Sorensen*, was sunk by a large whale. A schooner of eighty-four tons, the *Sorensen* had been purchased the previous year and retrofitted with a gasoline engine for use in whaling. It sank almost immediately, on 10 or 12 May, near Cape Addington, after being struck by the whale. Accounts of the incident vary; a contemporary newspaper story said the harpooned animal passed under the stern and gave the hull a slap with its tail, opening the seams and causing the schooner to settle quickly.[24] A later recounting by Tønnessen and Johnsen purveyed more romance; the whale, identified in their account as a blue, came around in a wide arc and rammed the side of the *Sorensen* repeatedly, as if its attack were deliberate and malicious, until the vessel sank.[25]

Such malevolent behaviour is more in keeping with the reputation of certain sperm whales than any of the baleen breed. Tønnessen and Johnsen's account perhaps benefits from the loss of Melville's *Pequod*. The true circumstances were probably similar to those which had attended the loss of the American whaling bark *Kathleen* in the Atlantic Ocean eight years previously:

> Instead of that whale going down or going to windward as they most always do, he kept coming directly for the ship, only much faster than he was coming before he was darted at. When he got within thirty feet of the ship he saw or heard something and tried to go under the ship but he was so near and was coming so fast he did not have room enough to get clear of her.
>
> He struck the ship forward of the mizzen rigging and about five or six feet under water. It shook the ship considerably when he struck her, then he tried to come up and he raised the stern up some two or three

feet so when she came down her counters made a big splash. The whale
came up on the other side of the ship and laid there and rolled, did not
seem to know what to do.[26]

Both accounts of the loss of the *Sorensen* relate how quickly the schooner
settled. The crew had little time to save anything but a few personal items. A
dead whale matching the description of their foe is said to have been found
the following night by a passing ship.[27]

Shown at Akutan about 1917, the cut-down bark *Fresno* was towed to the Aleutians
each year with men and supplies. It eventually burned at Bellevue, Washington in
the 1920s.
Special Collections Division, University of Washington Libraries

After losing the *Sorensen*, the Tyee Company struggled for two more years,
but in 1913 the firm ceased operations. Barneson sold the cut-down
American-built bark *Fresno* to the North Pacific Sea Products Company, and
they put it to use carrying men and equipment to and from Akutan.[28] By January 1916 Tyee's ten-acre parcel of land on Admiralty Island, with buildings
"suitable for a salmon cannery," were available for sale or lease.[29]

The United States Whaling Company struggled along under Norwegian
ownership for another decade before deciding that the Alaskan waters were
simply too rugged and too far away to exploit efficiently. By then, the Tyee
Company had long since disappeared, and the Akutan operation was now run
by Americans who were geographically better able to service it.

Alaska was the uppermost limit of Norwegian expansion in the northern hemisphere. Its failure was due to several reasons; before the Panama Canal was opened in 1914, the voyage to Alaska [from Norway] was twice as far as to the Antarctic. Excessive expenses were incurred before catching could commence, there were difficulties in marketing the oil in America, while high freight rates to and from the European market made competition difficult. It took a long time to train native hands in an attempt to meet the demand to employ as many of these as possible. The labor force was highly mixed and cosmopolitan, comprising Norwegians, Swedes, Finns, Russians, Germans, Newfoundlanders, Americans, Indians, Chinese, and Japanese. The working tempo was slack, while wages were on a higher level than for purely Norwegian crews.[30]

The withdrawal of Norwegian management and financing had little effect on the daily operation of shore stations in the Pacific Northwest, whose employees continued to depend upon Norwegian expertise and equipment at sea and upon the physical capabilities of a "mixed and cosmopolitan" labour force on shore. The result, particularly in the Canadian industry, was a rare and unique melding of workmen from three continents, struggling together to harvest from the North Pacific the best and largest whales.

Naturalist Roy Chapman Andrews, visiting Sechart in 1908, identified six "racial" groups at work there, including Norwegians, Americans, Newfoundlanders, Japanese, Chinese, and "Siwash Indians."[31] Perhaps because of the nature of the work, any man who was healthy and wanted employment could get a job in the shore stations. Racial expectations were present, however; even though men of different nationalities worked in the same crews, a pyramidal society existed, with management and shipboard jobs vested in whites while flensing and the more demeaning and demanding jobs belonged to others.

In British Columbia the large percentage of Japanese in the shore-station crews led to the formation of a sub-hierarchy. The first-cut flensers—those who made the initial incisions into the carcass as it was hauled up the slipway —were placed at this pinnacle and, just as the Norwegians obtained favored shipboard jobs, so did the Japanese exclude Chinese from this one. The Chinese were delegated to the more odious jobs in the sheds: mincing, feeding the dryers and rendering vats, and packing the product. Many of the Japanese spoke little English, and answered to a bilingual Japanese foreman who interfaced with white managers. This strengthened their solidarity but also their isolation.

Their acquisition of the best shore jobs was based in part on their methodical skill. The reputation of the Japanese for steady, good-natured work was soon established; when supervised by a Japanese foreman, managers consid-

The Rose Harbour flensing crew posed in the jaw of a whale for this 1918 snapshot.
Station manager Harold Duckitt stands at center; Japanese foreman Moichi Kosaka
crouches to his right.
Provincial Archives of British Columbia

ered them conscientious, hard-working employees: "I have never seen men
work the way they did," wrote Knut Birkeland of the Japanese flensers at
Akutan. "They went about [their work] in about the same spirit as that in
which boys play football."[32]

But if the white station managers paid tribute to the co-operative spirit of
their Japanese workmen, racial and occupational stereotypes inevitably per-
vaded the public's view of whaling. One reporter pictorialized a Canadian
flensing crew as "Norwegian, Japanese and Chinese ghouls" and compli-
mented the white managers as men "who have had experience in the Atlantic
whaling, and can be trusted with the more important positions."[33]

Behind the apparent goodwill and camaraderie, the difficult work and
lonely detachment from home life affected the Japanese workers like everyone
else. Many of the Asian workers, both Japanese and Chinese, had left families
in their homelands and sent most of their wages out of Canada. Boredom and
the deep separation from friends and society led to alcohol abuse, gambling,
and occasional violence, which were neither more nor less common in the
Japanese bunkhouses than in the others. Except for reading, fishing, and perhaps carving the ivory teeth of sperm whales, there were very few pastimes
beyond the daily routine. Missionaries called at the stations from time to

Even visiting entertainers did not stop the work of whaling. Here, in the summer of 1923, a saxophone sextette called "The Six Brown Brothers" perform at Akutan while the flensing of a humpback whale continues in the background.
Special Collections Division, University of Washington Libraries

time, but their arrival was generally considered an unwelcome intrusion upon the nightly poker game.[34]

Japanese and Chinese do not seem to have been so ubiquitous at the Alaskan stations. In the beginning, when these stations were owned in Norway and managed at long distance, the flensers and work crew were mostly Norwegian. Only after the initial collapse of the companies did the managers turn over their long-handled knives to Newfoundlanders and a few Japanese.[35] American government statistics compiled in 1916 indicate that only 39 of 233 men then employed in Alaskan shore-whaling were Japanese.[36] Further corroboration may be derived from the account of a man who worked at Akutan in 1922: "Sometimes in our bunkhouse we would put on a dance," he wrote. "Since most of the workers were Norwegians, the music usually consisted of Scandinavian waltzes and polkas, played by a fiddle and accordion."[37]

In British Columbia, however, the dependence upon Japanese and Chinese shore workers became complete. A visitor to Rose Harbour in the 1930s identified two working crews — sixty-three men in all — composed entirely of Asian workers. Of whites, there were only five.[38] These demographics eventually posed an insurmountable problem during World War II, when the Canadian government elected to remove residents of Japanese descent from the far western coast.

Native Indians appear to have been hired more in the early years than later, and primarily at Sechart and Kyuquot. Some may have been employed to tow carcasses from the nearby floats to the flensing slip.[39] Apparently, their skills as sea-hunters were not considered transferable to the new science of whaling. When they lived at the station, it was in shacks built for their use by the company. The Indian housing at Sechart was burned in 1914, for reasons not now clear,[40] and it is possible that native labour was not utilized in British Columbia after this date.

Permanent bunkhouses were provided for all other workers; whites, Japanese, and Chinese each had their own, usually a frame structure with sleeping quarters upstairs and a kitchen and mess hall on the ground floor. At Rose Harbour the white bunkhouse had a bathhouse complete with a wooden tank and a shower, with hot water provided by the boiler house. The Japanese built their own traditional communal tub at every station.[41] The bathing facilities were of more than a little importance. Flensing and processing often demanded hours of work amid the liquid residues on the slipway and the heavy dust of the guano-making sheds. Most contemporaneous descriptions of the stations disregard the fetid surroundings in praise of mankind's technological victory over the marine giants, but Roy Chapman Andrews's report accurately portrayed the grisly scene with an appropriate mixture of enthusiasm and pathos.

Andrews developed a fascination with whales and whaling and applied to his superiors at the American Museum of Natural History for a leave-of-absence to visit whaling sites around the world. In 1908 he toured British Columbia and southeastern Alaska, and became one of the first visitors from the scientific community to witness steam whaling in the Pacific Northwest. At Sechart he found none of whaling's supposed romance, but only "a group of large buildings at the end of a beautiful bay backed by a sheer wall of sombre forest. "From the water," he wrote, "it looked like what it was—a factory."[42]

Andrews studied intently, and sometimes participated, when whales were hauled up the slipway.[43] He watched as the stout wire cable was fastened to the narrowest part of the body—the caudal peduncle just ahead of the flukes—and described the large steam winch at the head of the inclined slip which drew the carcass ashore. Before the whale was half out of the water, Andrews said, the flensers severed the side flipper at the "elbow" and began the longitudinal cuts from head to tail which would allow the winchman to pull away long strips of blubber. These cuts were made with great speed; a whaleman who worked in British Columbia in the 1920s said that the men would start cutting as soon as the whale had begun to rise up the ramp. "By the time it got up to the stop," he said, "they'd have all the main cuts in it and would be all ready to go to it."[44]

Then they "went to it." A cable running from the steam winch to the whale's chin was brought up taut, and the blubber was slowly peeled from the body. Once the accessible surfaces were stripped away, a "canting winch"

A Japanese station hand oversees the reduction of intestines at Rose Harbour.
Photograph by Lewis L. Robbins, Kendall Whaling Museum, Sharon, Massachusetts

was employed to roll the carcass. The strips of blubber were cut into oblong blocks, fed into the slicing machine, and sent from there by an endless-belt conveyor to the rendering vats, which opened at the second-floor level of the adjacent building. In those vats the oil would be extracted by steam pressure and heat.

Meanwhile, other men worked on the carcass, chopping through the upper ribs, drawing out the organs and viscera, and then hauling the body onto the "carcass platform," built at a right angle to the flensing slip. There the flesh and bones were separated and the skeleton disarticulated. All this material would be boiled for oil by the Rissmüller methods and then pulverized for bone meal. Even liquids were saved; blood collected in troughs alongside the slipway was boiled, dried, and added to the fertilizer. The liquid remaining after the rendering process was sold to glue manufacturers, and the baleen was also set aside for cleaning and sale.

The salient feature of this operation was its smell. Neighbours recall the stench of blood and bone permeating the air for miles. Whaling could be felt on the face and the tongue, and the scent proved nearly impossible to eradicate from clothing. Captains and pilots of vessels would avoid tying up near the station, in the belief that the ship's paint would discolour. Piles of bones, stacked to await grinding, attracted flies beyond number, and the entire mess stank. "You couldn't go into the station," one chaser-boat crewman recalled. "The smell would take your skin off. I only went to see [the flensing] once."[45]

Skeletal bones were tossed on this pile at Sechart in 1905–6 to await transportation
to a grinding mill. By the autumn of 1907 there were four such mountains at the
two active stations.
Provincial Archives of British Columbia

Still, observers regularly remarked on the incredible energy of the flensers,
who seemed to revel in their work. The flensers formed a special brother-
hood, as Andrews came to learn when first he was welcomed among them:

> When all was set . . . one of the cutters called me over to the side of
> the whale. As I bent down to examine a spot he had pointed out, he
> thrust a huge knife into the belly. Out shot a stream of blood, almost
> black, and a horrible odor. Both struck me fair in the face. I went over
> onto my back, slid down the slip and into the water. . . . the station
> whistles blew and the bells rang.[46]

A few white workers joined the Japanese on the flensing platform; some of
these were from Newfoundland. A significant number of Newfoundlanders
were employed in the Pacific industry, either as ships' crew, flensers, or else
in various management positions. Like the Japanese, they arrived at the beginning of the industry in British Columbia, when men "experienced in Atlantic whaling", had to be imported. Many of the first arrivals had been affiliated with Ludwig Rissmüller's whaling operations and came to meet the

Pacific Whaling Company's contractual obligation to hire "consultants" from Rissmüller. They trained local men to run the plants.[47]

Thereafter they came readily and regularly, often working on seasonal contracts; many were recruited by an agency, James Murphy and Sons, in Placentia, who issued tickets and a per diem for travelling expenses before the start of each season.[48] Some worked as deckhands or shore workers and returned home in the autumn, while others, particularly among management, chose to move their belongings and their families to British Columbia. Among the latter group were Alfred Gosney, William Rolls, and George LeMarquand, all of whom accepted middle-management positions and in some cases brought their wives and children to live at the stations.[49] A few Newfoundlanders learned the fine art of whaling, and at least one such man, William Heater, made a long career for himself in the Victoria whaling fleet.

None but Norwegians, Newfoundlanders, and white Canadians crewed the chaser boats, save the Chinese cook, who could not expect to rise above his position on board and who was regularly vilified—in jest or for real—for his special brand of cuisine: "a curious hash," one critic said, "hard potatoes and fibrous meat . . . doughy, soggy bread, with corncake and the mud-colored coffee as dessert."[50] But that was the same grumble always heard aboard ship; as one nineteenth-century American whaleman put it, "The Lord sends the grub but the Devil sends the Cooks, no mistake."[51]

The Norwegians might be deckhands, firemen, or pilots, but the most coveted job was that of *skytteren*—"the shooter"—the man who stood behind the harpoon-cannon and pulled the lanyard or trigger at exactly the right moment. To him belonged the glory of the hunt, and to his skill accrued the bonuses payable to each member of the crew for every whale brought to the slipway of the whaling station. The gunners were the best paid, and the most heroic, of all the technicians employed in the whaling industry.

Without the learned and tested experience of the *skytter*, any whaling venture was doomed. A gunner who knew how long a whale might be expected to remain on the surface, how long it might stay underwater, and where it might be expected to rise again—his was the experience most in demand. It was he who counted the seconds as the chaser boat came alongside the great body, watching the whale's rising and falling until the one ideal moment—the moment the matador knows as *la toca de muerte*—when the harpoon might make a quick and successful kill.

Norwegian families whose young men pioneered the skill of harpoon-cannoneering tended to reserve the gun platform for their own, through an organization—a union, one might say—of whaling gunners. It was said that only a gunner's son, or a man who married a gunner's daughter, could earn the privilege of shooting. And this control extended through the entire colony of modern whaling stations from the North Cape of Norway to Tierra

The rugged mystique of the gunner was perfectly portrayed in a watercolour monochrome by Milton J. Burns, following his visit to Newfoundland just after 1900. Fully exposed to the sea, the *skytter* makes his one perfect shot.
Kendall Whaling Museum, Sharon, Massachusetts

del Fuego and north again to Akutan. "Whale shooters come from a small section around Sandefjord," a contemporary journalist observed. "Rarely does an outsider join the ranks."[52] Qualified Norwegian gunners held a total stranglehold on the best whaling jobs around the world for more than a half-century. Just before World War II there were perhaps 350 such men, "organized in a tight whaling unit which held a world monopoly in the business."[53] As late as the 1960s Norwegian whalemen continued to fire the cannons, even among the sophisticated Antarctic fleets assembled by Great Britain, Japan, and other whaling nations. The job was passed to sons and nephews and collateral relatives in very much the same way as the whaling rituals and tools had been handed down within a privileged few West Coast native whaling families.

To learn the craft of whale hunting was thus a delicate matter. For the novice attempting to begin, it was of particular importance to reach a politic agreement with *skytteren* about the shooting — in advance of the whaling season, and in writing. Such agreements usually stipulated a minimum number of shots which the gunner must allow his unwelcome student.[54] In the absence of such an arrangement, the entire year might elapse without the poor protégé being given a single opportunity to practise.

If the shooter gained all praise for a successful hunt, he also shouldered all the blame for any failure. When things were "skookum" he received the highest accolades from his crew. Shipboard morale soared. He became a Bunyanesque figure only slightly less virile than the great Svend Foyn himself. But if things went sour, he was castigated. Then, the frustrated gunner might throw tantrums or become morose, and he would start to squirm for some way to shift the pressure off his backside.

> The eyes of all the other men will be upon him, ready to commend or criticize, especially the latter. A bad shot will not soon be forgotten; and if it happens that the gunner misses several times in succession, the question will arise as to whether it is his particular fault or whether the blame should be laid to some other cause, as, for instance, perhaps the blacksmith did not adjust the harpoon properly when he had it in his shop, or it may have been the powder, or possibly something else. The members of the crew will put their heads together and discuss the matter pro and con, most of the time concluding that the gunner is no good.[55]

In fact, the gunner was asked to provide a good deal of courage and endurance. In the spartan environment aboard a chaser boat, a man's ability to stand up and do his job without complaint—no matter how rigorous—counted most toward his acceptance by the others in the crew. On the gunner was conferred the glory, but he had to earn every moment by standing up to the mighty head seas without rope or bulwark, somehow remaining poised and in full control of a volatile mass of iron and steel that literally weighed a ton. If he was skilled, and perhaps a bit fortunate, his first shot would send a whale toward paroxysm and death. If it was badly placed, he would be forced to fire again, perhaps a "shooting lance," which was nothing more than a barbless harpoon with an explosive head; after detonation the shaft could be pulled aboard and reloaded.[56] If the whale remained active, the gunner might go alongside in the boat's pram, an old-time iron lance in his hand to "rob him of his life."

Little wonder that a few missed shots occasioned a deep melancholy or sent the gunner into apoplectic fits. Superhuman he might be, but he was often tired and almost always wet. He dared not let the apprentice take his place at the gun, for *he* was *skytteren*. The only recourse left to an exhausted or frustrated gunner was to rant and roar and blame the gunpowder, which had somehow become damp and misfired; or blame the blacksmith, conveniently tucked away in his workshop ashore, who had not properly straightened the harpoon. And rant, and roar.

Whales were called everything from "rats" and "racehorses" to "black devils," depending on their size, speed, and behavior; also lots of names unfit for repetition. A short, mad spurt along the surface, during which the whales would fill their lungs and disappear ... just beyond shooting distance, would sometimes cause the gunner to throw his cap on the platform, jump on it and kick it over the side.

One thing that caused a gunner to blow up suddenly ... was to hear the report of a rival whaler's gun, on a calm day, when the rival would be just out of sight over the horizon. The more reports that were heard the higher the gunner blew, until he could stand it no longer, and would steam off at full speed to see what was going on. [57]

One crewman overheard his gunner tell the whales: "Go down, you cowards; go down and stay there; I hope you drown." After a minute the gunner screamed: "Come up, you wall-eyed brutes; come up and take your medicine." [58]

But the circumstances of the chase were often beyond the direct control of the gunner. Finding whales involved a substantial element of chance and depended also upon a pair of alert, knowledgable eyes in the lookout barrel atop the foremast. In a rare candid moment, a whaling captain at Naden Harbour once admitted to the importance of luck in his job. "Sometimes we see only one whale in a day—and get it," he said. "Another day we chase twenty and don't get a shot." [59]

Impatient men and unwilling whales have always collided. The *skytter* was not the first whaleman on the Northwest Coast to bellow over a missed opportunity or defective equipment. In the days of sail whaling, more than a few boatsteerers on the Kodiak Ground were "broken"—dismissed from their role as harpooner—only to be reinstated when the captain in turn "broke" their inexperienced replacements. The flaring tempers of an officer and his boatsteerer were sometimes a terror to behold:

at 8 AM lowered for whales   the waist boat struck and drew   the larboard boat struck and parted for which the second mate hove a bucket at the boat stearer but missing him ran forward and struck him with his fist the boat stearer said he would go no more in the boat with him came on board saw whales and lowered   the boat stearer kept back   the second mate came on board and the capt. came from mast head and began to beat him over the head with a board [until] they compelled him to get into the boat   when the boat stearer said he would take the capt. to [illegible] for flogging him the Capt then called him on board put him in the rigging and gave him 14 Lashes with mizentopsail brace. [60]

Even the great West Coast whaling chief Maquinna had his bad days, when a well-aimed strike was thwarted by the breaking of a line or the withdrawal of a mussel-shell point from soft blubber. John Jewitt, the American sailor held captive by Maquinna at Nootka Sound, recalled such occasions. "At these times," he said, "[Maquinna] always returned very morose and out of temper, upbraiding his men with having violated their obligation to continence preparatory to whaling."[61]

The personalities and the quirks of such hunters, from Maquinna's time forward, have outlived flesh and blood and expanded into folklore and legend. Former whaling men in British Columbia talk yet about the gunner Moses Erikson, who allegedly fell into a deep despondency and drank the alcohol from the standard compass. They talk of "Mad Harry" Anderson, who returned from a season of whaling to find his woman gone. Believing she had left him, he is said to have torn the linoleum from the floor of her apartment and shot her cow before learning she was in hospital, recuperating from an automobile accident.

The older whalemen also remember Knut Halvorsen, who once injured a whale in the waning light and followed it into the darkness with a large

Harry—sometimes "Mad Harry"—Anderson, posing with "Laddie" about 1936. His kind face and obvious affection for the dog contradict his wild reputation.
Photograph by Lewis L. Robbins, Kendall Whaling Museum, Sharon, Massachusetts

flashlight. He managed to tow the animal into Rose Harbour in the middle of the night, even though it was not quite dead.[62] Others are reminded of "Finn John," who served variously as mate and gunner in British Columbia. They speak comfortably of him, as if they had seen him yesterday, though none among them can recall his surname.[63]

The few non-Norwegian gunners in British Columbia kept pace and kept faith. Sprott Balcom's kin earned the right to shoot; his son Harry and nephew Willis both gained reputations for skill at the cannon. Willis Balcom was particularly well remembered. Born in Halifax—the son of Sprott's brother Sam—Willis spent most of his adult life in British Columbia as a whaling gunner. Among the boat crews he was known as a taciturn man, but "hell-bent on whales," the kind who rode by the gun in every kind of sea and spoke but little.[64] On shore he seems to have earned a reputation as his family's "black sheep." In 1910 he stood accused of "posing" as a sea captain— he was one—in order to pass a bad cheque.[65] But a disregard for the niceties of shore life held gunners in good stead, and it was said that Willis Balcom could catch whales when no one else could find any. A former whaling-station blacksmith explained this. He admitted, with a satisfied grin, that Willis would "sneak over to the American [Alaskan] side!"[66]

Harry Balcom also inherited his family's maritime interest and his father Sprott's savvy; he made a name as a seal hunter out of Halifax before moving west to join his father in the whaling business. A strong, handsome man, he seems to have taken on his father's style—rough and ready at sea but a well-groomed gentleman on shore—and he might well have inherited and amplified the Balcom fortune. Sadly, he was stricken with acute appendicitis while whaling in the *Green*, and died on 29 August 1918 for want of nearby medical facilities.[67]

The most vividly remembered British Columbian gunner—Norwegian or no—was the Newfoundlander William Heater. Bill Heater assumed command of the new chaser boat *William Grant* in 1910, and his whaling career aboard the *Grant* continued until the boat was thirty years old and Heater nearing eighty. He earned a reputation—depending upon whom you talk to—as a moody villain, street preacher, public relations agent, fine skipper, terrible skipper, or just a "wild old sucker."[68] Violent and mercurial, he seems to have been a man of two worlds who preached street-corner salvation during the off-season but raged sacrilegiously on the gun platform of the *William Grant*. Lewis L. Robbins, a pharmacologist, made several cruises with Heater in 1936–37 and photographed him on the whaling-station dock, hands in the pockets of his work pants and his hair a white mane under a flat cap. Heater's expression, apparently a typical one, suggests a gruff pleasure, and about as close to a smile as the wizened sea veteran could muster. Robbins pasted the snapshot in his photo album and added a cap-

Captain William Heater strikes a "formal" pose at the harpoon-cannon of his long-time command, the *William Grant*. Heater was already past seventy years of age and still actively shooting when this photograph was taken.
Photograph by Lewis L. Robbins, Kendall Whaling Museum, Sharon, Massachusetts

tion: "Capt. Wm. Heater, bible reading, hard swearing, moralist, 75 year old veteran of sealing and whaling."[69] That, efficiently put, was Bill Heater.

He was known for his common sailor's superstitions. "Heater always wore his sperm hat with the doo-dads on it!" said one former whaleman.[70] Whatever the "doo-dads" were, they caused the whales to come happily under the gunsights, though in later years Heater was considered a poor shot and the crew cited this whale or that whale which suffered a dozen harpoons before the "old man" finally despatched it. But the grumbling probably came from younger men who aspired to Heater's job at the bow of the *Grant*.

The company seems to have considered him an asset; management always sent reporters, scientists, and curious observers aboard the *William Grant*, and at least two generations of excited amateur photographers sailed with him. Pioneer cinematographer A. D. (Cowboy) Kean, who photographed the whaling in 1917, made a special excursion to the seal rookery at Cape St. James with Captain Heater.[71] And when Lewis Robbins arrived in the mid-1930s fresh from the University of Chicago, it was William Heater who introduced the young landlubber to the seaward side of the business.

Had the *skytter* been as observable—and as glamorous—as the cowhand in the valleys or the logger in the Douglas fir forests, Pacific Northwest folk history might have taken a more nautical turn, toward seafaring heroes who could hunt whales with a flashlight and harpoon three with a single shot, as Willis Balcom is said to have done in 1919. Writers and reporters visiting the whaling stations were quick to repeat the gutsy yarns concocted by the whaling men. Greenhorns have always been prey among nautical company, and never more so than in whaling camps, so these stories went around the world as truth; whales that gave milk on command, quizzical sea monsters, exploding harpoons accidentally fired over the deckhands' heads. The whimsical fraternity of whalemen might say anything, and since the outsider knew nothing and no harm was done, what did it matter? Whalemen took it upon themselves to make their occupation as glorious and dangerous as possible, though modern whaling was long in tedium and—because of the power of the harpoon-cannon—relatively short on heroism.

But to armchair sailors the cold cannon bolted to the bow suggested the virility of the men who used it. Firing a harpoon which itself equalled the weight of a bantamweight prizefighter, it symbolized modern whaling and gave rise to a genre of stories: of incredible shots, amazing near-misses, and fantastic adventures, stories now forgotten by the very men who created them. All because the harpoon-cannon waited there in its might, silhouetted against the sky and water. And among the greasy and sordid jobs of whaling stood the clean, brave seafaring hunter, brazenly defying the swell, legs braced apart and gun at the ready, waiting on a chance to do battle with the ocean king.

The harpoon-cannons employed by the Pacific Whaling Company were of a Norwegian design, muzzle-loaders fitted with stiff rubber pads on either side of the barrel to help absorb the shock and vibration of the recoil. This type of gun was obsolescent almost at once and completely outdated by 1926 when the leading cannon-makers, Kongsberg Vapenfabrikk, introduced a breech-loading model. The cannons used in the Alaskan fishery were a generation newer than those used in British Columbia—recoil was absorbed by an improved "shock absorber" chamber filled with a glycerine and water mix—but they were still muzzle-loaders of an outmoded design.

To load the cannon it was necessary to swing the muzzle toward the wheelhouse. A deckhand then introduced the charge into the front end of the barrel, followed by a wad of oakum to tamp it down, a cork patch, and a larger wadding made of wool to cushion the harpoon, which was added last.[72] Harpoons varied to local specification, five to six feet in length, their business end made of cast iron and the core filled with explosive powder. The appropriate charge varied; in the 1930s the time fuse ignited a three-quarter-pound charge of blasting powder a few seconds after impact with the whale.[73]

Inserting the harpoon was a two-man operation. It was placed in the barrel without its explosive tip and secured to the gun with yarn to prevent it from slipping out in heavy seas. The yarn fastened to small knobs around the circumference of the gun barrel. A sliding ring on the harpoon shaft connected the harpoon to the "foregoer" (or "foreganger"), a lightweight line which was spliced to the heavy whaling rope. In British Columbia in the 1930s, this foregoer consisted of 40 fathoms of 3.5-inch Manila, one end fed through the eye or ring on the harpoon, and the remainder coiled on a canted platform under the cannon, ready for immediate use. This platform could be folded upward to protect both gun and line in rough weather. The proximal end of the foregoer was spliced to 120 fathoms, give or take, of the best 6-inch Manila line, which was wound around the winch and fed into the locker just ahead of the wheelhouse.[74]

The final preparations before shooting involved screwing on the time-delay fuse and then the cast-iron bomb itself. The threads were customarily greased to maintain some waterproofing, since the performance of the black powder—as every cagey gunner knew—could be affected by dampness. The job of loading was relatively simple in the calm water of the harbour but could become torturous in rising seas. No wonder the circumstances gave rise to a series of hair-raising accounts of freak accidents, any one of which could have killed several men of the whaling crew but somehow never did.

Such yarns fit a folkloric mold, and the lot of them are somewhat suspect as fact. Living whalemen claim they never heard them. But writers and reporters published such stories, which all go in the same general direction: the gun is accidentally fired, and the wheelhouse or some other vulnerable target is suddenly imperiled by a 150-pound harpoon and explosive charge. In one version, a gunner swings the gun so far down that the outward-bound harpoon prangs the bulwarks of the high, flaring bow and rebounds over his head. The impact detonates the time fuse, and the harpoon explodes into shrapnel as it passes over the wheelhouse. As usual in such stories, no one is hurt.[75]

Whaleman-author William Hagelund recounted a similar accident which is said to have befallen the crew of the "Azure" (*Blue*) out of Kyuquot:

> They had almost shot a whale just at sundown, then in disgust over losing their shot, they supposedly secured the gun and headed into the anchorage. The next morning the sea was rougher as they headed out, and the ship began to roll and dive actively. The gun swung loose on its pivot, and suddenly the ugly snout, with its loaded harpoon and primed trigger, faced the men on the bridge. At that same instant the trigger was forced up against the foregoer rope and the snare released. The harpoon flew toward the frozen men, punched a hole through the weather board, knocked over the binnacle and plunged on past the stricken

helmsman, boring a hole in the funnel. As it soared on over the engine-room skylights and lifeboats, the foregoer pulled it up short just as the bomb exploded. The ship was sprayed by shrapnel, but surprisingly, not one man was hurt or killed.[76]

Such stories as these glorified the montonous pitching and rolling, un-comfortably wet quarters, hours and hours of useless steaming, and the inevitable frustration of missed shots. On shore, they provided some astonishing entertainment for visitors, and a case of nerves for new men about to venture after whales for the first time.

Certainly there was little romance left in the business. The lookouts, it was said, did not even bother to yell anything picturesque like "There she blows!" when a whale was sighted. "We all knew what we're after," one whaleman remembered, "so the men just gave the bearing."[77] And after a successful kill, it was all work. The towing cable would be thrown over the carcass, pulled to the tail, and retrieved aboard with a boat hook so that the flukes could be lifted up toward the deck by the winch. The tips would be cut off with a long-handled knife, to prevent the tail from "swimming" while under tow. Then the tail was lowered into the water again. A long compressor-tube spear would be plunged in, to inflate the body cavity with air. If the gunner intended to steam off after other whales, a mooring line would be inserted and a long flagstaff as well, to mark the position of the carcass in the water. It could be picked up later; chaser boats frequently returned to the station with two or three whales.

After picking up the carcasses, the captain would make a short notation in his logbook: "8.45 pm Towing homvards."[78] If the tow was impeded by rough seas, there might be an additional codicil: "6 pm Towing the whale in line astern. Engines going Dead Slow Blowing a hard S.E. and Werry Biig Sea."[79] This could make for some long work in the graveyard-shift hours:

2. a.m. arived at Station   2 sperms
4.30 a.m. Left station outward. Wind SE + rain
7. a.m. Hunting sperm 7.30 Shot whale
8. a.m. allongside towing to station   dirty weather S.E. strong.
10. a.m. arived at Station tied up to warfe day   ends dirty throughout.[80]

While on the hunt a captain might demand steam for as long as eighteen hours, or even longer. As the chaser boats from Rose Harbour and Naden Harbour began to hunt farther afield, it became commonplace to put into some hole-in-the-wall cove for the night or else to shut down the engine and drift through the darkest hours, perhaps between midnight and 4 o'clock. This method of fishing had been used in the sailing whaleships but found little favour among some of the newer generation of whalemen—one engineer complained of his captain: "I do not like this man's way of drifting in bad weather."[81]

Such was the life of the North Pacific steam-whalemen. But if their daily routine was a rocking tedium of long days, little warmth, less room, and nothing dry, at least the moments of chase and capture and the shock of the cannon's blast provided punctuation. Surely the *skytter* and his crew took heart in knowing they were heroes, the lifeline for the Japanese flensers and Chinese meat-cutters and for the Newfoundlanders who managed the stations, the irreplaceable source of the profits and dividends that satisfied investors and purchased fuel, food, and gunpowder for the season yet to come.

7

CONSOLIDATED WHALING

*The perfecting of methods of capture and prepara-
tion affords no solution to the problem of the continuance
of whaling. On the contrary, it leads straight to the ex-
tirpation of the stock.*

JOHAN HJORT
"THE STORY OF WHALING" (1937)[1]

The complications of a world war superimposed themselves upon the fiscal
mismanagement of the Canadian North Pacific Fisheries Limited and equally
so upon the troubled operations in southeastern Alaska. In Europe the
materiel of whaling was diverted into combat; shortages and delays affected
not only the shipment of new harpoons and whaling line, but manpower as
well. A volume of correspondence passed between the CNPF and Isak Kobro
regarding the hiring of gunners in Norway to replace some local men who
had not "given satisfaction." Kobro sent a dossier describing qualified men
just then "on the beach," and Lauritz Hansen seemed sufficiently expert.
Kobro represented him as a veteran of Japanese steam whaling, who had
most recently killed 169 whales in Mexican waters and 32 more off Ecuador
while working for a Norwegian company during 1913.[2]

Victoria agreed to hire him, and the *skytter* soon signed a bilingual contract—English on one side, Norwegian on the other—to come to work in
British Columbia for thirty-five dollars per month, plus bonuses of twenty-
five dollars for each sperm whale, thirteen for each blue whale, ten per finback, and five-fifty per humpback.[3] But when it came time for Hansen to
sail, he hesitated. The war had intensified; German submarines and surface
ships menaced the North Sea. Fearing for his safety on the customary route
from Liverpool to St. John's, Newfoundland, he opted instead for an oblique

crossing by way of New York in a neutral-flag hull.

"I have had to route 20 whaling men for Newfoundland this season via New York in stead," Kobro complained, "the risk of detention and loss being considered too great under English flag in the North Sea and on the West Coast of Brittain."[4] Kobro went to some trouble to find a neutral vessel, and by the time he arranged Hansen's passage, the CNPF had collapsed.

The war did not limit the operations of the beleaguered Canadian North Pacific Fisheries Limited. If anything, war increased the value of whale oils, particularly sperm oil, which could be used in the production of nitroglycerine for explosives. But the large whales were not being caught, and production statistics had slumped until the company fell into receivership and was sold at auction in June 1915. Its general manager, C. Rogers Brown, bid successfully for the assets and immediately declared a new firm, the Victoria Whaling Company. He petitioned Ottawa to reassign the CNPF licences to the new firm, and the Department of Naval Service, by now responsible for such matters, had only to follow the precedent set by Brodeur of Marine and Fisheries in granting the transfer. One slight complication threatened the business; before the transaction was carried out, the receivers learned there was a silent partner to the purchase.[5] It came to be known that Brown had done the bidding for an American insurance agent, William P. Schupp.

Like Ludwig Rissmüller, Schupp was of German descent, born in Saginaw, Michigan about 1866 of first-generation immigrants who had settled in the area two decades previously. After completing his education, he became associated with William Mackenzie in Canadian mining and lumber ventures. Later he entered the insurance business, and his brother, Henry Schupp, worked as a surveyor for Mackenzie and Mann during the building of the Canadian Northern. Schupp wrote bonds for the Canadian Northern, and later carried the insurance policies on Mackenzie and Mann's whaling vessels.[6] He clearly profited from his association with builders of empire; about 1910 he purchased a large estate, "Pawling Manor," near Staatsburg, New York—not far from the old whaling port of Poughkeepsie—and moved there with his wife Emily Kugler Schupp and their family. He continued to run his businesses from the estate, but as he became more and more involved with whaling, he spent increasing amounts of time in the Pacific Northwest.

It is not clear when he first became involved with whaling in British Columbia. He was probably working for Mackenzie and Mann during the reorganization of Sprott Balcom's old Pacific Whaling Company in 1909–10, and he was almost certainly a shareholder in the North Pacific Sea Products Company at the time it reorganized Lars Christensen's failed Alaska Whaling Company operation at Akutan in 1914. He acquired complete con-

Young William Schupp Lagen travelled to the Alaska whaling stations with his grandfather, William Schupp. The pair were photographed aboard the freighter *Gray* about 1933.
William S. Lagen

trol of the North Pacific Sea Products Company not later than January 1918, when shareholders approved its sale to the Victoria Whaling Company — by that time he kept a business office in Seattle — and he also acquired the American Pacific Whaling Company and its shore station at Westport (Bay City), in Gray's Harbor, Washington.

As soon as Schupp gained control of the Canadian assets, he sent the new Victoria Whaling Company to sea without waiting to pay for the purchase; it appears as though he withheld payment until the autumn of 1915, after the completion of the season, at which time the Victoria Whaling Company deposited $128,500 with the Canadian Bank of Commerce, payable to the receivers.[7] He retained some of the old guard to run the business, including Secretary Charles F. Munday, a veteran employee who had worked with the management of the original Pacific Whaling Company. C. Rogers Brown also stayed on until 1916. Then he transferred his activities to C. Rogers Brown and Company in Seattle and continued to act in the capacity of agent for Canadian whale oil shipped to the United States.

Maintaining continuity was surely a part of Schupp's strategy. He needed to cut away many of the excess ribs of the old Mackenzie and Mann umbrella, yet he recognized the whalemen for the temperamental specialists they were. Chances are he already knew some of the men personally, and he knew also that he would have to rely on them to return voluntarily each spring. To have them walk away in anger or frustration was to risk losing them for good and, once they were gone, where would he find skilled others to aim the harpoon-cannons and flense the blubber from the whales?

Schupp seems to have nurtured the men who worked for him season after season. It is said that he lent money to the whalemen if requested and gave advice if asked,[8] and it is difficult to find a man living today who will speak ill of him. He is remembered as a man who shot straight and supported his employees through the hard times; a man who kept the whaling industry alive. Some recall him as a dandy in tailor-made suits who arrived at the company dock in a chauffeur-driven long-sedan. Others equally qualified to remember say he preferred a conservative, tasteful blue coat. Still others claim he dressed less well than his employees. But one thing is certain: few speak badly of William Schupp.

The fact is that he set about to clean up shop, to rid the Victoria Whaling Company of marginal employees and marginal operations. Later, during the Depression, he would cut wages "by a half"[9] and exert his influence at the highest levels of Canadian and American government in order to maintain solvency. But his first task was to undo the ill-advised management decisions which had destroyed the Canadian North Pacific Fisheries. He also set about to expand nearby markets, particularly the exportation of whale oil to makers of cosmetic soaps in the United States.

He was acutely aware of the needs of the modern cosmetics industry. Whale oils were still used primarily in green and China soaps for industrial and crude household applications, but improved refining techniques had made it possible to use whale oils in higher-quality products. For the first time, the lesser grades of #2 and #3 oil, which were made from the tongue, intestinal fat, and the flesh and bones, could be processed to the equivalency of the best grades #1 and #0, which were made wholly from blubber. Through constantly improving techniques of hydrogenation, it had become possible to remove the residual "fishy" odour. Soap manufacturers began to purchase larger quantities of #0 and #1 as well as lesser grades. Schupp counted on this market in the United States. Unfortunately, he met with initial difficulty in selling Canadian oils there; wartime security measures enacted by the British Parliament included an embargo on whale-oil shipments to the States, the result of careless wording of an order-in-council passed to Ottawa early in 1915.[10] Its intent had been to restrict shipment of oil obtained by Norwegian firms under British "concessions" in the South Shetland Islands and South Georgia, to prevent its transfer to the German war effort. But the language of the order also prohibited domestic shipments across the U.S.-Canadian border.

Schupp argued that his company had never exported oil to Norway, nor did it intend to, but only wanted to exploit a new market close to the source of supply, that is, "leading soap makers in the United States."[11] Under the embargo, whale oils could not be transferred even from Schupp's Canadian shore stations to his own rendering plant at Bay City, and this was the key to the revitalization of the whaling, since Schupp planned to send all his oil production through the Washington plant and then wholesale it to agents across the United States. The first shipment was to include ten railroad cars of sperm oil from the 1914 catch to W. A. Robinson and Company, a major oil manufacturer and distributor in New Bedford; and four hundred barrels "odd lots," some to N. R. Allen Sons Company in Kenosha, Wisconsin and the rest to Atkins, Kroll and Company, Schupp's agents in San Francisco.[12]

Several major American purchasers stood ready to buy Schupp's oil, including Procter and Gamble of Cincinnati; Swift and Company in Chicago; Fels and Company in Philadelphia; and Galena Signal Oil Company in Galena, Pennsylvania.[13] Harvey Fitzsimons, who had formerly worked on behalf of the CNPF, now stepped in to mediate in Ottawa on Schupp's behalf. "They merely wish to be assured," Fitzsimons told Schupp, "that this oil will not fall into the hands of the enemies and when satisfied on that point there will be no obstacle in the way of any shipments to the States."[14] Before long, the matter was resolved, and Schupp began to make regular shipments to clients in the United States.

Wartime demands artificially elevated and supported the price of oil and assisted all producers, even the Norwegian-run United States Whaling Company, which continued to operate in southeastern Alaska. In 1916, for example, their production fell 25 per cent from the previous year, to 6,085 barrels of oil, but war pricing "redeemed" the operation.[15] That same year the Victoria Whaling Company produced 11,699 barrels, and the American Pacific Whaling Company station at Bay City turned out an additional 7,051 barrels. The Washington chaser boats had obtained mostly "small whales, giving an average of twenty-six to thirty barrels of Oil per whale," and Sidney C. Ruck, manager of the Victoria Whaling Company, had to admit that large whales were also scarce in Canadian waters. Nevertheless, he felt "sanguine" that the company had made "a nice little profit at normal Oil prices" and concluded that "there is still some life in the whaling industry."[16]

Schupp continued to pare down his operations. He sold the *Sebastian* and the *Germania* to the Canadian Northern Railway—another job perhaps simplified by his connections with Mackenzie and Mann—and delivered them to Port Mann on 23 March 1916 for conversion into towboats.[17] The *St. Lawrence* was also withdrawn, though not sold, and the *Orion* was converted early in the year for halibut fishing. The harpoon-cannon was removed from the bow, and $505 was spent to modify the vessel for longlining.[18] Veteran sealing skipper George Heater—brother of the feisty gunner in the *William Grant*—shipped a crew under terms set forth by the Halibut Fisherman's Union of the Pacific[19] and sailed with instructions to take both halibut and dogfish, since the latter could be used for liver-oil. The men would be paid twenty-five cents for every four-gallon tin of dogfish livers delivered to Naden Harbour or to Kyuquot. Of course, Heater was also instructed to report all their whale sightings, as well as those received from other vessels.

In April and May 1916 Heater and his crew gathered 37,000 pounds of halibut; a second load of 12,500 pounds was sold by 4 June. But the results were not sufficient to keep the *Orion* away from a good season of whaling. On 31 May Heater was advised to return to Victoria so that the vessel could be re-equipped as a chaser boat.[20] Schupp had closed the Rose Harbour station in 1915; now he was going to open it again, and he needed more vessels.

The whaling season of 1916 began in a regular maelstrom. The constant threat of war, the increasingly better-organized voice of the ordinary sailor, and the uncertain status of the whaling company contributed to frustration and tension on all sides. Disagreements among the boat crews came near to paralysing the effort. Most disquieting of these was the sudden challenge to the traditional authority and power of captains delivered by the engineering staff. Condoned by newly formed maritime labour unions, a struggle erupted between masters and chief engineers to determine who should have final authority over ship operation.

Animosity between "shellbacks" and the "black gang" in the stokehold originated in the earliest days of steam-auxiliary sailing ships. Sailors accustomed to working aloft in clean air quite literally looked down on the mechanical crew tending and feeding the steam engine. Stokers, they said, might understand their "contraptions," but they could not handle the halyards, buntlines, and reefing tackle that constituted the work of real seamen. Engineers thought it unnecessary to risk life and limb handling sail in all kinds of weather when ships could be run by steam. This separatism caused some steamship lines to install divided forecastles to house the two crews. No love was afterwards lost between the deepwatermen aloft and the engineers and firemen below.

Boys play at Rose Harbour, about 1916. Sam Kosaka (son of Japanese foreman Moichi Kosaka) and Nick Raine (son of station engineer John Raine), climbed to the lookout barrel, some forty feet above the deck, while the ship lay alongside the coaling wharf.
Douglas Raine

There was no space for such elitism in the chaser boats; everyone had to get along. But in 1916 the engineers decided to play out their hand, to see how far they could go in wresting traditional powers from masters and pilots. The captain's rule had always been supreme, of course, but the engineers saw a new power base in their massive cylinders and boilers filled with steam pressure. Without engineers, there could be no steam. Without steam, the ships could no longer leave the dock.

In British Columbia, the battle was joined by the National Association of Marine Engineers, Council No. 7 (Vancouver), whose representative attempted to force the Victoria Whaling Company to increase wages. The company refused to discuss the issue; manager Sidney Ruck contended that the men had received a wage increase in 1914 and one could not expect such a thing to happen every year. The association argued that the bonuses paid on individual whales had been reduced, thus negating the salary increase.[21] The situation reached a stalemate just as the whaling season commenced.

When the men began work, the chaser boat crews were rocked by a sudden rash of complaints stemming from an unspecified source. They seem to have begun with H. H. Shepherd, chief engineer on board the *Orion*. Oscar Hansen, the *Orion*'s gunner, complained that Shepherd "would not give him the right bells while hunting causing him to lose whales."[22] Hansen also claimed that the engine made too much noise and scared the whales. Furthermore, he said, Shepherd had failed to keep the whaling winch in good repair. Superintendent Engineer William N. Kelly made the rounds and discovered that there was indeed "serious trouble in the engine room which reflected on the gunner in such a way that he could not get near a whale through noise."[23] Kelly also found that one side of the winch was out of service; the shaft had seized in the bearings for want of oil.

Kelly tried to reduce Shepherd to second engineer but the chief refused the demotion. At the same time, Kelly tried to arbitrate another of Hansen's complaints, one regarding his pilot, J. H. McGregor, whom Hansen accused of being of "very little use either in whaling or making himself handy . . . outside of handling the vessel in or out of Port."[24] McGregor denied Hansen's charges, but Kelly determined them to be legitimate. McGregor had not bothered to keep the daily logbook, once forgetting this duty even when an unexploded bomb point, lodged in a sperm whale under tow, had pierced the hull of the *Orion*.[25]

Kelly then tried to convince William Heater to take McGregor as deckhand on the *William Grant*, but Heater refused. Seeing no further options, Kelly relieved McGregor of his job and demoted Chief Engineer Shepherd on a charge of incompetency, as provided in the terms of the shipping articles.[26] Shepherd again refused the demotion, and there was no subsidiary job available for McGregor; consequently, both were dismissed. McGregor was replaced by John Wells, the former pilot of the *William Grant*, and Shepherd's job was filled by the assistant engineer from the *Grant*, a man named Shannon. Shannon's former post was filled by "Joe Japanese"—also identified as "Joe Yamaska."[27]

William Heater was never known for a calm temper, and now he blew sky-high. In two weeks' time the managers had stripped his boat of its pilot and second engineer. And though undocumented, it probably did not sit well with him that the engineer's replacement was Japanese, a thing unheard of in the all-white atmosphere of the chaser boats. Heater began to make life miserable for everybody. He blamed everyone else for his inability to aim and concentrate, levelling most of his frustration on Rankin, his new chief engineer, who felt obligated to make a counterclaim against Heater's charges that the missed shots were the fault of the engineers. "The whole trouble is," Rankin told Superintendent Kelly, "the man is worked up so he cant hit anything, over loosing Jack Wells. 17 miss shots this month not [to] speak of the gear we have lost."[28]

Before Kelly could adjudicate this furor there was more trouble. On the day the freighter *Gray* was expected with supplies, the newly-promoted Chief Engineer Shannon concocted a story about a shortage of food aboard the *Orion* and forced the chaser to stay in port. When the *Gray* arrived, Shannon allegedly got into the "supplies," became inebriated, and fell overboard between the whaler and the freighter, striking the guard rail as he went.[29]

There was also trouble on the *Brown*. The details are not known, but its chief engineer returned to Victoria on board the *Gray* without giving notice. The second engineer was elevated to his post, and a new man, Sorenson, came aboard as second engineer. But Sorenson was not a certificated engineer, had no papers qualifying him for the job, and when the National Association of Marine Engineers heard of this—undoubtedly from one of their members at the station—they complained to the steamboat inspector at Victoria, charging that the whaling company had violated regulations prohibiting the employment of uncertificated engineers. Kelly had to explain that there were no more certificated men at the station. Furthermore, he accused the union of attempting to "disorganize" the whaling by writing constantly to engineers, even threatening them with suspension from the Association if they continued to work.[30]

There were other dissatisfactions. Gunner Hansen in the *Orion* took a dislike to Heater's former pilot, Wells, because of Wells's insistence upon shooting the harpoon-cannon. They had made no agreement about how many shots were entitled to the pilot, and the argument spilled over to the entire crew. Superintendent Kelly wrote: "the Gunner's crew of the 'Orion' seem to have lost all interest in the business. . . . Wells has been enthusiastic ever since he took hold but it appears that both Hansen and Miller . . . do not give him a show at the gun. As he puts it 'the man at the gun does not hunt it is the man at the wheel.' Wells had one shot & got his harpoon in, but the harpoon broke off at the neck. That is the only shot he ever got."[31]

If all that dissension failed to shut down William Schupp's struggling whaling concern, there was also an outbreak of beriberi in the Chinese bunkhouse at Naden Harbour. A man named Chow Wing died and two others fell gravely ill. Clothing had to be destroyed and the entire building disinfected.[32] At the same time, a general strike in Norway made it impossible for Isak Kobro to ship new harpoons. Kobro apologized, but the matter was out of his hands. The Victoria Whaling Company arranged for the shipment of Swedish steel in bars, and patterns were provided to local ironworkers, including Yarrows Limited at Esquimalt, who contracted to make harpoon legs for sixteen dollars each. Others were made at the Seattle Construction Company, and also in Vancouver. These locally made harpoons did not live up to expectations; Kelly returned at least some of the broken ones to Yarrows in hopes that they could determine what was wrong. The

Larsen, the famous whaling blacksmith, came to British Columbia from Norway in one of the chaser boats and stayed on for more than two decades. His skill in straightening harpoons bent during the chase was legendary.
Photograph by Lewis L. Robbins, Kendall Whaling Museum, Sharon, Massachusetts

matter was further complicated by individual preference: gunner Anderson used solid-shaft harpoons, for example, while Christian preferred those with hollow shafts.[33]

For want of approved shipping permits, the delivery of fine Italian hemp foregoer lines was also delayed.[34] Foregoers were subject to much strain during the tightening and slackening of the chase and became frayed with use; worn lines could lead to lost whales. At Akutan, Knut Birkeland gravely considered the circumstances: "Ten sets of foregoers were ordered from Norway," he wrote, "and I felt certain they would be brought to us on the first mail boat stopping at the station; but they did not come, and not a word about them either. What foregoers we have left will soon be rendered useless, and then good-by to the whaling."[35]

Normal operating expenses had to be met all the while. Coal ranked among the most expensive outlays, since all the chaser boats and the freighter required it for their power plants; the *Gray* consumed 12 tons every day to make 9.5 knots between stations. Coal for the single month of August cost $2,644.[36] To help defray costs, C. Rogers Brown began to charter the *Gray* for general freight. As early as 1915 it transported 120 tons of "best household lump coal" to the Port Simpson General Hospital, and in March 1916 loaded lime at the Pacific Lime Company plant on Texada Island for San Francisco.[37] The return cargo consisted of 800 tons of pyrites for delivery onto barges at Vancouver for use in munitions.

The charter plan soon backfired on the whaling company. In the middle of the 1916 season the *Gray* ran bone meal to San Pedro, California and arrived in the midst of a major waterfront strike. Unable to find a crew to offload, Captain Hawes had to wait for a settlement and did not clear the port until mid-July. In the meantime the whaling company had to spend some of the profits of the *Gray*'s long voyage to charter the *Prince Albert* to carry coal and supplies to its own employees.[38]

Freighting continued during the winter months, but these were soggy trips at best. During one passage from Victoria to Kyuquot the *Gray* shipped so much water that Hawes described his command as "more like a Submarine than a Steam boat."[39] Normally a run of less than twenty-four hours, the trip

A rare photograph of the *Gray* at sea, showing the general arrangement of the deck-houses and loading booms. Originally called the *Petriana*, the English-built ship served the British Columbia whaling stations for more than thirty years.
Matthews Collection, Vancouver City Archives

consumed nearly thirty-eight. On arrival at Kyuquot the crew pumped four feet of water from the chain locker, and, to add the proverbial insult to injury, every drop had to slosh through their forecastle before it could be pumped overboard.

The whaling season of 1916 eventually came to its arduous end, and successfully enough to warrant a return in 1917. But the faces and names among the whaling crews had begun to change. New men were ready to try their hand, while some of the veterans were looking at possibilities elsewhere. In 1918 Knut Birkeland would announce his retirement as manager of the North Pacific Sea Products station at Akutan. Others, including Ludwig Rissmüller, were dead.

Rissmüller seems to have parted from the CNPF about the same time as Sprott Balcom, but his departure from whaling is clouded by some of the same mystery attendant upon his entry into it. While William Schupp was hastily pruning the Victoria Whaling Company, Rissmüller followed the path taken by Thomas Roys four decades before. He left Victoria in 1914 and established residence in San Diego, California, at the Hotel St. James. It is not clear what problems may have beset him. Perhaps he felt threatened among the enemies of his native Germany. His wife apparently remained behind in Canada, and he was clearly in financial disarray. The value of his whaling stock had evaporated, since most of the subscribers to his patents had failed. Not long after his arrival in California, he was forced to surrender himself to the mercy of his former colleague, Sprott Balcom. In June 1915 he wrote Balcom: "I am very sorry that I am obliged to ask for an accomodation now and I never thought, that I ever could come in such a position. If only a few of the people, whom I loaned money at low interest and without any security would pay me, I would be easy, but as nobody pays me anything, it is absolutely impossible for me to raise a cent of cash."[40]

Like Roys, Rissmüller also suffered from failing health. A degenerative heart condition and arteriosclerosis, coupled with a two-week bout of flu contributed to his death on 16 April 1916; he was about sixty-four years old. Four days later his remains were interred in the Odd Fellows Cemetery in San Diego,[41] and Sprott Balcom spent part of his nine remaining years as executor of the Rissmüller estate, working through a legal tangle which involved real estate investments in Victoria and financial interests in the Lincoln Steamship Company and the Vancouver Dredging and Salvage Company.[42] If anything came to Rissmüller from the transfer of whaling patents — Schupp had quickly requested Ottawa to reassign the patents to the Victoria Whaling Company — it seems to have been too little and too late.

At the Armistice a certain sense prevailed that the slimmed and streamlined whaling outfits in the Pacific Northwest might actually survive. Oil now found a steady, if restricted, market among manufacturers of margarine and soaps. Bone-meal fertilizer filled a domestic farm demand, particularly in the Hawaiian Islands, where it proved beneficial as a soil additive on sugar-cane plantations. Some was sold to growers in California.[43] At home, meat shortages had propelled North American families toward one additional use of the whale: food.

Edible-meat production had been little more than a sidelight to the whaling business in the Pacific Northwest; certain cuts had been exported "in pickle" to Japanese consumers as early as 1907, but the production of fresh and canned whale meat in large quantities was not considered until 1917. At the height of trench warfare in Europe, most North American beef was diverted to the fronts. Protein-rich foods fell into short supply, and it was proposed to supplement home consumption with whale meat. Being mammals, many species of whales could provide a healthy substitute for beef and lamb. The meat of rorqual whales proved more digestible than beef, and, if anyone had known to care, it was lower in cholesterol, too.

Passengers aboard British Columbia coastal steamships had already tasted whale meat. Nothing was done to disguise it, and these diners reported it was like pork "although a little on the dry side."[44] Most others thought it weak in flavour, and much in need of a liberal hand with butter, garlic and onions.[45] Whaling crews ate the meat, too, since it was free and available in prodigious quantities. Their willingness to partake stands in clear contrast to the sail-whalemen of the nineteenth century, who do not seem to have eaten their prey, perhaps because of the relative inedibility of their preferred catch, the sperm whale.[46]

The reluctance of nineteenth-century whalemen to eat whale contrasts also with the reports of virtually all early explorers who visited the Pacific Northwest. Most of them sampled whale meat, beginning with James Cook and his colleague John Ledyard, who feasted upon roast whale and smoked bear given them by fur traders at Unalaska during the voyage of 1778. One hundred years later Charles M. Scammon found himself rushed to take measurements from a "sharp-headed finner [minke] whale" because of the local Makahs' interest in the food value of his find. And in 1911, Roy Chapman Andrews promoted whale meat as a staple food for the world's poor.[47]

North Americans' disinclination to dine on cetacean flesh hindered its sale until the meat shortages of World War I became acute. Then, a public relations campaign was devised in the United States to entice the public to buy whale meat. Whaling plants in the Pacific Northwest geared up to meet the anticipated demand; William Schupp purchased the Pacific Cold Storage Company of Tacoma, Washington, and also a refrigeration steamer, the *Elihu Thompson*, to bring meat from Akutan.[48] Additional cold storage facilities were added at Akutan, Kyuquot, and Bay City, and even more temporary cold-space was leased from Wallace Fisheries at Kildonan, on Vancouver Island.[49] At Kyuquot, special sanitized equipment was installed for canning whale meat, with storage for 50,000 cases.

The *Gray* also received special dispensation to operate between Canadian and U.S. ports in the transport of whale products.[50] Part of its cargo included whale meat. In 1917 a nineteen-ton load of meat from young whales was shipped in hundred-pound boxes to San Francisco restaurants. The following year, the *Gray* carried 550 cases of whale meat from the Grand Trunk Pacific Dock in Victoria to Seattle. Local reporters made it something of a *cause célèbre* that these particular cases of meat had been "lying on the dock for some time" before the crew of the *Gray* hauled them away.[51]

Whale-meat marketing began on the west coast in the spring of 1917, at ten cents per pound. In the New York market the following year, whale meat sold for sixteen cents per pound.[52] Some difficulties arose over where it should be sold. In Victoria butchers refused to handle it; since the whale lived in the sea, they argued, its meat belonged in the fish market.[53] On Vancouver Island the local press encouraged readers to take advantage of the low-priced "sea beef," and provided recipes for such delicacies as "Whale Meat Roll," "Whale Meat Shepherd's Pie," "Whale Meat with Sauce," and "Whale Stew." Editorialists argued, cajoled, prompted, and even promoted whale meat as *nouvelle* cuisine: whale steaks, they said, commanded a high price in chic New York restaurants.[54] The president of the prestigious American Museum of Natural History in that city—and Roy Chapman Andrews's employer—went so far as to host a banquet at which he served "Hors d'Oeuvre-Whale" followed by "Whale Pot au Feu" and an entrée of "Planked Whale Steak."[55] But it was difficult to overcome the old prejudice:

> the Whaling Company, in their endeavor to place their canned whale meat on the market, and to get it known, persuaded the Vancouver Board of Trade to allow it to be served . . . at the Vancouver Hotel, in 1917. . . .
>
> We were not told it was whale meat, and nothing happened except that several grumbled at the *rotten beef*. After the meat course was over, the chairman announced that the *beef* was whale-meat. Psychology is a peculiar influence; several people hurriedly left the room.[56]

Canned whale meat was demonstrated during Christmas 1918 in department stores in Winnipeg, Toronto, and the eastern United States.[57] At the Armistice, 29,585 cases had been prepared for shipment from Kyuquot,[58] but the return of beef brought a quick halt to the sale of all whale meat, as a bull-market sensibility returned to American life. The canned goods continued to sell in Japan and elsewhere; about 1923 a portion of a newly canned lot of

10,000 cases arrived in England, "presumably" one writer sneered, "for consumption in some of the colonies."[59] Another jingoist went so far as to link whale meat to a favourite racial stereotype. "They managed to sell off the stock they put up," he wrote, "to the West African market it went, I believe —but the West African nigger will eat anything!"[60]

The major Pacific Northwest supplier of "sea-beef"—and every other whale product—was William Schupp. By early 1918 Schupp possessed a controlling interest in every shore station between Bay City, Washington and Akutan, Alaska—with the single exception of the struggling United States Whaling Company on Baranof Island. On 2 May 1918 he incorporated a new umbrella company to oversee these various ventures, which he aptly called the Consolidated Whaling Corporation Limited. The legal offices on King Street, Toronto were not far from his old insurance firm. Operations remained headquartered at Point Ellice, Victoria.

The Consolidated Whaling Corporation was little different from previous companies in its form of organization; 25,000 common shares of stock at fifty dollars each and 7,500 shares of preference stock worth seventy-five dollars each provided a working capital of about $2 million.[61] Through a complex manipulation Schupp came into immediate possession of virtually all the stock. He did so by selling all his shares in the old companies to the board of directors of the Consolidated Whaling Corporation and then using the money to buy shares in the new company.[62] Naturally, he became its director.

The assets of the Victoria Whaling Company were transferred to Consolidated by a resolution of the corporate board dated 22 November 1918. The assessment clearly reveals the extent of the Mackenzie-and-Mann-era expansion: 40 acres of land with standing timber, buildings, and equipment at Kyuquot ($174,977.77); 164.7 acres plus buildings and equipment at Rose Harbour ($141,500.00); 20 acres with timber, buildings, and equipment at Naden Harbour ($200,500.00); and a 14-acre site at Sechart including buildings and equipment ($26,104.43). Additionally, there were four active whaling licences, patents, patent rights, franchise and development rights, the property at the Victoria docks, office furniture, and the vessels: eight steamers worth $880,000; the freighter *Gray*, valued at $225,000; and a small launch, the *Bachelor*, which added $1,900 to the final total of $1,811,062.10 (less depreciation).[63]

The low valuation of the Sechart site reflected the fact that very little remained of the province's first modern shore whaling station. Late in 1917 the Victoria Whaling Company had petitioned Ottawa for cancellation of that licence, saying it was not their intention to operate the plant. The cancellation became effective 20 February 1918,[64] and by that time most of the machinery had been shipped to Akutan.[65] The buildings at Sechart were leased to Vancouver Island Fisheries for use as a herring packing plant, with an option to purchase.

The business of the Consolidated Whaling Corporation began on 27 June 1918. On that day the new corporate name was typewritten onto the pages of the ledger of catch and production. For some years afterward, however, the ledger retained also the separate records of the North Pacific Sea Products Company and the American Pacific Whaling Company, much to the consternation of later statisticians and historians. In fact, Schupp tied his American and Canadian operations together, although legalities prohibited an easy transfer of equipment and supplies from one to the other.

Schupp's managers continued to run Consolidated from the Point Ellice dock in Victoria, while the work of the American Pacific Whaling Company and the North Pacific Sea Products Company was removed from Seattle to a new office and wharf at Bellevue, on the shore of Lake Washington. The lake had been linked to Puget Sound by a series of locks; the chaser boats and other floating stock of the firm could be sheltered comfortably in fresh water during the winter months. Now the whaling business began to demand more and more of Schupp's time. Early in the 1920s he would leave Pawling Manor behind; the death of his wife Emily about that time probably contributed to his decision. Though he would retain control of his "gentleman's farm"[66] near the banks of the Hudson River, he afterward considered Bellevue his home.

The economic advantages of the war economy had provided a benefit for the whaling companies in the Pacific Northwest, and the immediate postwar seasons of 1919 and 1920 continued to be among the most active in local history. Ten shore stations operated from California to the Aleutian Islands. But the staggering postwar recession of 1920–21 rocked the international whaling community. A glut of oil reached European markets from Antarctica, where newly exploited whaling grounds were rapidly giving up the largest and best of their prize; at the same time the depressed price of oil made it nearly impossible to sell at any gain. Norwegian whaling entrepreneurs had begun an oil cartel during the war years in an effort to establish price guarantees and sales contracts on a season's production,[67] but even this highly structured agency was ineffectual in shoring up the falling market for whale products.

Among the hardest hit was the United States Whaling Company. As of April 1920, 1,741 barrels of the lesser #2, #3, and #4 oils remained unsold from the company's 1918 catch, as well as nearly 8,000 barrels produced during 1919.[68] The company's *Star* chaser boats had been unable to catch large whales, a problem attributed to the long distance from the station site to the most active whaling grounds. But there was neither profit nor new stock sales to finance a relocation, and the managers concluded it was time to bring the company out of the whaling business. The United States Whaling Company finished the 1922 season, then the chaser boats were sold and the plant converted to herring meal and fish oil production.

The innovative Norwegian whaling captain Carl A. Larsen was just then preparing an experimental whaling expedition—the Hvalfangerselskapet Rosshavet—to explore the potential of the Ross Sea in Antarctica. Larsen outfitted the factory ship *Sir James Clark Ross* and purchased all three of the defunct company's chaser boats to accompany it there. The *Star I*, *Star II* and *Star III* steamed south to Seattle for a refit and were met by crews arriving from Norway via New York; two additional chaser boats, the *Selvik* and the *Pelikan* were brought from Norway and sympathetically rechristened *Star IV* and *Star V*. In September 1923 the flotilla set out toward an epochal and successful opening of the Ross Sea to commercial whaling, a long journey later recorded by Alan Villiers in *Whaling in the Frozen South*.

Plagued by the same poor market conditions that vanquished the United States Whaling Company, William Schupp elected not to open the stations during 1921. This was a risky move which threatened to undermine the traditional expectation of work, but Schupp may have expected to abandon the business in Canada. He already stood accused of such treason. In January 1920 his managing director, Sidney C. Ruck, together with Ruck's brother Cecil and Superintendent Engineer Kelly had applied to Fisheries Inspector Edward Taylor for a licence to "engage in the manufacture of oil and fertilizer and other commercial products from whales" under the name of the Vancouver Island Whaling Company. The application charged that the Consolidated Whaling Corporation intended to dismantle the whaling station at Kyuquot and send the equipment "to foreign territory"—meaning Alaska. As clear evidence of Schupp's intent, they cited the 1917–18 dismemberment of the Sechart station and the removal of its machinery to Akutan. "This enlarging of the field of operations in American waters and the consequent reduction of operations in Canadian waters," they said, "is a policy inaugurated by the President Mr. William Schupp, and the applicants feel that by introducing a new whaling industry at this time they will be preserving a Canadian industry."[69]

Sidney Ruck resigned his position with the Consolidated Whaling Corporation and travelled to England, where he entered into negotiations for the purchase of two Admiralty-built whaling steamers, the *Humpback* and the *Bullwhale*.[70] Unfortunately, Ottawa notified the partners that their proposed site at Winter Harbour, in Quatsino Sound, was exactly fifty-four miles from Consolidated's station at Narrow Gut Creek, Kyuquot Sound, and therefore forty-six miles shy of the law. Sechart was then considered, but little more transpired before the international economic collapse staggered the whaling industry. Kelly and the Rucks were forced to beg the chief inspector of

Fisheries in Vancouver to keep their application strictly confidential, for "obvious reasons,"[71] but their plan could hardly have been a secret. As early as March 1920 the *Colonist* had reported the formation of the new whaling venture and its option on two steel vessels "suitable for the business."[72]

Though the Rucks did not succeed, a few independent mariners of the old school did make an effort to profit from the large Alaskan whales. Such a man was Louis L. Lane, an American who earned his captaincy in the last days of sail whaling. In 1920 he took command of the three-masted diesel-auxiliary schooner *Carolyn Frances*, newly purchased for whaling by the Western Whaling and Trading Company, and in that vessel the following year he cruised from San Francisco to Kodiak Island and killed seventy-nine whales, all but four of them humpbacks.[73] In 1924 he returned again to the North Pacific in the large steamer *Herman*, cruised in the Gulf of Alaska and probably the Bering Sea, and again hunted humpback whales near Kodiak.

In 1925 he arrived at Kodiak in the *Gunner*, a small, "homemade" chaser boat converted from a U.S. Navy gasoline launch. This he fitted with a makeshift cannon, its barrel formed of cold-rolled steel and the breech block rebuilt from a sporting rifle. Captain Lane and his one crewman killed humpbacks in the bays of Kodiak Island and towed them onto nearby beaches for sale to fox farmers, who paid $500 per carcass. But his industry was cut short by two gun explosions. In the first instance, flying shrapnel imbedded itself in Lane's face and tattooed him for life. Later, the premature ignition of a whaling bomb injured a substitute captain. The *Gunner* was fast to a running whale at the time; after the crewman set the winch brake, the friction of the whaling line against the drum ignited a fire which quickly consumed the little launch. Luckily, a nearby towboat witnessed the scene and took the two men off before the *Gunner* exploded and sank.[74]

After the one-year layoff in 1921, William Schupp returned his whaling fleets to sea and did not abandon Kyuquot as Sidney Ruck had predicted. Instead, he turned his attentions farther south. After decreasing the capital of the Consolidated Whaling Corporation in 1923[75] he closed the large and costly reduction plant at Bay City, which was not processing enough whales to pay its way. Its new operators converted the facility for fish processing to capitalize on the vast runs of pilchards—California sardines—which had suddenly begun to congregate in northwest waters.[76]

The closing of Bay City freed men and chaser boats for whaling service in Alaska. In 1926 Schupp set about building a brand-new whaling station on Sitkalidak Island, just inside Cape Barnabas to the south of Kodiak Island. This new station would be sited at, and called, Port Hobron. A large barge was procured and the reduction machinery from the Washington station was loaded for the long haul to Alaska. Unfortunately, the towboat encountered

The *Orion*, in the foreground, had been sold by the time this atmospheric photograph was taken at the Consolidated dock in Victoria Harbour. The lookout barrels belong to the company's "colour fleet."
Provincial Archives of British Columbia

severe seas and the heavy cargo washed overboard into the domain of the leviathans it had once processed.

Now Schupp was faced with a difficult problem; he wanted a station in an unexploited area of southeastern Alaska, but to buy all-new equipment would be extremely costly. Then, belatedly, he considered the situation at Kyuquot. During the summer of 1925 Kyuquot had returned only 1,280 barrels of whale oil and 181 barrels of sperm oil from a meagre take of eighty-two animals. A company representative told Ottawa that whaling in the province had been "a pretty tough game in the last few years" and that its prospects did not look too bright.[77] Schupp was clearly ready to unburden himself of this weak link, so in September 1925 the buildings and wharves at Narrow Gut Creek were transferred to Orion Fishing and Packing Limited. Orion also obtained the historic chaser boats *Orion* and *St. Lawrence*, which were converted for pilchard fishing in the spring of 1926.[78] Not surprisingly, most of the reduction machinery was sent north to Port Hobron.

When the new Port Hobron station opened it was praised as the most up-to-date plant in Alaska; its equipment included a large dryer of a recent design, six blubber cookers and fifteen meat and bone cookers, at least some of which Schupp was probably forced to purchase new. Whaling began on 20 July 1926, and by mid-August more than eighty whales had been taken. William Schupp installed his brother Henry as station manager, a post which he held until his death in the 1930s.[79]

The chaser-boat fleet servicing Akutan and Port Hobron represented the entire history of modern shore whaling in the Pacific Northwest, for the American Pacific Whaling Company had long ago acquired the "eiendomelig" chaser boat *Tyee Junior* from the defunct Tyee Company and renamed it *Tanginak*; Schupp also operated the two Moran-built chasers *Unimak* and the *Kodiak* which originally steamed to Akutan for Lars Christensen's Alaska Whaling Company. The other four chaser boats were those commissioned for the American Pacific Whaling Company station at Bay City, Washington: the *Moran*, *Paterson*, *Aberdeen*, and *Westport*.

Schupp left Vancouver Island behind. Although the business of the Consolidated Whaling Corporation continued to be run from the office and wharf at Point Ellice, western Canadian whaling was now restricted to the two operational stations at Rose Harbour and Naden Harbour in the Queen Charlotte Islands. In 1927 Schupp again devalued the stock of the Consolidated Whaling Corporation[80] and eliminated the last vestiges of the North Pacific Sea Products Company. Its name disappears from the ledger books as of 1 August 1927; operations were brought under the aegis of the American Pacific Whaling Company, which had been without a whaling station since the closing of Bay City. After this devaluation and consolidation, Schupp made no more major alterations to the whaling programme. For the remainder of the decade the whaling business operated in a healthy fashion; enough whales were caught at two Alaskan stations and two in the Queen Charlottes to pay the bills.

At the same time, a California whaling company made a brief foray into Alaskan waters with a floating reduction plant. The California Sea Products Company had met with disappointing results at two shore stations in California, and in 1926 outfitted a former oil tanker, the *Lansing*, for offshore operation. This ship and three chaser boats operated during the summer of 1927 from Lazy Bay, a cove in Alitak Bay at the south end of Kodiak Island, and returned to San Francisco in October after taking 237 whales in Alaskan waters. The *Lansing* was subsequently employed off Southern California, but did not return again to the Pacific Northwest.[81] Innovations like these held promise for the future, but none of the Pacific Northwestern whalemen could have foreseen the Depression just a few years ahead or the particular problems which would attend upon the world's whaling industry. Their troubles were far from over.

The economic collapse in the autumn of 1929 threatened to topple William Schupp's unstable balancing of market prices against steadily rising operational costs. The restraints of a recession market were aggravated by unrestricted whaling around the world, and particularly in southern waters. The Antarctic catch in the southern summer of 1930–31 glutted the world market with whale oil at precisely the time when there were few cash buyers wanting it. The Norwegian oil cartel suffered a long string of cancelled contracts just as the Antarctic fleet returned with a 25 per cent larger catch than anticipated, and the overproduction—114,000 tons of it—sat in warehouses with no buyers on the horizon.

The oil market was so poor in 1931 that Schupp again idled all his stations and boats. Akutan, Port Hobron, Naden Harbour, and Rose Harbour all were closed; the *Tanginak*, *Paterson*, *William Grant*, the "colour fleet," and all the others remained in port. Schupp could not get a contract at any price high enough to offset operational costs at even one of the stations. He petitioned the Canadian government for assistance, calling for an increase in the duty payable on foreign whale oil entering Canada, but this request, though granted,[82] proved ineffective in increasing domestic wants.

To get his Canadian boats back to sea, Schupp even eliminated insurance —his old business—and "cut wages by a half,"[83] to the displeasure of whalemen who needed the work but were not prepared to absorb the loss. Even so, it was 1933 before Schupp was able to reopen a Canadian whaling station, and then he did so cautiously, sending the *Brown*, *Green*, *White*, and the *William Grant* to Rose Harbour.[84] Their crews brought in 418 whales despite the opinion of one Fisheries inspector that the vessels were in a poor state of repair.[85]

When he reopened Port Hobron, he sent his son-in-law, Marc A. Lagen to replace the late Henry Schupp as its manager. Lagen, a former concert singer, seems an unlikely choice for a whaling station manager; he had come to the business as the husband of Schupp's daughter Emily, whom Lagen had courted while she was an aspiring performer and he a theatrical manager. But the whalemen took a liking to young Lagen and apparently found him as sympathetic to their problems as his father-in-law. Some of them even came to call him "Cap," though he was no sailor.[86] Until World War II, Marc Lagen managed the daily affairs of the American Pacific Whaling Company, summering at Port Hobron and wintering at the company's office and dock in Meydenbauer Bay on Lake Washington.

Whaling in 1933–34 was in some ways very different from the whaling of 1929–30. The proof of Johan Hjort's thesis had come nearer to acceptance and finally, after a very long delay, the whaling countries had begun to see a need for controls upon the business of collecting whales. For the first time, international agreements attempted to protect the whaling industry by placing certain limits on whaling. An International Whaling Convention, sponsored by the League of Nations, had convened in Geneva in September 1931 to draft controls to regulate and supervise whaling. The convention document was based upon a 1929 Norwegian law which prohibited that nation's whalemen from taking right whales, or any pregnant female whale, or any cow travelling with a suckling calf in any waters of the world. The Norwegian legislation also protected immature whales—including blue whales less than sixty feet in length and finbacks less than fifty feet long—and demanded efficient and complete utilization of all carcasses.[87] No conservationist sponsored these regulations, nor was the morality of whaling brought into question. The sole function of the Norwegian law and the Geneva convention was to protect the hunt from its own irrepressible greed. By excluding from capture immature whales, pregnant females, and commercially-extinct species the signatory nations hoped to preserve the whaling industry into the foreseeable future.

The Norwegian example was strictly followed. Each nation would be expected to provide statistical data on each whale taken, including date of catch, location, species, sex, length, presence of a fetus, and—wherever practicable—stomach contents. The numbers of whales killed and quantities of oil and fertilizer produced were also to be reported. A statistical bureau to be established in Oslo (formerly Christiania) would require the data at "convenient intervals not longer than one year."[88] Gunners' bonuses were not to be based simply upon quantity; size, species, value, and yield of oil were suggested as alternative criteria. Minimum size limits such as those mandated by the Norwegian law were not incorporated into the new convention, and the sperm whale was excluded from its protections. But the killing of right whales—including "Pacific right whales"—was expressly prohibited.

This admonishment nearly came too late; the species had become so scarce in the North Pacific that a capture often made headlines. One writer, subscribing to an unfortunate misunderstanding, described a stone-dead right whale on a flensing platform as "one of nature's rarities."[89] He failed to perceive that man, not nature, had isolated the right whale in the seas. Even as late as the 1950s a maritime historian wrote mistakenly that the right whale was rare in Alaska because of its preference for Arctic waters.[90] Fisheries Inspector Richardson in 1934 knew a gunner at Naden Harbour, fifteen years in the business in British Columbia, who had never *seen* a right whale in the Queen Charlotte Island Fisheries District,[91] in the same waters where the American whaling ships had hunted them so diligently through the 1890s. The last few right whales brought to shore stations in the province were towed in during the mid-1920s; two, perhaps three, in 1924 and one in 1926. A few more were taken in Alaskan waters during the subsequent decade, but the total number of right whales taken by modern whalemen in the Pacific Northwest may not have exceeded twenty-five.[92]

Among the nations that signed the whaling convention at Geneva on 24 September 1931 were eighteen whose governments eventually ratified the

Badly exploited by intense hunting in the nineteenth century, right whales were rarely seen, and only a few were taken in the twentieth. This one was brought to Akutan, probably in 1916 or 1917.
Kendall Whaling Museum, Sharon, Massachusetts

document.[93] Canada and the United States did so, thus its conditions became binding on both the American Pacific Whaling Company and the Consolidated Whaling Corporation. Ottawa told Schupp that it would reaffirm Consolidated's whaling licences for the 1933 start-up only if the corporation agreed to eliminate bonuses paid on aggregate catch and install in their place a system based on one of the other acceptable priorities.[94] Schupp changed the bonus system to one which paid the premium on the largest examples of each species. Blue whales less than seventy feet long earned $9.50; larger specimens were worth $10.50. Finbacks and humpbacks were valued at $5.25 or $5.75, depending upon length.[95] Bonuses for sperm whales — not covered by the terms of the convention — were set at ten dollars per whale. The earlier system had been much simpler. "The seamen each got three dollars per whale," one former whaleman recalled. "The captain got ten. If he got a blue whale, it went up to fifteen."[96]

In 1934 Schupp's companies returned to something like normal operation after three years of recumbent waiting. That year he restarted Naden Harbour, so that all four of the Northwest stations were open for the first time since 1930. The combined fleets of the Consolidated Whaling Corporation and the American Pacific Whaling Company took 1,019 whales, and a new reciprocal trade agreement halved the U.S. import duty on oil entering from

Canada to 2.5 cents per gallon. But even this concession brought no significant improvement to the whaling business. When the boats went north for the 1935 season, Naden Harbour was once again closed.

Any means was employed to improve the sagging market for whale products. Schupp instructed Rose Harbour manager Alf Garcin to put up dried whale meat, which he hoped to sell as "jerky." Schupp's brother Henry had successfully made some at Port Hobron, but Garcin's subsequent attempt did not meet expectations. A five-hundred-pound sample lot, shipped to the American Pacific dock in the *Gray*, hung for two weeks in a shed but retained its offensively strong smell. Schupp told Garcin that he fed some to a dog, who seemed to like it, "but a few days later he got sick and did not eat any more."[97]

But William Schupp devised another plan far more grandiose than whale-jerky. He had seen that the vast catch of the pelagic whaling industry in Antarctic waters dominated the world market for oil. Fleets of factory ships and their attendant chaser boats, the majority of them Norwegian or British, had in one decade killed so many whales that Schupp's North Pacific contribution paled into insignificance. Whereas in 1919 the North American west coast provided about 10 per cent of the total world supply of whale oil, by 1929 Schupp and the few California producers accounted for only 2.5 per cent of production.[98]

Schupp also realized that the glut which accumulated after the 1930–31 season had eventually been sold, largely as a result of new chemical processes introduced in 1929 which made it possible to sell a margarine composed almost entirely of whale oils. Germany and Great Britain had become the leading fat-purchasing nations in the world,[99] and in preparation for the conflict yet to come, these two industrialized nations, as well as others, now moved ahead to build up their reserve supplies of edible fats.

Because he could not compete from the northern hemisphere, Schupp determined to send a whaling fleet to the southern. He had plans drawn for a factory ship to operate during the Antarctic season of 1934–35,[100] but the American Pacific and Consolidated firms could hardly pay for such a venture. So Schupp turned to the United States government's Reconstruction Finance Corporation, an agency established by President Franklin D. Roosevelt to assist financially-troubled businesses by providing long-term, low-interest loans. From the RFC Schupp hoped to obtain funding for the Antarctic venture as well as additional moneys for shore whaling in Alaska.

A fortuitous circumstance worked in his behalf; his sister-in-law worked in the White House. Victoria Henrietta Kugler had married Henry (Harry) Nesbitt, a former American Pacific Whaling Company sales representative. After the death of her sister Emily Kugler Schupp, Mr. and Mrs. Nesbitt assumed the management of the Schupp's "show place" farm at Staatsburg-on-the-Hudson,[101] and soon made the acquaintance of some very influential neighbours—the Roosevelts—in nearby Hyde Park. After Franklin D. Roosevelt became president, he asked Mrs. Nesbitt to come to Washington, D.C. as his personal housekeeper. So the Nesbitts moved, and the aging Harry Nesbitt found himself installed in a token job, sometimes described as "White House aide," but otherwise simply as "custodian."[102] Through the Nesbitts, Schupp gained personal access to President Roosevelt and other cabinet officials. His needs and ideas also reached the ear of Eleanor Roosevelt, the President's politically active wife.

Schupp probably petitioned for money in the spring of 1934 in a telegram to Harry Nesbitt. Although the text has been lost or destroyed—as have many RFC records—Schupp likely spelled out his idea in some detail and with some urgency, since the money was to be used for reconditioning vessels for both the Alaskan summer season and the Antarctic cruise to follow. Harry Nesbitt forwarded Schupp's plea to Secretary of the Treasury Henry Morgenthau, sending both the telegram and an explanatory cover letter under date of 14 April 1934: "Word has been received that the whales are coming thro the Arctic Circle, and will be feeding in the Bering Sea within the next two weeks. If there is anything . . . that you can do toward assisting this Historic and commendable Enterprise, I am sure that not the employees alone and their Families would feel deeply grateful, [but] the Business Interests would indeed appreciate it."[103]

Unfortunately, Nesbitt carried too little bureaucratic weight to bestir Washington. Hearing nothing, Schupp cabled again, this time more urgently, and Nesbitt answered confidently: "It was not five minutes after I received [the telegram] when the President was reading it, and then action just stepped in and things began to move, and from what information I can get you will get all the Tonnage that will be necessary to cope with the situation." Nesbitt also reassured Schupp that Secretary Morgenthau had promised to do all he could to get the "necessary funds to see the thing thro."[104]

But it does not appear that any money was forthcoming. A formal request for a $300,000 industrial loan to the American Pacific Whaling Company reached the Reconstruction Finance Corporation on 16 November 1934. One-quarter of the total was earmarked for the charter of a factory ship for Antarctic operation.[105] Local bureaucrats in Seattle were assigned to investigate the applicant company, but they were unimpressed by what they found. Net profit for the preceding ten years of operation had amounted to a slim $38,000, after depreciation and losses. The RFC examiners wondered if Schupp could pay back such a large loan. And what would happen if the company failed?

> Should the ship be chartered and operations extended to the Antarctic as requested it might change this showing, but whether it would be better or worse we cannot at present ascertain. . . . There is one factor re-

garding this security which is so uncertain that it should receive attention and consideration. This applicant and its affiliated Canadian Corporation are the only two known operators in this line on the North Pacific
Coast. Apparently neither of them is prospering. If it became necessary
to resort to the security it would certainly be hard to find a buyer. Therefore, the appraised value as a "Going Concern" vanishes and we have in
its stead the sale value of whaling plants and whaling ships. . . . it is
impossible for this Agency to approximate their realizable value under
necessary liquidation.[106]

The examiners recommended therefore that the loan be declined for lack of
proof of earning power. On 28 December it was done.[107] Undaunted, Schupp
again approached the government for a smaller loan, but when January 1935
passed and nothing more was heard, he became more insistent. Someone,
perhaps Harry Nesbitt, appealed directly to Eleanor Roosevelt; on 21 February the chairman of the board of the Reconstruction Finance Corporation received a note from the First Lady's secretary and a copy of one of Schupp's
cables: Mrs. Roosevelt wanted to know if he might possibly "hurry this matter along."[108] Five days later the RFC approved a loan of $175,000, and there-
after the American Pacific Whaling Company maintained an ongoing relationship with the corporation. Surviving records indicate that the company
filed nine loan applications, six of which were declined and one withdrawn;
the second of two loan approvals was given in 1939.[109]

Washington saw no reason to support Schupp's marginal enterprise. Almost immediately after granting the loan, the government repealed two major tariffs on imported whale oil. One, a 6-cents-per-gallon import duty, had
come into effect in 1930. The other was a new excise tax of 3 cents per pound
of oil, which, at 7.5 pounds per gallon, effectively raised 22.5 cents on every
gallon imported.[110] These rates had been imposed to protect farmers, whose
sales of vegetable and animal fats and lards had faced increasing competition
from the whaling industry.[111] But the tariffs had disadvantaged Norwegian
merchants, who wanted to continue selling some part of the Antarctic catch
to American soap manufacturers.

Norway protested circuitously, by placing an embargo on the importation
of automobiles from Denmark, where they were assembled with "80 percent
American work." Roosevelt was told that this "indirect retaliation" would
cease if the U.S. government took "favorable action on whale oil."[112] Secretary
of State Cordell Hull advised the president that the minimal U.S. involvement in whaling was "insufficient upon which to turn down a mutually
profitable trading arrangement on a substantial scale between this country
and Norway."[113] Hull also knew that the German government had begun to
offer favourable allowances to Norway, bartering automobiles and other

goods for oil to be used in the manufacture of edible margarine and other products.[114] Schupp protested in vain, as did two American-registry firms operating pelagic factory ships — the Western Operating Company (*Ulysses*) and the American Whaling Company (*Frango*).[115]

Schupp's whaling activities were further restricted under the terms of a second major whaling agreement, adopted in 1937 at a London conference. Although signed by fewer nations than had participated at Geneva six years before,[116] the International Agreement for the Regulation of Whaling applied more stringent controls. Statistics gathered since 1931 showed a fivefold increase in the numbers of whales killed, without a proportionate increase in the amount of product marketed. This suggested that much of the harvest consisted of immature whales.[117] The new agreement corrected the appropriate defects in the Geneva convention; factory-ship operations were specifically controlled for the first time, geographic hunting areas were established, and some species prohibitions were placed in effect. The taking of baleen whales in certain regions of the Pacific Ocean was specifically outlawed,[118] although factory ships were allowed in the western North Pacific — a concession to the Japanese, whose infant pelagic fleet was just then beginning operations.

Land stations could not be worked more than six months of any calendar year, and these six months had to be counted consecutively. A new provision required the completion of processing within thirty-six hours of the kill. And as before, cows with calves and suckling young were protected[119] and minimum size limits were imposed. Gray whales, hunted to virtual extinction in the North Pacific by Japanese whale hunters and American sail whalemen were included with right whales on the list of totally protected species. Other provisions from the Geneva convention were retained, including the proscription against bonuses based on numbers of whales caught. Importantly, the statistics of whaling were now to be undertaken by impartial inspectors — paid by the contracting governments rather than by the whaling companies — who would forward catch statistics to the International Bureau of Whaling Statistics in Sandefjord, Norway.

The mandate to examine carcasses brought biologists and other scientists into close contact with the whale hunt, and at least some government whaling inspectors pursued research interests while complying with the regulations. At Port Hobron, for example, a forty-eight-foot sei whale was brought in "solely for the purpose of inspection and study by the Whaling Inspector"; its processing was delayed "in order to give the inspector a reasonable time to photograph and take measurements of the speciman."[120] Flensing crews in the Pacific Northwest grew accustomed to doctors and university professors prowling the flensing decks. Roy Chapman Andrews's visit to Sechart in 1908 had paved the way for such interested observers;[121] by the 1930s scientists increasingly focused on specific and sophisticated data which might be of medical value to humankind.

The most regular of these visitors in William Schupp's domain was Dr. Eugene Maximilian Karl Geiling, research professor from the School of Pharmacology at Johns Hopkins University in Baltimore, and later of the Pharmacology Department at the University of Chicago. Geiling spent several summers at Rose Harbour, removing pituitary glands from whales in hopes of discovering the functions of that organ. Pituitary research had been impeded by the small size of the gland in humans and laboratory animals, but the whale pituitary proved to be large and easy to handle.[122] Geiling brought in several assistants and scholars engaged in similar research,[123] the most valuable of whom, from the perspective of whaling history, was his departmental aide, Lewis L. Robbins.

Robbins arrived at Rose Harbour in 1936, naive, untravelled, but with an artful eye and a real concern for the scene and its people. He brought with him a fine, large box-camera, with which he set about to record both. Through two summers he conducted scientific investigations, shipped biological specimens, and served as an *ad hoc* medical attendant to the station crew; he also created a photographic and written record of many otherwise undocumented aspects of life and work at a British Columbia whaling station.

Rose Harbour was a little-spoiled paradise, far removed from the mounting horrors of "civilization." In letters home the young Robbins wrote glowingly of Kunghit Island as a "beautiful setting . . . almost tropical in its wild profusion."[124] Given a small cottage all to himself, which he dubbed "Robbins's Roost," he found himself surrounded by giant trees, ferns, flowers, a garden, and a doorway bower made of the hanging vines of a honeysuckle. From Robbins's Roost he conveyed his enthusiasm for the romantic delights of the Pacific Northwestern frontier as seen by an unspoiled heart. Even while dissecting the giant carcasses on the slipway, he carefully memorized their beauty so he might convey it onto paper; he found his artistry capable of describing, in the manner of a painter, the intense palette of the internal whale:

> All the organs were a beautiful pale blue. The peritoneal lining shone gleaming white reflecting the sun's rays with a brilliant glare. The kidneys lay against the reddened back from which hung beautiful droplets of golden fat. The liver, smooth and glistening, was streamered with maroon. . . . The lungs looked like two long white sacs. But the greatest color combination was around the stomach. This was overlaid with a beautiful iridescent thin membrane, the omentum, which caught the light and returned it in many delicate hues from pink to indigo. The stomach itself was powder blue.[125]

Rose Harbour carpenter Iwabuchi was well known as a scrimshander. He was photographed by Lewis Robbins holding a pair of large teeth from a sperm whale that was brought to the station.
Photograph by Lewis L. Robbins, Kendall Whaling Museum, Sharon, Massachusetts

His medical knowledge was occasionally put to the test, with an unexpected and most satisfactory effect. After a chain broke and trapped the carpenter Iwabuchi inside a whale's jaws, Robbins treated his wounds and received in return a pipe, which Iwabuchi had scrimshawed from a single sperm whale's tooth.[126] Even though the long voyages of sailing whaleships were done, scrimshaw remained the folk art of the whalemen. Ivory cribbage boards made of two butted sperm-whale teeth were a popular item at the British Columbia whaling stations; pipes, bookends and even miniature harpoons were occasionally fashioned.[127] Robbins received presents of two canes carved from the jaw bones of sperm whales, one from Iwabuchi and another from "Russian Louie," an older deckhand in one of the chaser boats.[128] He was fascinated by the artistry of these men and touched by their kindness. After receiving the second cane, he wrote: "I shall have plenty of whale bone for support when I'm old and feeble. Can't hurt either man's feelings by turning down these gifts which I'm really very fond of."[129] When he returned to Chicago he brought many pieces of scrimshaw, including several cribbage boards, the pipe, a desk set, cigarette holders, ash trays, canes, napkin rings, lamp bases, and totem poles, the majority of which were created by Iwabuchi and "Russian Louie."[130]

After two summers at the whaling station, Robbins still retained his enthusiasm for the people and the land. But he also came to terms with the loneliness and displacement of the men who worked there season after season, far removed from the regular comforts of city life. "I have seen many fascinating things here," he wrote, "have performed an out-of-the-ordinary work; have conversed with and learned to know people who came to this isolated place from Norway, Sweden, Russia, Scotland, Ireland, England, Japan, China, and other distant places. My storehouse of memories is crowded with a kaleidoscope of impressions of vast beautiful scenery, of great leviathans whose homes are under the sea, of men whose homes are nowhere."[131]

Robbins must have seen also the advancing age and weakened condition of the chaser boats and the shoreside equipment, most of which had been in almost continuous summer service for a quarter-century. There was no denying that by 1937 the famous "colour fleet" had been through more than a lifetime of hard use. It was also beyond refute that the company had been less than meticulous in maintenance, adopting a necessarily penny-wise but pound-foolish approach during the grim days of the Depression. Since 1929 there had been little refitting at all; station manager Alf Garcin was supposedly told to fix one steamer each year and not look too closely at the others.[132] Bailing-wire repairs were the order of the day, tacitly approved to avoid an expensive visit to the Victoria Machinery Depot. Of course, purchase of replacement equipment was completely out of the question.

So Alf Garcin stopped looking too closely. The chief engineer in the freighter *Gray* later recalled that the boilers in the chaser boats were "just like a piece of lace," and a worker is said to have driven a hammer completely through the weakened keelson in one of them.[133] Metal fatigue and worn-out equipment began to cost dearly: first in lost whales, then in lost hunting days, and finally in crew injuries. "Shot one Finback," the gunner of the *White* reported, "line parted lost whale and 180 Fathoms main line, one Foregoer and a harpoon [The main line] rotton, it parted with no strain wathever."[134]

Occasionally the circumstances proved more severe than a parted line and a lost harpoon. Most damage occurred while fast to running whales, but the vessels were sometimes bent after a whale was killed, by the impact of the carcass against the hull during towing.[135] By the late 1930s the ordinary force of a whale's body slapping the side of the chaser boat threatened to open seams and sink it. On 30 May 1935 the logbook of the *Blue* notes: "Shot Sperm at 11 a.m. an he came up and bumped the boat very hard on the Starboard side doing damage to plating and framing starting rivets and [damaging propeller?]."[136] The next morning the *Blue* was found to be leaking, and extensive repairs were required.

Minor repairs had once been effected by canting the hull to one side, either by shifting coal or by beaching the vessel over a low tide; there was a soft bank at Rose Harbour admirably suited for this use.[137] But by the 1930s it was considered risky to rest the chaser's weight upon its weakened hull plates. Only the cost-saving argument mitigated in its favour; the president of the Consolidated Whaling Corporation might turn a blind eye to a dented hull and popped rivets, but he could hardly ignore the arrival of a crippled chaser boat at the VMD yard in Victoria. And the Victoria Machinery Depot's repairs were not cheap; an estimate given prior to the 1934 whaling season quoted $1,580 to replace five riveted plates in the hull of the *Blue* and $400 to make patches and replace some riveting under the stern of the *Black*.[138]

The condition of the boats caused friction among their crews, and plenty of stubbornness, too. In June 1934, for example, the propeller of the *Brown* was damaged and its shaft bent in a collision with a running whale. The crew managed to kill the animal and turned homeward with two giant bodies lashed alongside. Skipper Knut Halvorsen did not have full control of the chaser, and at 7:30 AM he condescended to accept a tow from the *White*. But when the two vessels reached the narrow tidal channel leading to Rose Harbour, Halvorsen dropped the tow and, as one observer said, "ruddered it in" to the float where carcasses awaited their turn on the haul-up slip.[139] He wanted to bring his whales to the float under his own power, perhaps to salve his ego but probably to retain for his crew all the bonus money due on them.

The *Brown* was beached on the next tide and examined while it lay ashore very like a dead whale on the flensing slip. But the damage was too great to patch; two blades were broken off the propeller and there was no replacement at the station.[140] Manager Garcin decided to have parts shipped from Victoria by the next trip of the *Gray*, but Halvorsen threatened to quit unless his boat was repaired at once.[141] So the *Brown* was duly towed south to the VMD, and its skipper no doubt received some kind of compensation for his enforced idleness among the midsummer delights of Victoria.

The potential for disaster grew with each succeeding year. When the *Brown* was once again damaged while towing whales in 1941, skipper Louis Larsen and mate John Fransen entered a condition report into the logbook and *signed* it, as if the entry were precious testimony: "Wall of starboard side bunker now in (dangerous) danger of collapsing," one of them wrote. "The ship in present condition considered hardly fit to carry on whaling."[142]

The *Brown* was again inspected, temporarily patched with cement,[143] and sent down to Victoria. Behind it on a long towrope followed the *Green*, which had been disabled by a boiler breakdown.[144] The trip from Naden Harbour consumed four days of the whaling season, and the *Brown* was subsequently laid up at Victoria for ten more days.

Parts gave way. Chief Engineer H. D. Hornibrook entered this succinct calamity into his engine room logbook on 29 May 1935: "Shot a sulphur bottom   when pulling it in the block on the mast came down and went through the deck. we lost the whale."[145]

Boiler problems were commonplace, and other engine failures were duly noted. Bilge pumps were inadequately maintained and unable to cope with rampant leakage: "We started at 3.45 AM. It is blowing very hard and a heavy sea   the bilge pumps are blocked up with dirt and we are having a hard time to keep the water out   the coal has shifted in the bunkers   we have a heavy list to port   I am sorry I came on this boat."[146]

Boiler tubes bursting, boiler walls shot through like lace, ancient rivets giving up to the incessant pounding of a fifty-ton corpse; these were the catastrophic possibilities. And then the cannons began to go awry.

> During the past few years three of the whalers have broken their gun forks in the following order: — "Brown," "Grant," and "White." In each case the gunners did not meet with any accident. H. Anderson, gunner of the "White" was in yesterday and he remarked that possibly the guns may be crystallized as well as the forks. . . . In 1920 it cost $1800.00 for a gun only, landed here from Norway; certainly since then the price has been considerably reduced.[147]

Whatever hints were dropped about replacing the vintage muzzle-loaders, Schupp would not take them; whatever the price, he could not pay it. So the aged guns began to exact their toll. "Mad Harry" Anderson fired at a whale during the 1934 season and the swivel mounting broke in two, sending the breech block and barrel crashing down on top of him. Unbelievably, he was extracted without severe injury, and the crew even managed to hang on to the wounded finback. A spare gun fork was installed at the station, and the *White* put back to sea the following evening, barely twenty-four hours after the incident.

Anderson's accident was a portent. One month later, Harold B. (Kris) Kristensen, gunner in the *Westport*, missed a blue whale when the harpoon fell short of its mark. Thinking the gunpowder at fault, he increased the charge and fired again. At this second shot the trunion fork and gun barrel snapped off flush with the top of the pedestal. The gun fell on Kristensen's feet and knocked him backwards so that his head struck the deck. The blow killed him instantly.[148] The *Westport*'s crew hastily returned to Akutan, arriving there at 4 AM to everyone's surprise, since no one had ever before witnessed the death of *skytteren*.

There were other deaths that year of 1934. Fireman Selover of the *Unimak* died while preparing the chaser boat for the start of the season. Chief Engineer O. F. Curry died of internal bleeding aboard the *Tanginak* while at

The chaser boat *Westport*, here fitted with a Bofors-type harpoon-cannon of Swedish design. In 1934 a crystallized mounting gave way and sent the vessel's cannon to the deck, killing the gunner.
Museum of History and Industry, Seattle, Washington

sea on 27 August.[149] Perhaps these, too, were portents; an engineer dying, a gun collapsing, a man killed just when the whaling seemed to have survived the doldrums of the Depression.

Two years later the *Westport* was lost altogether. For all the years of plying in and out of the rocky northwest coastline, finally with chaser boats barely solid enough to float, there had never been a shipwreck. Such bountiful providence ran out on 14 September 1936. Captain Schroder had just left the Akutan station to cruise in the Bering Sea. Conditions were poor, the darkness being compounded by fresh easterly winds, mist, and heavy seas.[150] The *Westport* was running along Reef Bight on the western side of Akutan Island and went aground before anything could be done to prevent it. None of the twelve men was lost—the U.S. Coast Guard cutter *Daphne* took them off and returned them to Akutan—but the chaser, valued at $59,000, was a total wreck. Without the *Westport* the men were supernumerary at the station, so the U.S.C.G. *Chelan* took them on to Seattle.

Local seamen's unions thought it a shame that sailors had to risk their lives in such old, rotting hulls.[151] As the voice of the unionized sailor became more voluble, a cry was raised for higher wages and improved conditions in the whaling fleet. Though difficult to document, William Schupp seems to have

A sorry scene. The *Westport* went aground on Akutan Island in 1936 and broke up in the surf. The crew were taken off but the vessel was a total loss.
William S. Lagen

faced pressure from various unions, including the Inland Boatman's Union and the Seafarers Industrial Union of Victoria, whose representatives demanded better wages and repaired vessels.[152]

But Schupp was still unable to maintain the whaling at full operational status. In 1939 only Akutan operated. No whaling was conducted anywhere in British Columbia that year, though a Japanese fleet made a good catch almost within sight of the coast; the halibut schooners *Kanaga* and *Liahona* reported the factory ship and four chaser boats just seventy miles west of the Queen Charlotte Islands.[153] Canadians were upset by Japanese whaling efforts so close to shore, but there were no local competitors.

Schupp tried again in British Columbia in 1940 with the *Blue*, *White*, and *William Grant*[154] and realized a fair return, so in 1941 he sent station crews to both Rose Harbour and Naden Harbour and used all six Canadian chaser boats.[155] War once again encouraged the production of whale meat and other products, but unlike in World War I, armed conflict now threatened to reach the very doorstep of the whaling stations. Not long after the Japanese attack on Pearl Harbor, the American government commandeered the Alaskan chaser boats and took Akutan for a strategic base. Before the end of the war the buildings were turned over to the Soviet Army on a lend-lease arrangement; the Russians allegedly looted the station of all its equipment before departing. Schupp and his descendants carried on an extended suit against the American government, and though they recovered ownership of the land, they were never repaid for losses incurred during the military occupation of Akutan Island.

Schupp's Canadian operation carried on in 1942, but only at Rose Harbour. Its activities were circumscribed by wartime regulations. Prohibitions against plain-language transmission of weather reports or the location of vessels impeded the hunt considerably. The aged boats, worn-out equipment, and a significant wage increase[156] unbalanced the operation, and the final blow was struck when the Canadian government relocated all residents of Japanese origin away from the West Coast defense zone. The internment of Japanese-Canadians robbed the Consolidated Whaling Corporation of most of its skilled station hands. The addition of men from Newfoundland and untrained local help could not make up for the loss, since there were few qualified men available for whaling work; most of them were already at war.

Schupp accepted an offer to lease one of Consolidated's chaser boats to Maritime Industries of San Francisco for use at the Fields Landing station on Humboldt Bay; the *White* was accordingly mustered out and sent south to work alongside that company's own chasers, the *Pedro Costa* and the *Hawk II*.[157] One of the lessees may have been Greek shipping entrepreneur Aristotle Onassis, who reportedly invested $15,000 in the California shore station. He wanted sperm oil to sell in the wartime market, but apparently found a better profit in providing whale meat to mink ranchers. The station also sold livers to the Borden Company for production of Vitamin A.[158]

The California season opened 1 April 1943 and closed 31 October, but during that entire time only twenty-nine whales were caught. The *White* returned to Victoria and did not hunt in the south again. In British Columbia that year the *Blue* and the *Brown* conducted all the whaling, and when they returned to the Point Ellice dock at the end of season their lines were turned around the bollards, engines rung to a stop, and crews sent home for good.

The stalwart freighter *Gray* had already departed the scene. Since World War I the whaling company had received regular offers for their stout coastwise cargo-carrier. Each offer was refused in turn until the ship became too uneconomical to run. In 1940 Frank Waterhouse and Company chartered the *Gray*, and in September 1942 Schupp sold it to them outright. Thirty months later, while flying the flag of the affiliated Union Steamship Company, the freighter met a huge sea off Nahwitti Bar. A rogue wave broke the *Gray*'s back. Captain Albert Cyr said that two-thirds of the 183-foot hull poised itself above water when the crest of the wave passed under amidships.

After that accident, the *Gray* remained at dock. Not long afterward its fittings were stripped and the bare hull joined the S.S. *Chilco* and other veteran British Columbia vessels to form the breakwater for the Iron River Logging Company's harbour at Oyster Bay, just south of Campbell River.

Fair go, everyone said, for the work of whaling was done.

8

IN THE MEAT BUSINESS

*Nous sommes maintenant bien loin de Melville et de
la marine en bois, et chaque année voit apparaître
quelque dispositif nouveau, quelque perfectionnement
meurtrier, ou quelque utilisation diabolique d'une partie
de ce cadavre démesuré.*

PAUL BUDKER
BALEINES ET BALEINIERS (1957)[1]

Near the end of the war, visitors began to poke around the Point Ellice dock
in Victoria Harbour to have a look at the five whaling steamers languishing
there. Sprott Balcom's "colour fleet" was thirty-five years old, every one
thin-walled in disrepair, used up by almost four decades of chasing, playing,
and towing whales in all kinds of weather. One veteran whaleman, asked by
one of those visitors to go to sea in them again, reportedly invoked Christ's
name in vain and said no, he didn't want any part of that.[2] Knowledgable observers concluded, sight unseen, that the idled machinery at the Rose Harbour and Naden Harbour whaling stations had likewise disintegrated.

Still, Marc Lagen made plans to put his father-in-law's company to sea
again, at least in Canadian waters. The Alaskan situation seemed hopeless,
now that Port Hobron was rusty with disuse and the Akutan station had been
ransacked. But Lagen thought the Canadian chaser boats might be made
seaworthy if pressing debts against the Consolidated Whaling Corporation
could be deferred. He contacted the major creditor, H. S. Hammill of the
Victoria Machinery Depot Company,[3] who not only agreed to suspend his
call for repayment but also offered a small amount of money toward the
restart, provided the company could be made operational by the summer of
1946. As a guarantee, Hammill required that a local management committee be named, its members to be approved by the whaling company's

creditors.[4] He also insisted that the Victoria Machinery Depot receive all contracts for the rebuilding of the aged boats. Consolidated's board of directors formed the management committee at a special meeting in September 1945; Hammill was nominated to it, and Marc Lagen was selected to manage the 1946 whaling operations.[5] Ottawa was approached about a loan to refit the chaser boats and one of the whaling stations, and a $200,000 loan application was filed in the United States with the Reconstruction Finance Corporation,[6] probably to repair the American Pacific Whaling Company's chaser boats.

Schupp ordered a full inspection of the "colour fleet," but the results were not impressive. He told one potential investor that the cost of placing them in operation would be "substantial."[7] Three of the American chaser boats, the *Aberdeen*, *Moran*, and *Paterson* were found to be in better condition—though the cost of preparing them for whaling was expected to reach nearly $25,000 —so Schupp petitioned Ottawa for dispensation to operate them in Canadian waters—an unconventional request rarely granted except *in extremis*. The Customs inspector at Victoria confirmed Schupp's contention that the Canadian whaling vessels were beyond repair.[8]

But before any whaling could be organized, Marc Lagen died.[9] The Reconstruction Finance Corporation rejected Schupp's loan application and subsequently turned down a second application as well.[10] Ottawa remained mute, except to deny permission to operate the American-registered chaser boats in Canadian waters. The combined blows seemed to take the wind out of Consolidated's sails. Nevertheless, Schupp scavenged behind the scenes for financial support, and the Canadian Bank of Commerce finally agreed to lend startup money, provided that one-half of any profit was earmarked for repayment of its loan.[11]

In addition to the money needed to refit the vessels, an additional $21,000 was required to buy supplies and pay crew advances, and $25,000 more was the price for retrieving the Rose Harbour station from verdant oblivion.[12] In Victoria, a rumour circulated that the whaling fleet was being spruced up for the coming season, while others whispered that certain investors, unnamed, were looking for money.[13] But nothing happened. No repairs were undertaken by the VMD; no whalemen were called back to work. On 13 November 1946 the Consolidated Whaling Corporation Limited entered into bankruptcy. Nineteen days later, G. O. Cumpston, senior partner of the accounting firm of Touche, Ross—and a member of Consolidated's management committee—was named trustee at a creditors' meeting. On that day the Schupp era of whaling in the Pacific Northwest came to a definitive close.[14]

Ten days into his trusteeship, Cumpston did the very thing that Sprott Balcom had conspired against during the early years of the Pacific Whaling

Company; he sold Schupp's assets to Lars Christensen of Sandefjord, Norway. Christensen, now in his early sixties, had long ago found his way blocked in British Columbia. Now, almost too late, he gained access to western Canada, and immediately requested a ten-day option in order to approach Ottawa for the necessary "concessions."[15] Christensen told the government that he would bring skilled Norwegians to rebuild the station at Rose Harbour and import two chaser boats of the 800-horsepower Antarctic type, vintage 1928–31, to replace the local fleet.[16]

But the licence was not forthcoming, and Christensen let his option lapse. In February 1947 the Consolidated Whaling Corporation dock and office at Point Ellice were sold to the Victoria Machinery Depot Company. The VMD bought the land only; it was directly adjacent to their own and could be used for expansion. They had no interest in, and did not buy, the chaser boats or the whaling equipment belonging to the defunct firm.[17]

On the last day of March the five chaser boats of the "colour fleet" were sold at auction. A. J. Maynard and A. R. Roberts of Victoria presided over the sale, knowing beforehand that the only bids would come from two competing scrap-metal dealers. When the last gavel had fallen, two of the steamers belonged to Capital Iron and Metals Limited; Morris Greene, Capital's president, bought the *Black* with a bid of $260, and a slightly higher offer of $285 won him the *Blue*. He might have bought all of them, except that he seems to have set his bid limit at $300 each. M. H. Kramer, from the Northern Junk Company, got the *Brown* and the *White*, but he had to pay $305 and $315, respectively.[18]

Miraculously, one of the five hulls escaped the wreckers' torches. The dented old *Green*, survivor of many a thumping from the whales and many a mechanical failure—the very vessel on which Harry Balcom had met his death in 1918—sold for exactly $300 to a fisherman and former sealer, Max Lohbrunner, in the name of his Deepsea Fish Company. He said he would refit it for company use.[19] Instead, the *Green* became an unfashionable relic near the foot of Bay Street; Lohbrunner moved his belongings aboard and brought another derelict fishing boat alongside. In his declining years he lived an eccentric life on and off the vessels, turning aside repeated requests to remove their rotten hulls from scenic Victoria Harbour. Eventually the old chaser boat let go its hull fastenings—with a little quasi-official encouragement—and settled stern down in the mud. The Navy and some volunteers used explosives to remove the harpoon-cannon, which remained visible just above the waterline, and it came in due course into the collection of the Maritime Museum of British Columbia in Victoria. The fragments of the hulk sank well out of sight of the Inner Harbour tourist trade.

At the time of the auction, the original British Columbian chaser boat was still in existence. Sprott Balcom's *Orion* had served a long term as a fishing

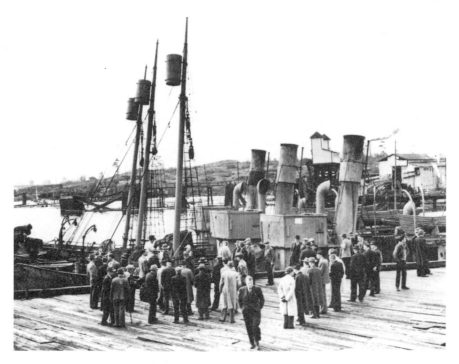

Faced with mounting debts and worn-out vessels, the Consolidated Whaling Corporation sold the venerable "colour fleet" at auction in 1947. Four of the five hulls were sold to scrap dealers.
Vancouver Maritime Museum

boat, and later as a fireboat for the city of Vancouver. First under its original name, and ultimately as the *Pluvius*, the old Norwegian outlived—but just barely—its newer brethren of the "colour fleet." The *Sebastian*, sold out of the whaling service in 1916, survived also, continuing its work as a towboat under the name *Saanich*. Strangely, its contribution to whaling was not yet complete.[20]

Almost immediately after the sale, the press heralded the formation of a totally new whaling company, incorporated in Vancouver with a capital of $500,000.[21] Management would be shared among British Columbia Packers; Nelson Brothers Fisheries; and W. F. Gibson & Sons—the Gibson Brothers—who held major interests in coastal logging.[22] This new Western Whaling Company was less a partnership than it seemed; Nelson Brothers was quietly affiliated with B.C. Packers, but this did not become public knowledge until a decade later, during an infamous combines scandal.

The Western Whaling Company was described as the "first all-Canadian-owned whaling company" in British Columbia.[23] Clarke Gibson was named

Near the end of a long working career as chaser boat, fishing boat, and private yacht, the *Orion* served briefly as a fireboat for the city of Vancouver, first under its original name and finally as the *Pluvius*.
Vancouver Maritime Museum

president; B.C. Packers' Robert E. Walker, vice-president; and Douglas Souter of B.C. Packers, general manager.[24] The consortium originally intended to rebuild the Rose Harbour whaling station, which would arise as a phoenix—or perhaps more like a moss-enshrouded Thunderbird—from the flaking tin-roofed sheds of the Consolidated Whaling Corporation.[25] Gordon Gibson took a work crew to Kunghit Island to begin reconstruction of the docks and buildings but they abandoned the effort when it became known that Crown Assets Disposal was preparing to sell the Royal Canadian Air Force seaplane station at Coal Harbour, on Vancouver Island near the head of Quatsino Sound. A third Gibson, brother John—M.P. for Comox-Alberni —made arrangements for Western Whaling to obtain the land and buildings,[26] and the Rose Harbour plan was happily abandoned.

The enclosed aircraft hangars at Coal Harbour offered significant advantages. They were large enough to accommodate a fully-equipped reduction plant, and the concrete ramp once used to launch seaplanes made a perfect flensing slipway. Living space was available in relatively new buildings nearby. A significant disadvantage was the great distance from the head of Quatsino Sound to the open Pacific, but this was overcome by contracting with outside parties to bring in carcasses from a convenient float established at Winter Harbour, near the mouth of the sound.[27]

Processing of the whales would follow precedent established by the earlier companies, but some new equipment was installed which promised improved yield. The main hangar was fitted with four Norwegian Kvaerner-type digesters of modern design, five Sharples "Super-D Canters," horizontal screen cookers, deep-bay cookers, a flame dryer, a "meal cyclone" meat press, four Sharples separators, a liver-oil plant, and a miscellany of settling tanks, blowdown tanks, and separating tanks. In the second hangar an evaporator plant was installed for the reduction of solubles, as well as a freezer for storing fresh meat. [28]

This intricate new equipment depended as always on the skill of the gunner and his crew, but the company owned no chaser boats. A motley squadron was assembled to begin the first whaling season in 1948; the largest of the flotilla, at 658 tons, was the *James Carruthers*, built in England in 1912. It had worked in the province as a salmon packer, halibut longliner, trawler, dory fisherman, and finally a log-boom towboat for the Gibson Brothers. [29] Now it was sent down to B.C. Marine Engineers and Shipbuilders for the addition of the harpoon-cannon and other conversions to make it into a whaling vessel. Louis Larsen took command. B.C. Packers contributed the *Nahmint*, a four-year-old San Francisco-built ex-minesweeper, under command of Johan "Big John" Bordewick. [30]

The third vessel, a ninety-three-foot towboat formerly used by the Coastal Towing Company, [31] must have given senior whalemen a sense of déjà vu, for it was none other than the Canadian North Pacific Fisheries chaser boat *Sebastian*, in its livery as the towboat *Saanich*. After a thirty-two-year hiatus it had returned to the whaling business. A fourth vessel, the *Tahsis Chief*, was equipped with the harpoon-cannon but was returned to Gibson's sawmill at Tahsis Island in disgrace after it failed sea trials. [32]

Ottawa granted the company its licence on 21 May 1948. [33] By that time Canada had subscribed to the terms of an important new whaling agreement, drafted between 20 November and 2 December 1946 at a meeting of the representatives of whaling nations in Washington, D.C. The International Convention for the Regulation of Whaling reinstated and amended the prewar accords, added new stipulations, and created a permanent regulatory body — the International Whaling Commission (IWC)—in which was vested "certain powers" [34] to recommend new regulations and amendments to the convention, as dictated by scientific findings.

The main duties of the new IWC were to gather catch statistics, designate protected species, set dates of whaling seasons, close certain waters to commercial whaling, limit total catches, define standards of measurement and specifications for catching equipment, and establish statistical and other record-keeping methods. The new convention also reinstated the use of independent observers, added sei whales to the species for which minimum lengths were specified, and permitted the taking of certain whales for research purposes. [35] Canada signed and ratified the new convention, and its regulations became mandatory for the new company.

It now became necessary to procure men for the chaser boats as well as for the shore stations. By 1948, union goals affected the ways in which maritime companies conducted their businesses, and the Western Whaling Company came to grips at once with this new power speaking for the working sailor, particularly the three-year-old United Fishermen and Allied Workers Union (UFAWU). The UFAWU had steadily collected members of small, often loosely knit fishermen's unions, and despite continuing allegations of Communist infiltration,[36] it successfully competed with the Seafarer's International Union to be the major spokes-organization for coastwise sailors throughout the postwar whaling era.[37]

UFAWU bargainers met at once with Western Whaling officers to work out a satisfactory wage and bonus agreement, establish a system of seniority in the seasonal rehiring of workers, and establish more prosaic benefits, such as the regular availability of clean bedding and towels. The bonus system grew more complex over the years and would form a major bargaining point during the annual renegotiation of union contracts. Western Whaling established a system based on whale "groups," organized by size and species, with the largest and most valuable whales earning the best premiums.[38] This system encouraged crews to seek out mature animals and minimized the random killing of immature and young whales. Because bonuses constituted an important perquisite over and above crew wages, the UFAWU argued constantly for increases, as well as for proportional payments of wages and bonuses to employees who became ill or were injured on the job.

Once a satisfactory agreement had been reached and men hired, the whaling began. The season opened 30 May 1948 and continued until 24 September; the chaser boats operated from Cape Cook to Cape Scott and killed approximately 182 whales—statistics vary—the majority of them humpbacks.[39] The short catch may be traced in part to green crews working together for the first time and in part to the incompleteness of the plant, which could not accommodate more than three whales per day. When more than three were brought to the float, chaser-boat crews were temporarily idled, just as they had been at Sechart in 1905–7. Even so, 5,679 barrels of oil were produced.[40]

The disappointing catch was also due, in part, to the makeshift chaser boats, which did not prove adequate to the task. After the first season the venerable *James Carruthers* and the reborn *Saanich* were dropped in favor of the Cummins-engined *Kimsquit*, a 104-foot ex-Navy vessel built in Nanaimo in 1944 and skippered by its owner, Jack Haan.[41] The *Tahsis Chief* was reconsidered in a better light, and this time it passed muster.

Another percentage of the shortfall may have stemmed from the awkward two-man leadership system; Western Whaling employed both a captain and a gunner—acting as mate—in each chaser. This division of power between two bosses led predictably to frequent and vociferous debates about who should have the say in matters of whaling. Most of the mates were Norwegians, but the captains and other hands might come from anywhere. The *Nahmint*, for example, set out for the whaling grounds in 1949 carrying skipper Einar Jensen, mate Erling Nelson and a complement of firemen, deckhands, and engineers which included two McLeods, Miskofski, Skog, Gonsalves, Holowaychuk, and Peterson.[42]

An immediate attempt was made to pack and ship edible meat. Protein shortages had again accompanied the deployment of troops during World War II, and when initial shipments of whale meat reached consumers in Victoria on 8 July 1948—at the fish counter of a downtown department store for thirty-nine cents per pound—the press rejoiced. Shoppers, they said, "stopped wailing about spiralling beef prices and started harpooning a 10,000-pound shipment of red whale steaks."[43] One reporter admitted he didn't care for whale meat, but suggested that one could probably "get used to it."[44]

Canned meat was also produced. Whale meat was cut into one-inch cubes and immersed for twenty minutes in boiling water containing 1.5 to 2 per cent soda ash. This process imparted "a more desirable flavour." The prepared cubes were then canned, 11.5 ounces of meat to 5.5 ounces of flavoured gravy in the "one-pound talls" familiar to every consumer of canned Pacific salmon.[45]

Before the close of 1948, William Schupp died at the age of eighty-two. His passing was marked by a volcanic eruption on Akutan Island, the first in more than a decade.[46] At a special meeting of the board of directors of the American Pacific Whaling Company on 14 July, it was resolved to sell the Akutan and Port Hobron properties to an anonymous bidder who tendered $500 for both. The money from the sale would be applied to the final debt owed to the Reconstruction Finance Corporation, and the RFC was asked to consider and approve also the sale of the *Aberdeen*, *Kodiak*, *Moran*, *Paterson*, and *Tanginak* to one R. D. Egge, who had bid $4,550 for the lot.[47] Schupp's attorney, W. L. Grill, announced that the company "would never engage in whaling operations again."[48]

The tendered bid for the station properties was accepted, and it was perhaps "tender" in several ways; the buyer was none other than Grill himself, who promptly returned title to Marc Lagen's widow, Emily. Schupp's grandson later described his action as "a bit of 'Beau Geste'";[49] Grill had grown fond of Emily Lagen and eventually married her. After the sale of vessels and land, nothing was left but the legalities; six years later, on 12 February 1953, the moribund Consolidated Whaling Corporation of Victoria was dissolved.[50]

"First, cut your whale into dainty pieces."

Whale meat provided a substitute for beef and pork during protein shortages in both world wars. Cartoonists and advertising copywriters tried every means to convince Americans and Canadians of its suitability as human food, but it was not generally popular.
Robbins Collection, Kendall Whaling Museum, Sharon, Massachusetts

Meanwhile, the Gibson Brothers chafed violently at high labour costs and the necessity of reaching agreement with the UFAWU at the start of each whaling season. Before the second season began they sold their interest to B.C. Packers,[51] who also bought out — officially — the interests of the Nelson Brothers and placed their own J. M. Buchanan as president of the whaling company.[52] New wage and bonus negotiations were settled with the union, and the men returned to work on the 1948 wage scale. The catch increased to 250 whales, but a glut of vegetable oil in the world market severely inhibited the company's ability to sell the oil; B.C. Packers was forced to store it against a time of higher demand rather than letting it go at a loss in a "soft" market.[53]

The Western Whaling Company, now wholly owned by B.C. Packers, continued to upgrade its fleet, in 1950 adding a proper chaser boat, the 246-ton *Bouvet III*. The *Bouvet III* had seen considerable Antarctic service since its construction in Middlesboro, England in 1930,[54] and had most recently worked in Labrador. Most of the makeshift chasers were discarded, except for the *Nahmint* and the *Tahsis Chief*, the latter being reworked into a suitable whaling vessel. A small auxiliary, the *Speedmac*, may have been utilized for whaling during 1950, but it and another, the *Towmac*, were primarily used to bring whales from the Winter Harbour float to the whaling station.[55]

B.C. Packers contemplated new whaling gear as well. Manager Doug Souter travelled to Europe to examine the new electric harpoons which were being promoted as an efficient and humane alternative to the time-consuming struggles that usually accompanied killing with the grenade-point harpoons. The concept of catching whales with electricity was not new, even in British Columbia; a British patent on such a device had been filed as early as 1868,[56] and in 1904 Ottawa had received a petition on behalf of the Gulf of Georgia Fish and Curing Company, which intended to "catch whales and species of whales in the Gulf of Georgia and kill them by electricity."[57]

Modern electric harpoons were little more than an ordinary harpoon fired from a conventional harpoon-cannon. In place of gunpowder, however, the head carried the distal end of a conductor-wire connected to the generator aboard the chaser boat. Immediately the harpoon was secure in the blubber, a switch on the master control panel allowed a predetermined amount of electricity to pass into the whale. The first recorded success with the electric harpoon was the collection of four finback whales by German whalemen off Norway in 1929. Further experiments conducted by the *Sir James Clark Ross* fleet in Antarctica in 1932–33 resulted in the electrocution of perhaps 250 whales, and Dr. Albert Weber, who pioneered the German electric-harpoon effort, continued to assist with research until the collapse of Nazi Germany, when he destroyed his papers and committed suicide.[58]

Tests were resumed after the war, by RCA Victor in the United States, the General Electric Company Limited in England, and Elektrohval in Norway, among others. But the story got around that two gunners had been electrocuted in Antarctica,[59] and whether because of fear of dying or fear of unfamiliar new equipment, the concept earned no favor within the gunners' union. Some of the gunners claimed they could feel the residual charge in the harpoons,[60] but proponents speculated that they were feeling only concern over losing their specialized jobs to new technology.[61]

Sea trials pointed to some practical problems. If improperly placed, the electric harpoon might stun the whale without killing it, a dangerous and embarrassing situation for a boat's crew who discovered that the carcass lashed alongside was a live eighty-foot animal in no mood to go to the whaling station. Japanese investigators discovered also that the blubber sometimes served as an insulator, allowing the charge to trickle off into the sea. The flexible electrical wire often stretched and broke, and sometimes the insulation was stripped away by the shot and impact, thus rendering the electrical component useless. Even more important, electrocution broke down the blood tissue, congealing the solution and discolouring the meat. More edible meat was lost in this way than was ruined by shrapnel damage when using the conventional harpoons. Carbon dioxide gas-injection harpoons were also tried, but these, too, turned the meat black.[62] Finally, cost mitigated against the new harpoons. One electric model marketed in the 1960s by RCA Victor cost about $1,700 each. Six or seven were eventually provided to whalemen in eastern Canada in 1962, but they were not adopted.[63]

At Coal Harbour the "experimental ship" *Speedmac* was used to test "improved methods of hunting and despatching whales" — probably including the electric harpoon.[64] But even without electric and gas-injection harpoons, the Western Whaling Company processed 314 whales during 1950, which the government touted — erroneously — as the largest catch in the province since 1936.[65] Chaser-boat crews must have been pleased, especially since the bonus system had been augmented by a "whale pool," which operated in this way: the company kicked in money for each whale legally caught and processed, as much as $150 for a blue whale. At the end of the season, the accumulated cash was divided among the union crews on a share basis. Mates received twenty shares each; chief engineers eleven; second engineers nine; cooks, deckhands, and firemen each drew six.[66] The pool reduced the inevitable competition and jealousy, since each man received a share of the total catch of all boats proportionate to his rank. It was continued unchanged until 1956, when the mates were split off into a second pool all their own; management then paid into both pools on each whale caught.[67]

The total of whales taken each season and the other requisite statistics were forwarded to the International Bureau of Whaling Statistics in accordance with the terms of the whaling convention. Responsibility for this bookkeeping fell to the Pacific Biological Station in Nanaimo, and more specifically its young junior biologist Gordon Chesley Pike, who was assigned to monitor the first postwar season and continued to do so until the company ceased operations two decades later. Pike tallied the catch by size, sex, species, stomach contents, and other determinants according to the IWC mandate, but he also obtained material for biological study, documented the maturity of harpooned whales based on the development of their reproductive organs, and conducted parasitological examinations in hopes of unravelling some of the mysteries of the whales' migration into warmer waters.[68]

Several gray whales, protected since 1937, were taken by the Western Whaling Company on research permits. This specimen was brought in during April 1953.
Photo by G.C. Pike, courtesy E. Mitchell

The maturity study proved to be of particular significance; Pike discovered that about 50 per cent of the finback whales and 35 per cent of the humpbacks among the first catch of 1948 were immature. This meant either that the local waters had remained depleted since the prewar overhunting or else that the area was not regularly frequented by the larger and older whales.

Pike's experiments were usually conducted in co-operation with the whaling company; the tests often sought commercial as well as scientific validation. In one instance, he assisted the Pacific Fisheries Experimental Station in Vancouver in a study of the "technological utilization" of certain cetacean organs,[69] and ongoing tests were devised to determine the feasibility of recovering vitamins, insulin, and other pharmaceutical products from the flensed carcasses. One such experiment studied the extraction of insulin in commercial quantities from the islets of Langerhans tissue dispersed throughout the pancreas of the whale.[70] Other experiments also supported potential commercial results; the Pacific Fisheries Experimental Station tested a chlortetracycline-and-water solution which was believed to impede meat spoilage,[71] and one programme used locator balloons to mark harpooned whales. These were fastened to the whale by a small dart fired from a .303-calibre rifle.[72]

The IWC's decision to allow the taking of small numbers of endangered whales for research purposes came into play in British Columbia in 1951; ten of the totally protected gray whales were taken on such a permit, and an eleventh was also killed, apparently in error.[73] The study continued for several years; at least ten gray whales were taken for research during 1953.[74]

Two ex-Antarctic chaser boats were procured for the 1951 season, and the plant was expanded by 25 per cent to handle the anticipated catch.[75] One vessel was the *Polar V*, a 278-ton, 119-foot Norwegian chaser which had been constructed in 1931 by Nylands in Oslo—the same firm that built Svend Foyn's *Spes et Fides* and Ludwig Rissmüller's *St. Lawrence*. Western Whaling purchased the *Polar V* directly from Antarctic service, rerouting it from Cape Town, South Africa just in time for the season in British Columbia.

The other purchase, the *Globe VII*, seemed an entire generation newer. A 251-ton Antarctic type built in Moss, Germany in 1935, it represented Western Whaling's first step toward obtaining the most modern chaser boats.[76] Both vessels were sparsely appointed, with few amenities added since the time of the *Orion* and the *St. Lawrence* and little improvement in their notorious sea character. Of course, life at sea was still preferable to the noxious odours of the shore station, but sometimes only barely:

> It was the first vessel I had ever been on that seemed designed for discomfort. The heat in the galley turned the butter liquid, and I found it hard to enjoy even a dry cracker while sweat ran down my face. Sleeping quarters for the crew were located in the dark and fusty bowels of the ship where my bunk pressed so closely to the underside of the deck I brushed it with my shoulder each time I turned over. Except for the overheated galley, there was no place at all to sit down; even the lavatory was a damp cubicle of streaked iron whose door clanged to and fro with each roll of the ship.[77]

In 1954 the company prepared a sixth chaser boat, the *Lavallee*. Built at Tacoma, Washington in 1942 as a navy YMS-class auxiliary minesweeper,[78] it was diesel-powered and fitted with sonar and steel cable for whaling. Both the diesel and the steel cable were unusual; gunners eschewed diesel engines, claiming that their noise scared whales away before the vessel could reach shooting range. But the Western Whaling management must have appreciated the *Lavallee* because it operated without the two firemen required in the steam-driven chasers, and statistics eventually proved it capable of competing against the silent steamers even though, aesthetically speaking, it lacked the handsome and rakish flared bows and low bulwarks of the chaser boats built to the job.

To assure a smooth continuity of operation, it was necessary for management to continue its bargaining sessions with the UFAWU, whose representatives regularly restated the need for improved living conditions and wage increases for both shoreside workers and boat crews. A flat bonus of

$100 was added to all other moneys earned by union men during the 1955 season. The following year this bonus was staggered according to total catch, and amounted to $100 for every eligible crew member on a total of 450–74 whales; $125 if 474–99 were caught, and so forth up to $225 for 575–99. The company offered one dollar for each whale exceeding 600.[79]

But there were other objections besides wages. The union claimed that management had reduced the number of men working ashore; now there was one employee moving meat from the grinder to the dryer where there had once been three. Now three worked the meat shed which once had been a four-man operation. Seven handled meat whereas nine had been at this job in preceding years.[80] Management responded with a bleak profit picture, but union representatives refused this justification.

Another source of dissatisfaction during 1956 and 1957 stemmed from a management decision to hire veteran Antarctic gunners and eliminate the two-man captain-gunner hierarchy on board the chaser boats. The availability of skilled, experienced gunners was completely fortuitous, owing almost entirely to the collapse of a large and controversial pelagic whaling fleet which had been sponsored by Greek shipping magnate Aristotle Onassis.

Following his fledgling whaling experience in California, Onassis decided to make a serious entry into the industry. Through the Olympic Whaling Company, founded with the assistance of American interests in New York and registered in Uruguay, Onassis acquired a second-hand T-2-class tanker, the *Herman F. Witton*.[81] Lengthened and rebuilt at the Howaldt-Kiel shipyards in Germany as the factory ship *Olympic Challenger*, Onassis then converted a fleet of navy corvettes into chaser boats and set out to hire the expertise necessary to make a success of whaling. He registered the vessels in Panama and sent them to the Antarctic for the whaling season of 1949–50, much to the dismay of Norwegian and British whaling firms, who had already agreed to divide a fixed IWC catch limit among their eighteen companies. The addition of the *Olympic Challenger* group meant a loss to Norway of 64,620 barrels of oil, at that time valued at £1,370,000.

Among some Norwegians the matter was more personal. Onassis had selected a German yard to outfit his ships, and the director of outfitting was a former administrator of Norwegian whaling under the Nazi regime. Lars Andersen, Onassis's whaling manager, was likewise suspect; he had belonged to the Norwegian Nazi Party and learned some of his trade in the German factory ship *Walter Rau*.[82] At the conclusion of World War II, Norway had passed a law prohibiting its citizens from whaling under any foreign flag. Violators were threatened with fines and loss of property,[83] perhaps even revocation of citizenship.[84] At the same time, the number of available jobs was limited, since much of the whaling equipment had been damaged or sunk during the war. Onassis created a class of outlaw whalemen by offering hundreds of jobs with the new fleet, jobs which the unemployed Norwegian seamen were almost entirely unable to refuse.[85]

Norway complained, alleging that the *Olympic Challenger* crews killed immature animals, out of season, and in geographically restricted areas. Onassis in turn accused Norwegian whaling managers of obstructing his operation "by all means in their power,"[86] by threatening to boycott German shipyards, by undercutting the price of oil agreed upon in advance of the season, and even by refusing to provide fuel to his fleet tanker *Aristophanes* when the tanker had been damaged and was allegedly in distress. Finally, he accused the Norwegian government of offering immunity to its outlawed whalemen —no property to be confiscated, no fines levied—in exchange for declarations that the *Olympic Challenger* had violated the several provisions of the IWC convention.[87]

The fleet continued to operate for three years, and in mid-1954 Onassis ordered it to the west coast of South America in advance of the Antarctic season. But by this time Panama had ratified the IWC convention, thus making Onassis's ships subject to IWC regulations, and three South American nations—Chile, Ecuador, and Peru—had formulated the so-called "Santiago Convention"[88] which established a 200-mile sovereign territory including rights to fishing and whaling. On 15 November Onassis's ships became the test case for the Santiago Convention. Peruvian warships and planes encountered the *Olympic Challenger* and its flotilla taking sperm whales within the protected zone and forced the ships into Callao at gunpoint.

Crews were charged with the capture of 2,500 to 3,000 whales in Peruvian territory, and the vessels were impounded until Onassis's insurers grudgingly paid a $3 million fine. The fleet steamed belatedly to the Antarctic whaling grounds, but Onassis chose to disband the operation. Most of the ships were quickly sold to the Japanese, including the *Olympic Challenger*, which was transferred to the Kyokuyo Hogei Company and renamed *Kyokuyo Maru No. 2.*[89]

It was then that the fate of Onassis's skilled whalemen hung in the balance. Those who did not want to work in Japan, or who had no offer there, may have believed that they could not or should not return to Norway. They had violated the 1945 Norwegian law as well as a strict, if unspoken, ethical promise to support Norway's whaling industry. Rumours circulated that the Norwegian government intended to strip away their citizenship if they returned.[90] A few of these men solved their immediate problems by accepting employment in British Columbia.

One of them was Arne Borgen. One of the most influential and skilled of the *Olympic Challenger* gunners, Borgen was asked to revamp the Canadian operation and substitute some of his deepwater colleagues for local men who

The older chaser boat *Bouvet III* (far right) and the converted fishing boat *Nahmint* were matched with the newer chaser boats *Polar 5* and *Globe VII* to form the Western Whaling Company catching fleet in the mid-1950s. A fifth chaser, the *Lavallee*, is not shown.
Vancouver Maritime Museum

had not proven satisfactory.[91] Borgen's reputation attracted skilled gunners from the corners of the earth; Sofus Haugen came from Chile, and Edward Karlsen travelled to British Columbia from South Africa. Others came also; Bjarne Andersen, Karl Bordal, Fyelling, and Dahle. All of these men were at Coal Harbour by 13 September 1956, the day when crew and boat changes went into effect.

The two-man system which pitted the captain against the gunner-mate was abolished, since "it was just like any other job with two bosses."[92] The captain-gunner was installed as sole commander. Borgen assumed the role of overall fleet captain in the *Polar V*. Karlsen took the *Bouvet III*, and Dahle the *Globe VII*; Fyelling went in the *Tahsis Chief*, and Bordal had the *Lavallee*.[93] Only one Canadian captain remained in place: McPhail, in the *Nahmint*. Members of the United Fishermen and Allied Workers Union were shaken by these changes. They decided to write to the Norwegian Whaling Association for information about Borgen, Karlsen, "and the Olmpic Challenger situation generally,"[94] and they passed a motion calling on Western Whaling to give hiring preference to former mates when selecting captain-gunners, rather than to Onassis's rebels. But when the 1957 season began, only Ed Holloway had been so promoted,[95] and he replaced McPhail, so the UFAWU made no real gain toward recognition of Canadian gunners.

Among the newcomers there was much skirmishing, as crews got acquainted with gunners and gunners got to know the boats. Plant Manager Hec Cowie transferred Andersen from the *Tahsis Chief* to the *Bouvet III*, and from there to the *Lavallee*. Bordal moved from the *Lavallee* to the *Globe VII*, and Karlsen—who had originally driven the *Lavallee*—went to the *Globe VII* and then to the *Tahsis Chief*. Haugen left the *Tahsis Chief* for the *Bouvet III*,[96] in an ever more complicated season of "musical boats" which seemed to disrupt everyone except Holloway in the *Nahmint*—none of the Norwegians wanted that ship—and Arne Borgen. As fleet captain, Borgen certainly had first choice of boats, and he did not move from the *Polar V*.

Some of their reputations were less than sanitary. Karlsen, a white South African, is said to have been called "Killer Karlsen." A ruthless hunter who displayed complete disdain for the terms of the IWC convention, legend holds that he was twice reprimanded in British Columbia for shooting lactating whales. "But they wouldn't fire him," one employee said later. "He was too good."[97]

But if there was ever a man capable of amassing unto himself all the folkloric reputation of the *skytter*, the "Iron Man" reputation of Willis Balcom or the zany insistence of "Mad Harry" Anderson, then that man was Arne Borgen. His prowess was unrivalled, and his accomplishments in the Antarctic had already become legend by the time he arrived at Coal Harbour. He was not a mean or shallow man, nor someone necessarily to be feared.

Arne Borgen was most often seen on the bridge, beret on his head and binoculars in hand. He hunted whales in the Maritimes after 1967, where this photograph was taken.
Edw. Mitchell

Most knew him as an "urbane, dark-slacked, beret-wearing professional."[98] In the prime of his life—he turned forty in 1957—Borgen retained a calm pragmatism; whaling was only that, killing whales. If he had any qualms, they may have concerned the nobility of providing whale meat to mink ranchers, but even in that he had to shrug his shoulders. "Nothing will stop a woman," he was quoted as saying, "who wants to have a fur around her neck."[99]

Arne Borgen would be remembered as a cool competitor, larger than life—a man who once celebrated his birthday by killing twelve sperm whales between noon and 9 PM, a feat locally accepted as a world record.[100] Yet his legend is suffused with some of the light-hearted humility and comic absurdity which breathes life into all Norwegian whaling heroes beginning with the great Svend Foyn. At Coal Harbour you might have heard that Arne Borgen was frightened only once, and that this happened on a clear day, un-

der a beautiful cloudless sky, on a sea broken only by the rising and spouting
of the great whales. Arne was at the wheel of the chaser boat, steaming full
ahead. He had already been at sea for some time, and, unknown to him, an
inexperienced engine-room crew had expended all the fuel in the central
tanks while ignoring the wing tanks mounted in the sides of the hull. Suddenly, whales spouted broad on the beam. Borgen threw the helm down, and
the weight in the wing tanks so unbalanced the ship that it heeled over at full
speed, pinning the skipper against the wheelhouse door, out of reach of either the wheel or the engine-room telegraph.

The men below, one of them a passenger, gripped the handiest solid object and turned to the porthole glass to find not the azure blue of sea and sky
but the ugly green of the undersea. Some of that green had begun to enter
the cabin and was sloshing about their ankles. Luckily, the mate had been
caught on the upward-facing deck. He climbed down into the wheelhouse,
centred the rudder, and rang down for full stop. As the chaser came to rest it
slowly righted itself. As usual in such stories, no one was hurt.

Borgen is said to have preferred another story, about running aground on
an Antarctic iceberg. But that time out of Coal Harbour, when his chaser
went on its beam ends at full speed, that, he used to say—at least others *said*
he said—was the only time he had ever been *scared*.[101]

He also became a victim of the wayward-harpoon legend. Whether told on
Willis Balcom or John Fransen or a great Antarctic gunner like Arne Borgen,
the premise was always the same: someone or something would accidentally
trigger the harpoon-cannon at a time when it was not aimed seaward. True or
not, the excited retelling of these near-catastrophies always exaggerated the
danger of the business while simultaneously making light of it. No one ever
suffered an injury, except to his pride. The version told on Arne Borgen is
this: his chaser boat was lying at the pier at Coal Harbour while the crew
readied it for the next trip out. A deckhand had test-fit a shell into the
breech of the harpoon-cannon and—as he must for the benefit of the story—
walked away from the gun on some emergency or other. Meanwhile, a second
crewman unknowingly came along to test a harpoon for straightness and
pushed it deep into the barrel of the cannon, while a third crewman sat
mindlessly under the muzzle, coiling the foregoer.

Arne Borgen happened along just then, and while inspecting the gun he
brushed against the trigger. The shell fired, and the harpoon sailed over the
head of the rope-coiler, across the dock and through the roof of the rendering
plant, slid across the ceiling beams, emerged through the opposite wall, and
came to rest just a few feet from the front window of the station manager's
cottage. Luckily, there was no powder in the grenade and the manager was
away; when he returned he found a pile of sawdust and wood chips on the
plant floor, and the story soon got out. Arne Borgen had some explaining to
do.[102]

The introduction of the Norwegian gunners did improve the catch. In 1957 a total of 635 whales was obtained, and this figure rose to 774 in 1958 and to 869 in 1959.[103] The 1957 catch was already the largest in the history of Western Whaling, and though it included immature animals, the annual report of the Biological Station at Nanaimo found no evidence that the stocks of whales were being overtaxed.[104] Gordon Pike pronounced the future of provincial whaling "optimistic," believing that adherence to IWC regulations and "conscientious" biological study would make it possible to recognize depletions before any serious damage could be done.

The increased catch of 1957 also encouraged the UFAWU to press home its suit. They had backed off on the issue of Norwegian gunners when it came to pass that none of the former captains was prepared to make an issue of the question.[105] But now the company seemed to be on solid ground, or so the statistics implied, and it was time to improve the lot of union brothers in the whaling business. B.C. Packers disagreed, claiming a slow world market for whale products; further, the company charged that union men were committing malpractices against the company. Cooks from the chaser boats, for example, were accused of buying supplies from chandlers who had not been approved by the company and paying too much for the goods.

In this instance management may have been correct. The excessive cost of food seems to have had less to do with the per-item price than with quantity purchased. One newly hired cook, visiting the ship chandler to place his first order for supplies, allegedly was asked: "How much of this do you want sent home?" The supplier had become accustomed to charging off the cook's personal account to the company as a kind of perquisite.[106]

Wages and hiring practices remained the core issues. Union representatives claimed that chaser boat crews worked longer hours and earned less pay than seamen in comparable maritime jobs elsewhere. Management also came under fire for continuing the traditional practice of hiring Newfoundlanders; the union argued that the "500 seamen out of work on this coast"[107] should receive preference in hiring. Discontent festered throughout the late 1950s and came to a head at a heated bargaining meeting in January 1959, when the company refused to increase wages beyond the 1958 rate and insisted that all contracts be signed by month's end in order to avoid delay in preparing the vessels for whaling. B.C. Packers threatened to close the Western Whaling Company unless the terms were immediately accepted. Now, more than ever before, the union and the whaling company were locked in a standoff; the union representatives did not know the true margin of profit, and management did not know how far they could forestall union demands.

The UFAWU representatives took the ultimatum to the membership for a vote. Members were told that the company might make good its threat to close Coal Harbour. On the other hand, this was not the first time the whaling management had threatened to quit the business. "We cannot tell,"

UFAWU representative Alex Gordon admitted, "whether the Company ultimatum is a bluff until it is tested, and past experience of the UFAWU certainly proves that testing is required."[108] A vote was hastily taken. Newfoundlanders—still at home—voted unanimously to reject the proposal to whale on the 1958 agreements, and the local membership agreed; twenty-four of twenty-nine union men in the chaser boats refused the company offer, and 71 per cent of the shoreside workers voted against it.[109]

Federal and provincial mediators stepped in to solve the dispute, but they were unable to break the deadlock and soon withdrew. The union then cut its wage demands by one-half and dropped some peripheral issues. Despite mediation efforts and concessions, the whaling officers remained obdurate.[110] In February 1959 Plant Manager Cowie drafted a letter to all personnel who had worked at Coal Harbour the previous year: "In view of the United Fishermen and Allied Workers' Union's refusal to accept the Company offer to commence a 1959 whaling operation on the basis of last season's contracts, the whaling operation has been discontinued. We have no other employment for you." The termination notice was prepared as a form letter; there was no other text.[111]

The mediators again stepped in, this time proposing a two-year contract which increased boat crew wages by twenty dollars per man per month, with an additional fifteen-dollar increase to be added monthly after 1 April 1960. Station crews would return to work on 1958 wages, but would be guaranteed a 4 per cent increase on 1 April 1960.[112] Both union and management accepted these terms, but almost immediately the company announced its intention to bring a seventh chaser boat from Newfoundland, its crew to share in the whale pool and the various bonuses. The vessel would be skippered by Johan Borgen—Arne's father—one of the senior gunners of the much-reduced Newfoundland whaling fleet.[113] UFAWU representatives rose in objection. The added vessel would cut into crew earnings by further subdividing the pools, and the employment of Johan Borgen revived the Newfoundland issue since the union again had to ask why local men were not given preference.[114] The proposal seems to have been withdrawn.

The catch of 1959 was topped by an eighty-three-foot blue whale, and included 369 finbacks, 259 sperm, 185 sei, 28 blue, and 27 humpback whales. Even from these 868 whales, the margin of profit proved to be very small. Whale meat had fallen to $1.20 per unit in the world market, where it had once obtained $2.10; sperm oil fetched only 7.6 cents per pound in London, the lowest price since World War II.[115] In addition, the chaser boats were reportedly in need of major overhaul and supply and wage costs remained high. When the UFAWU again pressed suit for improved conditions and pay for the coming 1960 season, B.C. Packers' President, J. M. Buchanan, announced that there would be no whaling. The union protested,

Shore whaling was revived briefly at the mouth of the Columbia River about 1960. Bioproducts, Incorporated, converted two fishing boats to chasers and brought several humpback and sperm whales to the flensing platform at Warrenton, Oregon. The spectacle was sufficiently unusual to attract a crowd of bystanders.
Richard Carruthers, Bioproducts, Inc.

but to no avail; management determined that the market for the product was unsatisfactory.

In 1960 shore whaling on the west coast of North America was conducted from two stations located at Point San Pablo inside San Francisco Bay and by a marginal operation near the mouth of the Columbia River at Warrenton, Oregon. The Warrenton firm, Bioproducts, Incorporated, outfitted the sixty year-old dragger *Tom and Al* on behalf of the Oregon Fur Producers Co-operative to catch whales for mink food and for the manufacture of protein meal, fertilizer, and vitamins. A former U.S. Navy gunner supervised the 90-millimetre harpoon-cannon on the bow of the vessel; the half-day or day-long cruises were conducted along the Washington and Oregon coastline. A second small fishing vessel, the *Sheila*, was also fitted with a harpoon-cannon. A separate organization, established by Bioproducts but called "Hvalfangst-Oregon" ("Whaling-Oregon"), landed and flensed the carcasses. Whalemen were paid by the size of the whale; eighteen dollars per foot for sperm whales, twenty-two dollars per foot for finbacks.[116] The Bioproducts venture lasted just a few years, from 1960 to about 1965, with minimal result. Tønnessen and Johnsen have suggested that only thirteen whales were taken during this period.[117]

Paradoxically, it was just when British Columbia whaling had come to this standstill that the International Whaling Commission selected Vancouver as the site for its 1961 meeting. Not long after this event, prospects for whaling from the Canadian Pacific coast were improved during a series of talks between B.C. Packers and representatives of Taiyo Gyogyo Kadushki Kaisha, Japan's largest whaling and fishing combine.[118] An agreement was reached to collaborate in the collection of fresh-frozen whale meat from provincial waters for export to Japan. The Coal Harbour station would be reopened under the name Western Canada Whaling Company Limited.

Like Japan's newly regenerated pelagic whaling industry,[119] the Coal Harbour operation would provide edible meat to a country both philosophically and logistically indisposed to the production of domestic livestock as a food source. The Packers-Taiyo agreement called for a fully cooperative whaling venture to produce meat to stringent Japanese standards; frozen and canned products would be exported in refrigeration ships. Taiyo agreed to send two chaser boats of a modern Antarctic type as well as meat-packing specialists. B.C. Packers offered to contribute vessels, plant, and crews, while retaining a controlling 51 per cent interest in the firm.[120] These arrangements were approved by all parties, and, as one manager later remarked, "We were in the meat business."[121]

Western Whaling had processed some meat throughout the 1950s, but it was intended for chicken fodder and mink food. In 1953, for example, about five hundred tons of fresh and frozen meat had been bagged in fifty-pound sacks.[122] Whale meat had also been canned, mostly for pet food; these tins were trucked to a warehouse where a label was glued on to identify the contents as whale meat for "non-human consumption."[123]

A major offensive was set in motion to obtain the goodwill of the United Fishermen and Allied Workers' Union with respect to the arrival of the Japanese nationals. In December 1961 the union agreed to include the Japanese as "teachers" under UFAWU agreements; two mates and eight meat-cutter teachers would be accepted as "temporary workers." But the union reserved its right to reconsider the situation at the end of the initial whaling season in the autumn of 1962.[124]

The wage agreement was similar to the 1958 contracts. Though management balked, chief engineers would earn $390 per month; mates $310; cooks $300; deckhands, firemen and oilers, $287. A $53 monthly board allowance was included, and the whale pool was reinstated. Into it the company would contribute $140 for each blue whale used for fresh meat, $90 for each finback similarly utilized, and $30 for each sei whale.[125] Gunner-skippers were not covered by union contracts and, as usual, negotiated their own arrangements with management. Some union men were disgruntled that jobs were being given to Japanese, but some retraining was required to ensure proper hand-

Westwhale 5 was the first chaser boat imported by Taiyo Gyogyo K.K. to work in British Columbia. It was one of Arne Borgen's commands. Eventually, five Japanese vessels were integrated into the whaling fleet.
Vancouver Maritime Museum

ling of the meat. Any dissatisfaction was mitigated by a strong sense of optimism around the docks, the first in a long time, and a tentative truce was declared in the union-management war. In true team spirit, the chaser-boat fleet was renamed in the style of an Antarctic pelagic fleet; Arne Borgen's flagship *Polar V* went back to sea as the *Westwhale*, and the other chasers followed in numerical sequence. Bjarne Andersen's *Bouvet III* became *Westwhale* 2; Sofus Haugen's *Globe VII* was renamed *Westwhale* 3. The diesel-powered *Lavallee*, under Per Hvatum, began the new industry as *Westwhale* 4.[126]

The two new ships from Japan received new names in turn. The larger of the two, a 1,600-horsepower, fourteen-knot chaser boat called *Katsu Maru*

No. 5 became *Westwhale* 5; the other, the 307-ton *Seki Maru No.* 2, was soon rechristened *Westwhale* 6. Japanese captain-gunners and mates arrived with the vessels, as well as engineers to oversee the operation of their engines. Aboard *Westwhale* 5 came an energetic gunner named Ishii, a large and strong man about six feet tall, who would soon give Arne Borgen some serious competition in the whale hunt.

The Japanese boats carried modern electronic detection gear,[127] as well as a new type of harpoon fitted with a blunt or concave grenade which had been developed under the supervision of the Japanese Committee for the Improvement of Equipment of Whaling Vessels. The blunt point significantly reduced ricochet which, since the times of Svend Foyn and Thomas Roys, had turned many a clean shot into a frustrating miss.

B.C. Packers also surmounted delicate negotiations with their star performers, the whaling gunners. This was accomplished in large part at an informal meeting held on board the *Westwhale* 2 on the morning of 10 April 1962. It is not known if Arne Borgen attended, but the other five gunners — Andersen, Haugen, Hvatum, Ishii, and Kitayama — were there, together with their respective mates: Bigset, Noringseth, Samson, Ikoma, and Miyahara.[128] Plant manager Lorne Hume stressed the need for co-operation among the gunners, as well as prompt delivery of carcasses, since the new product would be spoiled by excessive delay. There were some touchy points of diplomacy, particularly when Hume had to explain why the "Red Ensign," rather than the "Rising Sun," had to fly at the stern of the Japanese boats. He countered this potential indignity by requesting Captains Ishii and Kitayama to ask their mates to explain the proper way of attaching the lightweight bamboo waif poles, which were unfamiliar to the Norwegian and Canadian men. And it was decided that the Japanese mates would shift around the fleet to teach the Canadians how to protect the freshness of the whale meat while at sea.[129]

Communication difficulties were exacerbated by the arrival of the Japanese, who had little fluency in English and could not understand the Norwegians at all. Hume later recalled that a linoleum floor of the old seaplane hangar proved useful as a giant chalkboard; from the first meeting at Coal Harbour, the men drew pictures on the floor whenever language barriers impeded discussion. "This is the way we got [the Japanese and the Norwegians] to start communicating," he said.[130]

The changeover to fresh-frozen edible meat brought some subtle changes to the island communities of Coal Harbour and nearby Port Hardy. Japanese men began to appear in the village streets, and a new item — whale sandwiches — became popular in the local cafés. Cooked in butter, the sandwiches enjoyed a brisk sale, particularly at a well-remembered community fundraiser for the local hospital. The alternative, barbecued salmon — much prized in British Columbia—is said to have sold poorly because of the novelty of the whale meat.[131] Increased cleanliness at the station, mandated by the production of edible meat and a decrease in bone meal and blood-meal production, reduced the stench of dead whale which once permeated the area. In former years, nearby residents knew about the whaling work from the smell: "When [the men] went to Port Hardy to a dance they'd take about three showers," one woman remembered. "After about three turns around the floor nobody would dance with them. They smelled of whale."[132] One man, employed just before the switch to edible meat production, said it was then customary to take four sets of clothing for the whaling season; one for the work, one for the dining hall, one for casual wear in the bunkhouse, and a fourth for going home. "You burned the first three at the end of the season," he said.[133]

The addition of the foreign workers caused some rumblings but no trouble. As in Sprott Balcom's day, the particular closeness of the Japanese workers amplified whatever resentment existed. The flensing crews worked in a paramilitary style, wearing identical garments which to some Canadians resembled a uniform. They seemed to accept orders without question, as a soldier would,[134] and they appeared to accept their own immediate supervisor as an officer. "All [the Japanese manager] had to do was walk through," one Canadian remembered, "and production went up like crazy."[135] "The Japanese had a kind of frantic approach to everything," another said. "[They were] totally gung-ho."[136]

Some whites thought they were being pushed around. One former employee later claimed that the Japanese managers "came down pretty heavy on some of the local guys." The white workers "mumbled," but they voiced no objections, since "they were doing pretty well on the thing."[137]

Racial tensions were not so pronounced at sea, and in at least one case union solidarity and the fellowship of seamen outweighed chauvinism. During the first month of whaling, the mate of the *Westwhale 6* suffered a temporary paralysis from the waist down and was at least three times hospitalized for examination by company doctors. They found no specific ailment. Unable to obtain relief, the mate asked to be taken to a larger hospital downcoast, and in this he was supported not only by his Japanese skipper but also by the Canadian deckhands who worked with him: "The skipper and Taiyo engineer are very concerned," a shipmate wrote, "as they have both sailed with the mate previously and see a great change in him. . . . Apparently the skipper has had words with [the station manager] over this matter . . . and said he won't take the boat out again after the trip unless the mate is given proper attention & examination. . . . As far as we're concerned on deck," he concluded, "we'll back the old man [Captain Kitayama] up. The mate is a Union brother, and is at all times an extremely amiable guy and excellent shipmate, and is now obviously unwell."[138]

Job stress, both on board and on shore, seems to have been high. It was known that Coal Harbour could process eight whales during a normal work day, but there were sometimes twenty carcasses requiring immediate attention lest the meat spoil. Men worked whatever hours were necessary to fulfill the requirements of prompt processing. One man remembered a ninety-hour work week; he would leave the plant at 3 AM only to get up at 7 for the next day's labour.[139] The shipping of frozen meat proved to be a particular trial. Chunked as it was stripped from the carcass, blocks of red meat were packaged and retained in a freezer shed[140] and then moved twice a year to Taiyo's refrigeration ships. The cargo holds in these ships were pre-chilled at sea, perhaps to -40° Fahrenheit. Under the midday sun, the shore crew would ride with the pallets to the bottom of the hold and transfer the cargo by hand. "We all hated that," one man said. "It was summer. When you came out of the hold, it was like being in the Tropics." After much protest, the captains of the refrigeration ships agreed to turn off the refrigeration units three days before arrival at Coal Harbour. Even so, temperatures in the hold rarely exceeded 0° F.[141]

Dried meat was shipped out of Quatsino Sound on railroad barges, each of which carried ten to twelve railroad cars. Employees loaded hundred-pound sacks of meal with a conveyor belt, but these sacks would have to be lifted manually from the forklift truck to the belt and again from the belt to the deck of the car. The worst job, however, belonged to the men in the meal shed. They stood amid the sticky solids that came from the drum-dryer, breathing the dusty air and shovelling the material into an auger, which carried it away to be mixed with regular meal.[142]

The catch of the first combined season was 713 whales, but the product was not immediately sold. Currency restrictions in Japan made it necessary to store the frozen meat in British Columbia, since it was impossible for the Japanese to export yen to pay for it. This situation was ameliorated the following year and normal trade relationships were resumed. When the *Banshu Maru* called in 1963 for a load of frozen meat, the shipment was hailed — wrongly — as the first cargo of edible meat ever exported from British Columbia.[143]

In 1963 only 548 whales were taken, including 220 finbacks, 154 sei, 147 sperm, 30 blue, 24 humpbacks, and 3 Baird's beaked (bottlenose) whales. But total production increased because of the large blue and finback whales in the catch. Chaser boats had to travel a long distance offshore even to find this many animals, and some of the meat spoiled even though carcasses were cut open at sea to allow the cold ocean water to circulate through the body. "Today . . . we caught 18 sei and 1 fin," a crewman wrote his union representative, "but sad to say that some of the boats had to go

100–200 mile out to get them so leaving them pretty rotten by the time they get in and this owing to the shortage of whales to make up the meat bonus, tends to cause much dissention all around."[144]

There were other competitors for the meat. Sharks were always present, so the crews began taking potshots at them with .22-calibre rifles. Other, more worrisome competition came in the form of a Soviet factory-ship fleet which was sighted by a halibut fisherman 360 miles west of Cape Flattery.[145] On 20 July 1963 the Coal Harbour chaser boats met the Soviets in person, just 35 miles due west of Cape St. James. Manager Hume recalled later that one of the Western Canada Whaling Company chaser boats raced against a Soviet chaser for the same large blue whale. Neither of them caught it, although another Canadian vessel later harpooned it.[146] Canadians charged the Russians with taking undersized whales and females with calves, but their main concern was the potential extermination of local whale populations by the large and efficient Soviet factory-ship fleet.

B.C. Packers and Taiyo soon upgraded their own fleet to improve the catch. The oldest boats were set aside, and Taiyo sent larger, modern vessels to take their place. *Westwhale 3* (ex-*Globe VII*) was the first to be retired, in 1962. The hull went to the shipbreakers, but its triple-expansion steam engine was saved for the movies; in 1965 the gigantic power plant was transported by rail to Hollywood, where it served as a prop in the Twentieth-Century Fox production of *The Sand Pebbles*.[147] It was followed into retirement by Borgen's steady *Westwhale* and then by *Westwhale 2*.[148] The *Westwhale 4*, ex-*Lavallee*, was removed from whaling service in 1964 and refitted as a purse-seiner; B.C. Packers sent it to Harbour Breton, Newfoundland, where the company intended to establish a herring fishery.

The remaining chaser boats underwent a refit at Celtic Shipyards in Vancouver during the winter of 1964–65 and were joined at sea by two additional Japanese chaser boats, the largest ever sent to work in British Columbia. One, *Toshi Maru No. 21*—promptly renamed *Westwhale 7*—had been constructed in 1952 as the Norwegian *Suderøy XVI*. Before serving the Taiyo fleet it had been in Antarctica for the Suderøy Hvalfangstaktieselskapet, and was for a short time named *Kos 51*. On arrival in Canada it was turned over to Arne Borgen. A near-sister ship, *Toshi Maru No. 22*, ex-*Suderøy XVII*, was renamed *Westwhale 8*.

Unfortunately, the addition of these sophisticated vessels did not significantly augment the 1965 catch. Johan Hjort's contention that improvements in whaling technology do nothing to improve the lot of the whale seemed to be borne out. The large cetaceans were scarce. Whether from local hunting or Soviet pelagic whaling or both, the numbers of finback, blue, and sperm whales had declined drastically. The smaller sei whales now took first place in the annual statistics, and the superstitious among the whalemen were

Whaling gunners appeared smaller as chaser boats grew larger. Bjarne Andersen is almost hidden on the bridge of the *Westwhale* 7. Radar and radio equipment improved the hunt, and the ramp to the gun platform may be seen at left.
Lorne Hume

quick to point out that seis always appeared wherever a whaling ground had been commercially exhausted. In 1965 sei whales accounted for 604 of 865 carcasses processed. Of sperm whales there were only 151, and of finback fewer still — 83.[149] These returns seemed to validate research opinion that the once-prominent humpback and blue whales had been exterminated. Concerned with such scientific findings, the International Whaling Commission enacted a one-year ban on the killing of humpback whales in the North Pacific — renewed for a second year in 1966 — and five years of protection for blue whales.[150]

As the Western Whaling chaser boats steamed out to begin the 1967 season, the search became one of frustration. Even with the newest spotting and detecting equipment, the crew of the *Westwhale* 7 saw no whale during fifteen of the first twenty-three hunting days in April;[151] this despite the fact that weather conditions were moderate.

A sophisticated chaser boat, the 1950-vintage *Fumi Maru No.* 8, was prepared for service as *Westwhale* 9.[152] But at the close of the season, only 496 whales had been caught, and management made the long-feared announcement: Coal Harbour would not reopen in 1968. The market for oil had fallen 50 per cent, they said, owing to competition from vegetable and other oils.

The steadily rising standard of living in Japan meant a shorter sale of whale meat. Its price had fallen, and the large percentage of sperm whales in the recent catch was of little interest to the Japanese, since the meat was inedible. Even the mink industry had come onto hard times, so that it was more difficult to sell meat for non-human consumption.[153]

Again the UFAWU protested. Members voted an emergency resolution calling upon Prime Minister John Diefenbaker and Minister of Fisheries Angus MacLean to absorb and underwrite the whaling operation. A copy was sent also to Premier W. A. C. Bennett in Victoria.[154] The union pleaded for government intervention to keep the work flowing, but none was forthcoming.

When whaling came to an end, the *Westwhale* 7 was streamlined for its new duty as the *Pacific Challenge*, the largest steam towboat in the Pacific Northwest. The steam engine proved unprofitable, and had to be replaced.
Vancouver Maritime Museum

At the close of business in 1967 the men dispersed for good, and the vessels went after them. Arne Borgen took a gunner's job in eastern Canada, and later worked in the Philippines, where he died of natural causes. *Westwhale* 7, was sold to Pacific Towing Services Limited, and in 1970 it was converted to a towboat at the McKenzie Barge and Derrick Company shipyard in North Vancouver.[155] Its potent trawler-type hull emerged as the province's only oceanic steam-tug, the *Pacific Challenge*. In its new guise, the former

Antarctic "killer boat" earned more publicity than it ever had as a whaling vessel, though its newsworthy steam power proved uneconomical and had to be replaced.

Westwhale 9 was sold to new owners in New Orleans.[156] Only the *Westwhale* 8 remained in the whaling business. Arctic Fishery Products Company bought it and Arne Borgen sailed it from a shore station at Dildo, Newfoundland, thus reversing the path of Ludwig Rissm ̈ ller's *St. Lawrence* and making a complete and full circle of the business from the Canadian Maritimes to the west coast and back again. At Dildo, whales were processed alongside the station's other commercial harvests: groundfish and blueberries. The full circle of industry was continued also by Taiyo Gyogyo K. K., which soon entered into partnership with Fishery Products Limited to form the Atlantic Whaling Company. They reopened an abandoned whaling station at Williamsport, Newfoundland, and used one of the *Fumi*-class chaser boats there until the plant burned down in 1971.[157]

Once more the conclusions of Johan Hjort had been verified. The whaling industry was never subject to the same natural economic laws as other businesses. Instead, every improvement, every increase led only to the further reduction of the very animals whose presence encouraged incentive. It was for this reason that the Norwegian law of 1929 and subsequent whaling conventions attempted to regulate the whale hunt. The demise of the last whaling operation in the Pacific Northwest followed the same principles which beset and eventually bested the Yankee and French right-whalemen of the nineteenth century, and after them Thomas Roys, Sprott Balcom, Mackenzie and Mann, and William Schupp in turn.

Gordon Pike, the research scientist who kept the last records, closed up shop at the Pacific Biological Station in Nanaimo and accepted his new assignment in eastern Canada. In Montreal not two years later he committed suicide, just as though the decline of whaling in his favourite British Columbia had signalled his personal decline. His ashes were scattered over the calm waters of Departure Bay, near the place where Sprott Balcom and *Capitaine Morin* had hunted whales not so many decades before.

In June 1972, at the United Nations Conference on the Human Environment in Stockholm, the endangered whale was adopted as the symbol of mankind's abuse and potential destruction of the natural resources of the planet. The assembled delegation voted to support a strengthening of the International Whaling Commission and called for a ten-year moratorium on commercial whaling.[158] But at the annual meeting of the IWC two weeks later, the commission's scientific committee advised that there was no basis on which to call for such a moratorium. Instead, the commission proceeded with its newly-adopted stock management scheme, which called for individual catch quotas for each species within discrete geographic sectors of the world's seas.[159]

Both the Canadian and American governments went ahead with plans for unilateral moratoria; by the end of the year the Canadian Department of Fisheries announced a total ban on commercial whaling and mandated the closing of the few whaling stations operating in Nova Scotia and Newfoundland. Whalemen cast out of work by the termination of their industry were paid a reimbursement.[160] The United States government also passed legislation which has helped to ensure that whales are not exploited in American territorial waters.[161]

Finally, a decade after the Stockholm meeting, a test moratorium on commercial whaling was passed by a vote of the IWC general assembly. The few remaining whaling nations and their commercial allies cast the seven dissents: Japan, the Soviet Union, Brazil, Iceland, the Republic of Korea, Norway, and Peru.[162] The moratorium called for a reduction to zero of all commercial catch limits from all whale stocks, beginning with the southern 1985–86 pelagic season and the summer 1986 coastal whaling season. By the time of the vote, Canada had withdrawn from the IWC. That withdrawal, announced 26 June 1981, came as a result of complex political pressures, some of which revolved around an IWC proposal to regulate the catching of smaller cetaceans—particularly the narwhal and beluga—which live entirely within Canada's 200-mile territorial limits. Ottawa's public decision, however, was justified on the strength of Canada's withdrawal from commercial whaling. In 1982 the Whaling Convention Act was repealed, and the Cetacean Protection Act was added to the Fisheries Act; it placed the regulation of whaling back in the hands of the Canadian government and virtually prohibited commercial killing of large whales.

It seems unbelievable, in this century of steamships and harpoon-cannons, refrigerated meat and preference-share stock certificates, that native peoples would continue their traditional whale hunt along the shallows of Swiftsure Bank and Umatilla Reef. But the Makah hunt did continue, at least until 1928, and perhaps a few whales were taken just before World War II during a revival of Makah culture.[163] Soon thereafter, however, a student of native affairs concluded confidently that no Makah would eat the whale.[164] In truth, there had been rapid changes, an almost sudden leavetaking from the old ways. Compulsory schooling of native children had forced the Makah from their isolated village at Ozette on the outer coast of the Olympic Peninsula. They came in to Neah Bay to live next door to cousins who had always lived there. At Neah Bay they found a school, automobiles, telephones, radios, and communication with the larger society around them. Their acculturation continued at an ever more rapid pace, so that by 1950 the mention of whale meat, once a delicacy, might drive supper guests out of a Makah house.

At mid-century, fewer than a dozen native West Coast and Makah whalemen were alive. They were then in their fifties and sixties and seventies. None is living today.

Appendix A

THE JOURNAL OF THE *CAROLINE*
ON THE NORTHWEST COAST, SEPTEMBER 1843

The ship Caroline *of New Bedford, Daniel McKenzie, master, arrived at Sitka, Alaska in the late summer of* 1843. *There the men encountered the aggressive stance of the Russian American Company managers toward the growing threat of trade intrusion from American and French whaleships. From Sitka the* Caroline *proceeded southward to Nootka Sound and the Strait of Juan de Fuca. A journal of that voyage kept by George W. R. Bailey, a "green hand," is one of the earliest surviving accounts of whalemen landing on the Northwest Coast. The holograph manuscript is preserved in the collection of the Kendall Whaling Museum, Sharon, Massachusetts, and reproduced with permission.*

Sunday 3d Light wind at N.N.W. & pleasant weather: Course by the wind L^d tack: at 1 PM. saw a sail ahead: large ship standing to the South'ard & West'ard: at 2 PM. hauled up the M^n Sail & F^r Sail to allow the sail to windward to come up with us: showed English Colours; at 4 PM sent the Mate on board the strange vessel, for the purpose of gaining information concerning the Bays: at 5 the Mate returned, accompanied by Capt. C^{hs} Humphries of the English trading Bark Columbia just out from Sitka & bound to Columbia River: reported the Wm Hamilton of New Bedford with 2100 bbls. & that she was forced to leave the Harbour of Sitka, on Tuesday last, by Russian Gun Boats: the Russian Authorities will allow no whale ships to enter their Ports or Bays, on account of their Fur Trade: at 7

PM Capt. returned to his vessel: kept Ship S.E. by 1/4 S: saw Humpbacks & Finbacks: M[iddle] P^t calm, until 2 AM. when a light breeze sprung up at W.N.W.: at daylight saw Land, all along on our L^d beam, high & covered with snow: saw Humpbacks & Finbacks & two very large Sulphur bottom Whales: L P^t light wind at W.N.W. & clear pleasant weather: Course E.S.E. 1/4 S: land covered & concealed by a dense fog:

> Latt by Obs. 55°49' N
> Long by Chr 135°37' W

Monday 4th Calms, & small winds at N.W. weather clear, warm & pleasant: at sunset the Columbia was 5 on our S^d beam: fog floated off from the land & surrounded us: M P^t wind light at N.N.W.: weather foggy: Course S.E. 1/4 S: at daylight kept Ship E.S.E.: saw the Columbia in the same position as last evening: sent up the Fore & Mizzen To G'llnt Yards & set the sails: bent the Flying Jib & set it: at 7 AM fog cleared away: saw Land all along abreast of us: luffed ship up to E by S 1/4 S: L Pt. moderate wind at N.N.W. & clear pleasant weather: running in for the Land: saw great numbers of Humpback, Finback & Sulphur bottom Whales: at 9 AM sunk the Columbia on our S^d quarter: broke out a new Fore Topsail: at Noon we were off the South end of King George's Islands.

> Latt by Obs 55°07' N
> Long by 45 Miles off the Land

Tuesday 5th Moderate wind at N.N.W.: & clear pleasant weather: Course E.S.E.: unbent the M^n To Gllnt S'l & repaired it: bent & set it again: at 5 PM kept Ship off S.E. by E: at 6 saw land ahead: kept Ship off S by E: M Pt. wind at N: Course S 1/4 W: under all sail: set whole watches: running across Dixons Channel: at daylight we were off the North Cape of Queen Charlottes Island: luffed ship to S.E. 1/2 E. saw a Sail on our lee bow, standing to the southard & East'ard: watch employed in making Cartridges for our Muskets & bags of Langrags [?] for our "swivel": shot a Hawk, with a Fish in its claws: fell into the water: L Pt. wind subsided gradually into a calm: weather clear & pleasant:

> Latt by Obs 53°46' N
> Long 24 Miles off Land

Wednesday 6th Calms & occasional light airs of wind: at sunset we wore off Fredericks point: M Pt. at 7 PM a light breeze sprung up at N.N.W.: kept Ship S.E by E: wind veered to E & increased: at daylight it blue fresh at E.S.E. took in the To G'llnt S'ls & Fly Jib: at 6 AM tacked Ship, running N.E.: saw a very large Suphur bottom, or Razor back close to the Ship: L Pt. fresh winds at S.E. & thick rainy weather: Course by the wind S^d t'k: running in for the Land: wind increasing: at 9 AM. single reefed the Topsails: at 11 double reefed them: saw large numbers of Humpbacks & Razorbacks:

> Latt by Act 53°10' N
> Long 4 Miles off Land

Thursday 7th Strong gales at E & thick, rainy weather: Course by the wind, S^d tack: at 1 PM. being within 4' of the Land we wore upon the L^d t'k: bent & set the M^n Spencer at 4 PM: at 5. furled the M^n Sail: at 6 furled the Jib: M Pt. strong gales at E.S.E. & foggy: at 9 PM. furled the Fore & Mizzen Topsails: wind moderating a little, at 1 we set them again: wind veered to S.S.E.: at 5 AM wore Ship on to the S^d t'k: secured the B'ts: L Pt. strong gales at S.S.E. & thick rainy weather: at 9 AM furled the Jib: at 12 furled the Fore & Mizzen Topsails & wore Ship upon the L^d t'k:

Latt by Act 52°55' N
Long by Act 15 Miles off Land

Friday 8th Wind moderating fast. weather foggy: at 3 PM wore ship up on the S^d t'k & set the Fore & Mizzen Topsails (reefed): saw the Land ahead: at 6 PM. being within 4' of the land, we wore Ship upon the L^d t'k: set the Jib & M^n sail & turned one reef out of the M^n Topsail: M P't. wind veered to E.S.E. & became moderate: weather foggy & rainy: at 5 AM put Ship upon the S^d t'k: at 6 saw the Land ahead: L P't weather thick & rainy: at 10 AM, being about two leagues off the Land, we wore Ship upon the L^d t'k. heading S: saw plenty of whales, but none of the *right* kind.

Latt by Act 52°50' N
Long " 15 Miles off Land

Saturday 9th Moderate wind at S.E. & thick rainy weather. Course by the wind L^d t'k: at 2 PM. the wind died away & the weather cleared off pleasant: heavy swell running from the S.S.E.: turned the reefs out of the Topsails & sat the M^n To G'llnt S'l; braced round the Yards upon the S^d t'k: at 4 PM. a breath of wind springing up at N. we wore Ship upon the L^d t'k: fell calm again: at 7 caught a light breeze from the S.W.: M P't. wind moderate at S.W. with occasional light squalls of rain: at 1 AM. put Ship upon the S^d t'k: clear moonlight night: at 5 AM. set all sail: L P't. wind veered to S.S.E.: weather clear & pleasant: course by the wind S^d t'k: Mate employed constructing boarding Pikes.

Latt by Obs 52°37' N
Long " Obs 15' off Land

Sunday 10th Fresh wind at S & fair weather. Course by the wind S^d tack—Sail on weather bow—at 3 P.M. Tacked Ship—at 7 took in our light sails—Chief Mate employed constructing boarding Pikes for the destruction of the in^{n}ocent Nootkains—M Part—clear moonlight night—at 12 wore Ship upon S^d Tack & set all sail—at 3 saw a light on our weather bow: at daylight made the strange sail out—French whaler: at 9 AM being within a mile of the Frenchman, her Master came on board of us—Ship Reunion of Havre, 20 Mths out 1700 Whle 100 Spm. lost her M^n Sail & Mizzen Topsail, the last gale, in beating out from land, when she had got embayed in endeavoring to reach some place where she might recruit, they being on short allowance of Grub & Water—at 11 AM Capt Smith returned to his Ship, having been persuaded to accompany our Capt to Fugos Straits under promise of mutual protection.

  Latt by Obs 52°29' N
  Dist of Land 30 Miles

  Monday 11th Light breeze at S & clear pleasant weather. Course by the wind L^d tack —at 5 PM hove Ship too for our consort —at 6 he came up to us & Capt S---came on board — Reunion busily engaged in repairing her sails —M Part —wind fresh at S.S.E. & increasing weather thick —at 10 PM Capt S---h returned to his Ship —at 1 AM furled our Top Gallt Sails & double reefed the Topsails —by 3 AM it blew a gale at S.E. accompanied with thick rainy weather —furled the Jib & secured the Boats —at 3 1/2 AM close reefed the Mizzen Topsail & clewed up our Fore Topsail to close reef it but the gale increasing unfurled it —rain falling heavily —at daylight saw the R--- astern under close reefed M Topsail & Fore Sail —clewed down Main Topsail to close reef it by order of our Capt. who was lying in his berth, not having showed his nose on deck as yet —while engaged furling the Topsail the wind suddenly died away nearly calm & shifted to S.W. —at 6 AM our brave Capt appeared on deck & ordered all hands to be called —unbent the M Topsail & bent a new Fore Topsail in its place. L Part wind fresh & variable between W.S.W. & S.S.W. —sea rough & irregular —at noon the R--- was 4' off our larbord.

  Latt by Obs 52°06' N
  Long by Chr 132°22' W

  Tuesday 12th Moderate wind at S.W. & fair weather —Course S.E. —M Part —clear moonlight night —at 10 PM the Ship was felt to tremble violently accompanied by a strong kind of rumbling grating noise — waked some of the crew in the Fore Castle & up they came upon deck in their Shirts to know what was the matter —shock was of about 10 seconds duration —supposed it to be a shock of an earthquake —L Part —wind hauled to S.S.E. weather cloudy & lowering —shortened sail for the R---.

  Latt by Obs 50°27' N
  Long " Chr 133°05' W

  Wednesday 13th Wind fresh at S.E. & increasing weather cloudy —Course by the wind S^d tack —watch employed repairing M Topsail —at 4 PM furled the Jib & Main Sail & doubled reefed the Topsails —secured the Boats —by sunset the wind had increased to a gale, accompanied with rain — took in all sail but a double reefed M^n Topsail, F Staysail, M [Spanker] & Mizzen —wore ship upon L^d tack —R--- 1 mile off weather beam —M Part — Strong gale at S.E. & thick rainy weather —sea breaking over the Ship continually at 12 PM wind abated a little sea very rough —at daylight R--- not in sight —at 6 AM we Shiped a sea which filled our decks & set everything afloat & men & Pigs were swimming about together —L

Part — Fresh gales at S & fair weather — at 8 AM saw our consort 3' to leward — run down to the Reunion.

        Latt by Obs 50°39' N
        Long " Chr 132°08' W

        Thursday 14th Strong gales at S & pleasant weather — Course by the wind S^d tack — sea very heavy — at 5 PM close reefed the Topsails — our Consort being 4' astern we wore Ship & run down to her — M Part: clear moonlight night: L Part: at 10 AM saw the Reunion 8' on our lee Beam — bore away from her.

        Latt by Obs 50°12' N
        Long " Chr 130°58' W

        Friday 15th Moderate wind at S & pleasant weather — Course by the wind, S^d tack — heavy swell from the S.E. — M Part — Variable winds & calms — heavy falls of rain occasionally, towards morning became steady at N.N.W. — Course S.E. by S — L Pt wind Strong at W.N.W. with occasional showers of rain — Reunion 6 Miles off S^d Beam

        Latt by Obs 49°08' N
        Long " Chr 129°52' W

        Saturday 16th Fresh gales at W.N.W. & fair weather — Course E.N.E. at 1 PM shortened sail & yawed ship one way & other to allow the Reunion to come up with us. M Part — clear moonlight night with occasional showers of rain — saw Humpback & Finback Whales — saw a Rainbow this morning from the Fore Top Gallant Cross Trees of our Ship which appeared as a complete circle very distinct & brilliant — L Part — light wind at W & pleasant weather — at 10 Capt Smith came on board of us accompanied by his Doctor, at 12 AM saw the Island of Nootka on our S^d bow bearing N.E.

        Latt by Obs 48°46' N
        Long " Chr 126°40' W

        Sunday 17th Moderate gales at W.S.W. & fair weather — occasional light showers of rain. Course E.N.E. unbent our new Main Topsail & bent the old one again — at 7 PM Capt S---h returned to his Ship — M Part: wind at S.W.: Course E 1/2 S: clear moonlight lnight — L Part: Light wind at S.E. & pleasant weather.

        Latt by Obs 48°35' N
        Long " Chr 125°31' W

        Monday 18th Moderate gales at S.E. & cloudy weather: Course by the wind S^d tack — cut a new chain hole, clear of the galley — bent the Larbourd Chain Cable: at 5 PM saw the entrance to Fugus Straits bearing E.N.E.: at 7 PM Tacked Ship, heading S.S.W. — Reunion one mile astern. M Part: wind variable, between S & E. & squally one minute calm & the next blowing a gale, accompanied by thunder & lightning and heavy falls of rain: night dark as a pocket: sea remarkably phosphorescent: at 8 PM double reefed the Topsails: calm from 12 to 2: bunged [?]

up at N.: set all sail: Course S.E.: at 5 AM luffed to the wind on the L Tack: at daylight saw the Reunion 10 miles off our weather beam: at 6 AM double reefed the Topsails again, in a heavy squall of rain: L Part variable winds & fair weather: at 8 AM caught a steady breeze at N.W.: tacked Ship & made all sail: saw large numbers of the Humpback Species of Whale:
       Latt by Obs 48°18′ N

Tuesday 19th Moderate breeze at N.W. & fair weather: Course by the wind L^d tack: at 2 PM bent our S^d Chain Cable & at 3 PM we were in the entrance of Fugus Straits: Reunion one mile astern, standing directly in: our brave Captain had just discovered that there is great danger to be apprehended from the Natives: kept Ship off for the passage—at 3 1/2 PM our Hero, not daring to venture further, hove Ship too & made signal for Captain Smith to come on board: Capt S---h boarded us: saw a Canoe crossing channel inside of us. Second Mate of the R---n came on board & reported in Sight to be a Whale Boat. our Captain immediately decided that a Ship had been taken by the Natives—the Reunion is short of water & provisions & Capt S---h is very anxious to go in here & anchor, if not for Whales, at least for recruit. at sunset Capt S---h & Mate returned to their Ship & both Ships stood out to sea: M Part. light wind at S.E. & fair weather. standing off the Land: heavy swell heaving in from the S.W: at 9 PM tried to luff the Ship to the wind with the Main Topsail aback: could not succeed: Ship got stern way on her: after hauling out the Spanker twenty or thirty times & trimming the Yards we at last got the Ship before the wind: at daylight we were off the passage: R---n 1 mile off L^d beam: L Part, dead calm: saw a canoe crossing the Bay inside of us: very heavy swell setting in from seaward: at Noon Cape Flattery bore S. distant 4 miles:

Wednesday 20th At 1 PM a light breeze sprung up at W. weather pleasant—saw a Canoe approaching the Ship—Cap't imagined that the Natives were going to attack the Ship—ordered all hands to be called & nobody to leave the deck—six other Canoes soon followed the first one—Whaling Canoes, all contained a crew of 10 or 12 persons, men & women, their dress is composed of Bear Skins & shirts or pieces of cotton obtained from the fur Traders, thrown loosely over the shoulders & a kind of chip or Bark cap or Hat of a conical form. They appeared to be of an inoffensive & peacable character & much disposed to traffic. The complexion of the generality of them is a dark Copper colour, although to some, the Chiefs & their wives or the higher class air very light coloured & some of the women are even handsome. They offered for sale a few fine salmon, demanding in exchange, Iron Files. our Captain obtained four large ones for an old Rasp File, & four others for two Hand Saw Files. They offered for sale the Oil which they had taken from Whales. Their manner of taking the whale is as follows. He is first struck with a kind of Harpoon made of Pearl Shells fitted with a socket on to a stout pole of hard & polished wood, some eight or ten feet long, which is withdrawn or falls out of itself as soon as the harpoon has entered. These Poles air carefully preserved—Some six, eight or more drugs, made of the skins of the seal inflated with air, air made fast by a stout rope of sinew to the harpoon: these act as a great impediment to the progress of the

Natives of Noothka Island

tortured animal through the water & serve as a guide to the place of the whale's next rising: as soon as he is *up* again, he is transfixed by as many more harpoons as possible, until at last he is fairly worried to death.

We ascertained from the Natives that there had been two trading Vessels lately, but no Whale Ships, that there would be plenty of Whales in the Straits in the course of three or four weeks. Our Cap't does not like the appearance of things & has decided to sail for St Francisco in Calafornia & from thence to the Line and cruise for Sperm Whales —on the contrary Capt S---h is very anxious to remain here & wait for the season for whaling here but doesnt like to do so alone, & is justly angry at being enticed to come here & then deserted. Made all sail on the Ship & stood out to sea, made signal for Capt S---h to come on board, but his 2ᵈ Mate in his stead, (american) Mate returned to his Ship & brought back with him a Native Chief who spoke English: would not do. our Capt not to be persuaded: Mate went back to his Ship in a passion: Wind veered to N. W. M Pt wind N. W. Course S. W. at daylight Cape Flattery bore E. distant 40′ —R---n off Sᵈ beam —at 6 AM luffed to the wind & bore up for her —L Pt made signal for Capt S---h —sent his 2ᵈ Mate with Capt Smiths compliments: sent 2ᵈ Mate back, requiring Capt S---h to come on board & get some *slops for his men* (who are sadly in want of them) —Capt S--- came on board —broke out for Slops.

Thursday 21st Light wind at N. W. & fair weather —Course by the wind Lᵈ tack —our Capt had the generosity to exchange Slops which cost about $50 in the U States for a French Gold Lever Watch which cost $130 in Paris —Capt S---h sent his 2^d Mate to bring [?] a nine pound Cannonade for which our Capt gave two bolts No 1 Canvass — toward sunset the weather thickened & looked threatening & a dense fog sent Capt S---h off for his Ship, in some haste, when he was but twenty yards from our Ship he could not be seen — furled the Top Gallant Sails & double reefed the Topsails — M Pt wind light at N. W.: weather foggy & rainy — Course S. W. — at AM 8 turned reefs out of Topsails — at 9 AM saw the R--- half a mile off our lee bow — at 10 tacked ship heading N.N.W.

        Latt by Act 48°05′ N
        Long ″ ″ 125°25′ W

From here the Caroline *proceeded southward along the coast to northern California.*

THE ROSE HARBOUR WHALING STATION, 1928

The Rose Harbour whaling station, constructed in 1910 by Canadian North Pacific Fisheries on Kunghit Island, Queen Charlotte Islands, is the most fully documented of the Pacific Northwestern shore-whaling stations, with the arguable exception of the Akutan station built by Norwegian interests in the Aleutian Islands. This ground plan of Rose Harbour shows the layout in 1928, when the station was owned by William Schupp and the Consolidated Whaling Corporation. It is similar in configuration to the others constructed in the Pacific Northwest during the period 1905–26, and most closely resembles the whaling stations at Naden Harbour, B.C., and Bay City (Westport), Washington. The plan was redrawn and corrected by Joan Goddard from an appraisal drawing prepared for the Consolidated Whaling Corporation and now preserved among the Lagen Papers, University of Washington Libraries, Seattle. It is included here courtesy of William S. Lagen.

Whales were brought up channel between Kunghit and Moresby Islands to a float just out from the end of the haul-up slip. When ready for processing, a carcass would be brought in and winched onto the haul-up slip [1]. Flensers immediately removed the blubber, which was taken to the blubber slip [2] for further attention, and from there the smaller cuts of blubber were sent to the rendering building [3]. The hot oil was cooled in the cooling tank [4] and stored in the oil-storage building [16]. Baleen was prepared and stored in gill bins [5], gill tanks [6], and the gillbone (baleen) drying shed [7].

Meanwhile, the flensed carcass was moved by winch onto the carcass slip [8]

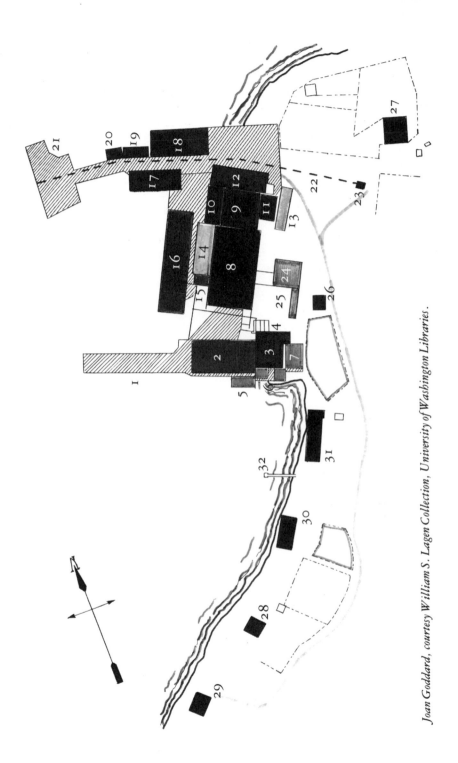

Joan Goddard, courtesy William S. Lagen Collection, University of Washington Libraries.

where the bones, meat, and viscera were utilized in related processes conducted in the meat pounds [9], press [10], glue tanks [11], drier [12], and bone mill [13]. Ground bone was stored in the bone shed [14]; guano was kept in two guano sheds, No. 1 [17] and No. 2 [18]. The sperm addition [15] and oil-storage building [16] housed the oil and the special products of the sperm whale.

Other buildings at Rose Harbour served as administrative offices, residences, and outbuildings, including the company store [19], office [20], boiler house [24], lighting plant [25], blacksmith's shop [26], manager's residence [27], engineer's residence [28], white men's bunkhouse [29], Chinese bunkhouse [30], Japanese bunkhouse [31], and finally the privy [32], which was located on the end of a rickety dock extending over the channel.

Vessels visiting the station, including the supply freighter *Gray*, tied up at the main wharf [21]; a track [22] ran from the coal hopper [23] to the end of the dock, to permit easy refueling of the freighter and the chaser boats.

SOURCES AND NOTES

Repositories holding manuscripts, unpublished documents, and newspaper citations are abbreviated throughout the notes according to the following key:

Balcom	Graeme Balcom Collection (private)
BT	Board of Trade Records, Public Record Office, Kew Gardens, London, England
CRMM	Columbia River Maritime Museum, Astoria, Oregon
CSH	Cold Spring Harbor Whaling Museum, Cold Spring Harbor, New York
de Lucia	Raymond de Lucia Collection (private)
Dukes	Dukes County Historical Society, Edgartown, Massachusetts
FDR	Franklin D. Roosevelt Library, Hyde Park, New York
FNA	French National Archives (microfilm), Old Dartmouth Historical Society Whaling Museum, New Bedford, Massachusetts
FRC	United States Federal Records Center, Waltham, Massachusetts
Harvard	Harvard University, Houghton and Baker Libraries, Cambridge, Massachusetts
HBC	Hudson's Bay Archives, Provincial Archives of Manitoba, Winnipeg
Huntington	Huntington Library, San Marino, California
IMA	International Marine Archives (microfilm), Old Dartmouth Historical Society Whaling Museum, New Bedford, Massachusetts
Kelly	Dr. Richard Kelly Collection (private)
KWM	Kendall Whaling Museum, Sharon, Massachusetts
Lagen	William Lagen Collection (private)
Lucas	Robert Lucas Collection (private)
Maury	Matthew Fontaine Maury Abstracts (microfilm), Old Dartmouth Historical Society Whaling Museum, New Bedford, Massachusetts
MIT	The MIT Museum, Massachusetts Institute of Technology, Cambridge, Massachusetts
MMBC	Maritime Museum of British Columbia, Victoria
MSM	Mystic Seaport Museum, Mystic, Connecticut
NARS	National Archives, Civil Archives Division, General Branch, Suitland, Maryland
NBFPL	New Bedford Free Public Library, New Bedford, Massachusetts
NHA	Nantucket Historical Association and Whaling Museum, Nantucket, Massachusetts

NHT	*Norsk Hvalfangst-Tidende* [Norwegian Whaling Gazette]
NLCHS	New London County Historical Society, New London, Connecticut
NMM	National Maritime Museum (Greenwich), London, England
ODHS	Old Dartmouth Historical Society and Whaling Museum, New Bedford, Massachusetts.
PABC	Provincial Archives of British Columbia, Victoria, British Columbia
PAC	Public Archives of Canada, Ottawa, Ontario
PMS	Peabody Museum of Salem, Salem, Massachusetts
PRO	Public Record Office, Kew Gardens, London, England
Richmond	Joshua B. Richmond Collection (private)
Robbins	Mrs. Lewis L. Robbins Collection (private)
RP	Providence Free Public Library, Nicholson Whaling Collection, Providence, Rhode Island.
SAG	Sag Harbor Whaling and Historical Museum, Sag Harbor, New York
Starbuck	United States, Commission of Fish and Fisheries, "History of the American Whale Fishery From its Earliest Inception to the Year 1876," by Alexander Starbuck. In *Report of the Commissioner for 1875–1876*. Part 4, Appendix A (Washington, D.C.: Government Printing Office, 1876; reprint ed. New York, 1964)
UBC	University of British Columbia Libraries, Special Collections, Vancouver, British Columbia
UW	University of Washington Libraries, Seattle, Washington
VCA	Vancouver City Archives, Vancouver, British Columbia
VMM	Vancouver Maritime Museum, Vancouver, British Columbia
Wood	Dennis Wood Abstracts, New Bedford Free Public Library, New Bedford, Massachusetts (also Old Dartmouth Historical Society, New Bedford; and Kendall Whaling Museum, Sharon)
WSL	*Whalemen's Shipping List and Merchants' Transcript*
Yale	Yale University Library, New Haven, Connecticut

Other abbreviations used throughout

fo.	folio
n.d.	no date cited
n.p.	no publisher cited

NOTES TO PREFACE

1. Kenneth Brower, *The Starship and the Canoe* (New York: Harper Colophon Books, 1983), 174.
2. Johan Hjort, "The Story of Whaling: A Parable of Sociology," *Scientific Monthly* 45 (July 1937): 19–34; Johan Hjort, "Human Activities and the Study of Life in the Sea," *Geographical Review* 25 (October 1935): 529–64; and Johan Hjort, Gunnar Jahn, and Per Ottestad, "The Optimum Catch," *Hvalrådets Skrifter: Scientific Results of Marine Biological Research* 7 (1933): 92–127.

3. Johan Hjort, J. Lie, and Johan T. Ruud, "Norwegian Pelagic Whaling in the Antarctic, VI: The Season 1935–1936," *Hvålradets Skrifter: Scientific Results of Marine Biological Research* 14 (1937): 33.

4. Bark *Pantheon* of Fall River, Frederick L. Crapser, keeper. MSM (769).

5. The Russians first sent the *Aleut* in 1933. Japanese factory-ship expeditions began in 1934; many whaling vessels were sunk during World War II, but were rebuilt with the assistance of the American occupation government. Discussion of modern pelagic whaling may be found in J. N. Tønnessen and Arne Odd Johnsen, *The History of Modern Whaling*, trans. R. I. Christophersen (Berkeley and Los Angeles: University of California Press; London: C. Hurst and Company; Canberra: Australian National University Press, 1982). The Greenpeace protest is detailed in Robert Hunter, *Warriors of the Rainbow: A Chronicle of the Greenpeace Movement* (New York: Holt, Rinehart and Winston, 1979), 209–35.

6. Samuel Eliot Morison, "History as a Literary Art," Chapter 13 in *By Land and By Sea* (New York: Alfred A. Knopf, 1953); reprinted in Emily Morison Beck, *Sailor Historian: The Best of Samuel Eliot Morison* (Boston: Houghton Mifflin Company, 1977), 389.

7. Edmund Morris, *The Rise of Theodore Roosevelt* (New York: Coward, McCann, and Geoghegan, 1979), 388.

NOTES TO CHAPTER 1

1. Jules Michelet, *The Sea*, trans. W. H. Davenport Adams (London: T. Nelson and Sons, 1875), 209.

2. James Cook and James King, *A Voyage to the Pacific Ocean...Performed under the Direction of Captains Cook, Clerke, and Gore, in His Majesty's Ships the* Resolution *and* Discovery, *in the Years 1776, 1777, 1778, 1779, and 1780*, 3rd ed. (London: H. Hughs, 1785), 2: 321–22, 328–29.

3. G. W. Steller, *Reise von Kamtschatka nach America* (1793): 42; quoted in Frederick W. True, "The Whalebone Whales of the Western North Atlantic," *Smithsonian Contributions to Knowledge* 33 (Washington, D.C.: Government Printing Office, 1904), 33.

4. Francisco Antonio Maurelle, "Journal of a Voyage in 1775, to Explore the Coast of America Northward of California, by the Second Pilot of the Fleet, Don Francisco Antonio Maurelle, in the King's Schooner called the Sonora, and Commanded by Don Juan Francisco de la Bodega," in Daines Barrington, trans., *Miscellanies* (London: A. Hamilton, 1799), 496, 502.

5. Jean F. G. de la Pérouse, *A Voyage Round the World Performed in the Years 1785, 1786, 1787 and 1788 by the Boussole and Astrolabe* (London: A. Hamilton, 1799), 1: 358.

6. Etienne Marchand, *Voyage Autour du Monde Pendant les Années 1790–92*, 3 vols. (Paris, [1798–1800]).

7. Cook and King, *Voyage*, 2: 421.

8. *The Northwest Coast: A Century of Personal Narratives of Discovery, Conquest & Exploration from Bering's Landfall to Wilkes' Surveys 1741–1841* (New York: Edward Eberstadt and Sons, [1941]), 43.

9. Personal communication, A. C. S. Payton, London, 9 January 1985.

10. John Meares, *Voyages Made in the Years 1788 and 1789, from China to the North West Coast of America; To Which are Prefixed, an Introductory Narrative of a Voyage Performed in 1786, from Bengal, in the Ship Nootka*, 1st ed. (London: Logographic Press, 1790), 240.

11. Nathaniel Portlock, *A Voyage Round the World; but More Particularly to the North West Coast of America: Performed in 1785, 86, 87 and 88 in the King George & Queen Charlotte,*

Captain Portlock and Dixon (London: Printed for John Stockdale and George Goulding, 1789), 123.

12. Warren L. Cook, *Flood Tide of Empire: Spain and the Pacific Northwest, 1543–1819* (New Haven: Yale University Press, 1973), 138.

13. George Dixon, *Remarks on the Voyages of John Meares, Esq. in a Letter to that Gentleman* (London: printed for the author, 1790), 27.

14. John Meares, *An Answer to Mr. George Dixon, Late Commander of the Queen Charlotte, in the Service of Messrs. Etches and Company; in Which the Remarks of Mr. Dixon on the Voyages to the North West Coast of America, &c, Lately Published, are Fully Considered and Refuted* (London: Logographic Press, 1791), 17.

15. John Ehrman, *The Younger Pitt: The Years of Acclaim* (London: Constable and Company, 1969), 350.

16. Pitt, quoted in Charles Enderby, *Proposal for Re-Establishing the British Southern Whale Fishery, Through the Medium of a Chartered Company, and in Combination with the Colonisation of the Auckland Islands, as the Site of the Company's Whaling Station* (London: Effingham Wilson, 1847), 14.

17. Duke of Leeds to William Pitt, 2 June 1790, PRO 30/8, 151; quoted in Ehrman, *The Younger Pitt*, 350.

18. Dundas, quoted in Gordon Jackson, *The British Whaling Trade* (London: Adam & Charles Black, 1978), 105. Pitt agreed that the trade and fishing privilege granted by the convention, "though no new right, was a new advantage."

19. [J. Debrett], *The Errors of the British Minister in the Negotiation with the Court of Spain*, no. 11, 22 September 2790 [sic] (London: printed for J. Debrett, 1790), 38.

20. "The New South Sea Fishery or A Cheap Way to Catch Whales," by I. Cruikshank (London: H. Humphreis [Hannah Humphrey], 4 January 1791), 14 11/16" x 21 1/2" (plate measurements). The plate also contains a poem: "The Hostile Nations veiw with glad Surprise / The Frugal Plans of Ministers so Wise / But they the Censure of the World despise / Sure from their faithfull Commons of suplies / Convinced that Man must fame immortal gain / Who first dare fish with Millions in the Spanish main." The second print is anon., "Billy and Harry Fishing For Whales off Nootka Sound" (London: W. Holland, 23 December 1790), 8 3/16" x 13" (image size).

21. "An Act for the Encouragement of the Southern Whale Fishery," 26 Geo. 3, ch. 50 (1786), 955–75. The bounties became effective 15 June 1786, payable to ships clearing out between 1 May and 5 September 1786 provided they sailed south of 7° North latitude, took not less than twenty tons of oil or head matter, and returned to a British port by 1 July 1787. The bounties were payable to fifteen vessels; the first three to return earning £500 each; the second three, £400; then £300, 200 and 100 for the final three arrivals. For an additional five ships fitted out and sailed within the time period—provided they passed south of 35° South latitude and returned not less than eighteen nor more than twenty-eight months from 1 May of the year they sailed—a bounty of £700 was paid to the first ship to return, £600 to the second, and so on, until the last of the five received £300. Master and three-fourths of crew had to be British subjects, or else "sovereign Protestants intending to establish themselves."

22. Ehrman, *The Younger Pitt*, 347.

23. By 26 Geo. 3, ch. 50 (1786) whaleships were prohibited beyond 15 degrees of the Cape of Good Hope, and could not pass north of 30° South latitude in the Indian Ocean. The 1788 extension in 28 Geo. 3, ch. 20 (1787) opened the Indian Ocean eastward to the 180° meridian, as far eastward as 51° East, and northwards; "An Act for Amending an Act Made in the Twenty-Sixth Year of His Present Majesty's Reign, *For the Encourage-*

ment of the Southern Whale Fishery; and for Making Further Provisions for that Purpose," 28 Geo. 3, ch. 20 (1787).

24. A tun ordinarily constituted a cask of 252 gallons capacity. Shields's cargo, then, totalled about 35,028 gallons, or, by the American measuring standard—a 31.5-gallon barrel—approximately 1,112 barrels.

25. Samuel Enderby, Jr. to William Pitt, Paul's Wharf, London, 30 August 1790, PRO 30/8/133, fos. 32–33.

26. Lord Hawkesbury to the whaling merchants, 20 January 1791, BT 6/95, 195. The Champions were partners in an operation with shipping interests in London. John St. Barbe was one of a family of merchants affiliated with the East India Company.

27. "Vessels Employed in the Southern Whale Fishery Sailed in the Year 1791," BT 6/95, 249.

28. So skilled was Brown that he had once been sought out as an adviser to a planned, but later aborted, North Polar exploration; see Derek Pethick, *The Nootka Connection: Europe & the Northwest Coast* 1790–1795 (Vancouver: Douglas and McIntyre, 1980), 122. Curtis's ownership is documented in "Vessels Employed," 249. *Butterworth*, a former French ship built in 1778, is listed in *Lloyd's Register* for 1792 under the ownership of T. Pritzler. Priestly owned the fifty-six ton sloop *Prince Lee Boo* as of 1792, and also the eighty-six ton sloop *Jackall*, as of 1796; British whaling historian A. G. E. Jones identifies Theo. Pritzler as a sugar refiner also involved in Greenland and Davis Straits whaling; little is known of Priestly; A. G. E. Jones to Author, Tunbridge Wells, Kent, 12 July 1986.

29. Joseph Ingraham, *Journal of The Brigantine "Hope"* (Barre, Mass., 1971); quoted in Pethick, *Nootka Connection*, 122.

30. *Ibid.*, 100–1.

31. George Vancouver, *A Voyage of Discovery to the North Pacific Ocean and Round the World* 1791–1795, ed. W. Kaye Lamb (London: Hakluyt Society, 1984), 3: 983.

32. William Brown to William Curtis, Whittette Bay, Whaoo [Waikiki Bay, Oahu], 29 December 1792, PRO 30/8/127, 222r.

33. *Ibid.*, 222r-223v.

34. "Vessels Returned from the South Whale Fishery in the Year 1795," in Samuel Enderby Papers, Chalmers Collection, Mitchell Library, Public Library of New South Wales, Sydney; quoted by Rhys Richards to Author, Paremata, Wellington, New Zealand, 5 September 1985. The returns explicitly state no sperm oil or "whalebone"; Richards has suggested the "whale oil" may have been seal or sea-elephant oil, but the quantity is curiously large.

35. "An Act to Allow Ships Carrying On the Southern Whale Fishery to the North of the Equator, the Same Premium as they are Now Entitled to if they Do Not Pass the Equator," 30 Geo. 3, ch. 58 (17 June 1793), 1091–92. Whaleships were eventually permitted westward of 180° in the South Pacific by 43 Geo. 3, ch. 90 (1803).

36. Enderby, quoted in Barry M. Gough, *Distant Dominion: Britain and the Northwest Coast of North America*, 1579–1809 (Vancouver: University of British Columbia Press, 1980), 85.

37. James Colnett, *A Voyage to the South Atlantic and Round Cape Horn into the Pacific Ocean, for the Purpose of Extending the Spermaceti Whale Fisheries, and Other Objects of Commerce, by Ascertaining the Ports, Bays, Harbours, and Anchoring Births, in Certain Islands and Coasts in Those Seas, at Which the Ships of the British Merchants Might Be Refitted* (London: W. Bennett, 1798).

38. Haswell, quoted in Frederic W. Howay, ed., *Voyages of the "Columbia" to the Northwest*

Coast 1787–1790 *and* 1790–1793 (Boston: Massachusetts Historical Society, 1941), 77.

39. Frances Densmore, *Nootka and Quileute Music*, Smithsonian Institution Bureau of American Ethnology Bulletin 124 (Washington, D.C.: Government Printing Office, 1939), 1–2.

40. United States, House of Representatives, "Report on the Population, Industries, and Resources of Alaska," by Ivan Petroff. In *Seal and Salmon Fisheries and General Resources of Alaska* (Washington, D.C.: Government Printing Office, 1898), 4: 446–47. Another Tlingit legend tells how the bird named *Khunnakhateth* seizes the whale with each talon and takes it from the water, making lightning and thunder by its gigantean struggles.

41. John R. Swanton, *Haida Texts and Myths: Skidegate Dialect*, Smithsonian Institution Bureau of American Ethnology Bulletin 29 (Washington, D.C.: Government Printing Office, 1905), 281.

42. Hilary Stewart, *Looking at Indian Art of the Northwest Coast* (Vancouver: Douglas and McIntyre, 1979), 96.

43. For comparative accounts, see Franz Boas, "Tsimshian Mythology," *Thirty-First Annual Report of the Bureau of American Ethnology*, 1909–1910 (Washington, D.C.: Government Printing Office, 1916), 714f.

44. Swanton, *Haida Texts*, 281.

45. Story told by Jackson, "late chief of Skidegate," in *ibid.*, 356–57.

46. Densmore, *Nootka and Quileute Music*, 109–10.

47. *Ibid.*, 112.

48. *Ibid.*, 48.

49. It has been noted that the two-pronged bone assembly which held the shell-point blade had a male and female component; if the male barb broke during the hunt it showed that the hunter had committed adultery. If the female side broke then the charge could be levied at the wife; see Edward Sapir and Morris Swadesh, *Nootka Texts: Tales and Ethnological Narratives* (Philadelphia: Linguistic Society of America, University of Pennsylvania, 1939), 223.

50. Erna Gunther, "Reminiscences of a Whaler's Wife," *Pacific Northwest Quarterly*, 33 (January 1942): 67.

51. *Ibid.*, 66.

52. Billy Mason, quoted in Ronald L. Olson, *The Quinault Indians* (Seattle: University of Washington Press, 1967), 47–48.

53. Peter S. Webster, *As Far As I Know: Reminiscences of an Ahousat Elder* (Campbell River, B.C.: Campbell River Museum and Archives, 1983), 24.

54. United States, Department of the Interior, "Tribes of Western Washington and Northwestern Oregon," by George Gibbs. Part 2 in *Contributions to North American Ethnology* (Washington, D.C.: Government Printing Office, 1877), 1: 175.

55. James G. Swan, diary, 6 April 1866. UW.

56. See Herbert Taylor and James Bosch, "Makah Whalers," *Carnivore*, 2 (Part 3 Special Supplement, November 1979): 10–11. Whaling canoes were traded downcoast, perhaps as far as the Columbia River and the Tillamook people on the Oregon coast; see Olson, *Quinault Indians*, 68.

57. Personal communication, John Angus Thomas, Neah Bay, Wash., 1 August 1983.

58. Densmore, *Nootka and Quileute Music*, 53.

59. Ivan Doig, *Winter Brothers: A Season at the Edge of America* (New York: Harcourt Brace Jovanovich, 1980), 145.

60. Haswell, quoted in Howay, *Voyages of the "Columbia,"* 78.

61. Densmore, *Nootka and Quileute Music*, 64.

62. T. T. Waterman, *The Whaling Equipment of the Makah Indians*, University of Washington Publications in Anthropology 1 (Seattle: University of Washington Press, 1920; reprint ed., 1955), 47.

63. United States, Department of the Interior, "Tribes of Western Washington," 175.

64. Waterman, *Whaling Equipment*, 30–31.

65. Personal communication, Greig Arnold and John Angus Thomas, Neah Bay, Wash., 1 August 1983.

66. Waterman, *Whaling Equipment*, 32.

67. Densmore, *Nootka and Quileute Music*, 51.

68. David R. Huelsbeck, "The Economic Context of Whaling at Ozette" (draft of paper presented at Eleventh International Congress of Archaeological and Ethnological Science, Vancouver, 1983), 13.

69. John Meares, *Voyages Made in the Years 1788 and 1789, from China to N.W. Coast of America: With an Introductory Narrative of a Voyage Performed in 1786, from Bengal, in the Ship Nootka*, 2nd ed. (London: Logographic Press, 1791), 2: 22.

70. United States, House of Representatives, "Resources of Alaska," 392–93, 434. Poison harpoons were also tested by European and American whalemen beginning as early as the 1830s, when W. and G. Young of Leith, Scotland, used a small amount of hydrocyanic (prussic) acid in a glass tube which was attached by a heavy copper wire to the harpoon point. But crews expressed fright that they might be sickened or killed by handling blubber of poisoned whales. One such case is reported from the crew of the Nantucket whaleship *Susan Swain*, which sailed in 1833. The Frenchman Thiercelin continued tests with strychnine and curare in poison-capsule bomb-lances. He eventually tested his equipment on a whaling cruise and is said to have killed ten whales and rendered them without ill effects. For discussion, see Robert F. Heizer, *Aconite Poison Whaling in Asia and America: An Aleutian Transfer to the New World*, Smithsonian Institution, Bureau of American Ethnology, Anthropological Papers 24 (Washington, D.C.: Government Printing Office, 1943), 454–55.

71. Martin Sauer, *An Account of a Geographical and Astronomical Expedition to the Northern Parts of Russia...Stretching to the American Coast, Performed by Commodore Joseph Billings, in the Years 1785, Etc. to 1794* (London: A. Strahan, 1802), 181.

72. Gawrila Sarytschew, *Account of a Voyage of Discovery to the North-East of Siberia, the Frozen Ocean, and the North-East Sea* (London: J. G. Barnard, 1807), 2: 16. These were likely humpback whales, which among all the larger cetaceans are most likely to demonstrate the kind of behaviour described by Sarytschew, although it is conceivable they were breaching gray or right whales.

73. *Ibid.*, 1: 70.

74. See F. von Wrangell, "Statistische und Ethnographische Nachrichten Übir die Russischen Besitznugen an der Nordwestkuste von Amerika," pp. 54–55, quoted in Edward Mitchell, *Magnitude of Early Catch of East Pacific Gray Whale* (Eschrichtius robustus) (Washington, D.C.: Center for Environmental Education, 1981), 4.

75. F. H. von Kittlitz, *Denkwurdigkeiten einer Reise nach dem Russischen Amerika* (Gotha, 1858), 266–69; quoted in Heizer, *Aconite Poison*, 430. Kittlitz thought it petty of the Russian American Company not to provide the Aleuts with modern whaling lines, and speculated openly that the firm might have made a profitable new branch of the business had they paid more attention.

76. For analysis of the "lay" system and other financial characteristics of the American whaling industry, see Elmo P. Hohman, *The American Whaleman: A Study of Life and Labor in the Whaling Industry* (New York: Longmans, Green and Company, 1928; reprint ed., Clifton, N.J.: Augustus M. Kelley, 1972).

77. Charles Romontar shipped at Payta, Peru, on board ship *Seconnet* of New Bedford (1855—60 voyage); Joseph Zafouk, aboard bark *Hecla* of New Bedford (1869—70); David Kanaka, aboard ship *Eliza Adams* of New Bedford (1863—67); George Kanaka, aboard bark *Benjamin Franklin* of New Bedford (1866—67); Levy Kanaca, aboard bark *Prudent* of Stonington (1850—53); Spooner Kanaka, "of Savage Island," aboard bark *John and Winthrop* (1885—88). In 1854—55, for example, the ship *Fabius* of New Bedford carried among its crew Harry Ratomah, Jim Kanaka, "2nd Jim" Kanaka, Jack Whylootacko, George Whylootacko, Moses Kanaka, Bob Kanaka, Ben Kanaka, and Joseph Kanaka; crew list, ship *Fabius*, KWM. This renaming did not remain exclusive to the Sandwich Islanders; in 1871, the ship *Navy* sailed to the Northwest Coast and the western Arctic with two crewmen named "Frank and Joe Guam"; ship *Navy* of New Bedford, 29 May 1871. KWM (156). In 1843 the ship *Friends* of New London had sailed from that port with "Thomas New Zealand"—a Maori by description—among the crew; crew list, ship *Friends*, FRC, RG 36, Records of the Collector of Customs, District of New London.

78. See Frederic W. Howay, "Early Relations Between the Hawaiian Islands and the Northwest Coast," *Publications of the Archives of Hawaii* 5 (1930): 11—38.

79. Gardner eventually brought the two Kanakas to New Bedford, "the first brought to this place," and later returned them to the Sandwich Islands; see Daniel Ricketson, *The History of New Bedford, Bristol County, Massachusetts, Including a History of the Old Township of Dartmouth and the Present Townships of Westport, Dartmouth and Fairhaven, From Their Settlement to the Present Time* (New Bedford: published by the author, 1858), 370; also John M. Bullard, *Captain Edmund Gardner of Nantucket and New Bedford: His Journal and His Family* (New Bedford: published by the author, 1958), 35.

80. This voyage of 1808—10 mixed whaling and sealing; Captain Spence reported having been at Guadaloupe Island on the coast of Mexico and there taking 150 tons of seal oil and 11,000 seal skins. The 523-ton ship was fully outfitted for whaling, and was reported as a whaler in Sydney, Australia, in 1809. A passenger later recounted how the crew lowered for two whales, unsuccessfully, after leaving the Sandwich Islands; Archibald Campbell, *A Voyage Round the World, from 1806 to 1812; in which Japan, Kamschatka, the Aleutian Islands, and the Sandwich Islands were Visited*, 4th American ed. (Roxbury, Mass.: Allen and Watts, 1825), 113—15, 165—66; also A. G. E. Jones to Author, Tunbridge Wells, Kent, 10 April 1987.

81. The *Syren* sailed in 1819 under the house flag of the Enderby family, run by Samuel, Jr. (d. 1829), George (d. 1829) and Charles (d. 1819) since the death of patriarch Samuel in 1797. The logbook of the voyage contains a transcription from a journal kept by Captain James Carge, dated 14 July 1806, in 34°52′ N., 158°22′ E.: "Good weather and a plenty of Whales all the Way [?] 120 Miles of Lat^d & 600 Miles of Long^d Also saw Sperm Whales Lat^d 32=29 North & Long^d 163=00 West & Blackfish & Porpoises in great Numbers"; ship *Syren* of London, Frederick K. Coffin, master, 8 August 1819 - 18 April 1822, end matter. CRMM 1982.8.30.

82. Thomas Pennant, *Introduction to the Arctic Zoology*, 2nd ed. (London: Robert Faulder, 1792), 51.

83. Ship *Tiger* of Stonington, Mrs. William E. Brewster, keeper, 26 July 1846. MSM (38).

84. Bark *Pantheon*, 12 July 1843, on the Northwest Coast.

NOTES TO CHAPTER 2

1. Herman Melville, *Mardi: And a Voyage Thither* (Evanston, Ill.: Northwestern University Press and the Newberry Library, 1970), 6. Melville deserted from the whaleship *Acushnet* of Fairhaven at Nukahiva, in the Marquesas Islands, in July 1842; had he remained aboard, he would have spent the summers of 1843 and 1844 hunting right whales on the Northwest Coast. The *Acushnet* narrowly escaped disaster in those waters, very nearly fetching ashore on the "Ischericows Isle" (Chirikof Island), southwest of Kodiak Island, in a dense fog on 25 June 1844. By deserting, Melville missed the opportunity to learn firsthand about the "Hartz logs" he disparaged in 1849 in *Mardi*; ship *Acushnet* of Fairhaven. Maury 12, no. 3.

2. Both vessels noted in Charles Lyon Chandler, "List of United States Vessels in Brazil, 1792–1805, Inclusive," *Hispanic American Historical Review*, 26 (November 1946): 606.

3. Worth, quoted in E. C. Williams, *Life in the South Seas: History of the Whale Fishery; Habits of the Whale; Perils of the Chase and Methods of Capture* (New York: Polhemus and DeVries, 1860), 7.

4. "Career of Captain Benjamin Worth," Nantucket *Inquirer*, 28 February 1825; noted in Edouard H. Stackpole, *The Sea-Hunters: The New England Whalemen During Two Centuries 1635–1835* (New York: J. B. Lippincott Company, 1953), 399. Captain Gardner returned the whaleship *Balaena* to New Bedford in 1823 with a report on the activities of a Russian gunboat on the far Northwest Coast, but it is not known if he got this information personally, or from a fur-trade captain in more southerly waters; see Chapter 3.

5. "Certificates of Registry Issued at Nantucket, Massachusetts 1815–1870," NHA (67), 1832, fo. 3a. Later historians claimed that Folger took his first Northwest Coast whales in 1835; they were probably misled by the notes of Frederick C. Sanford, a part-owner of the *Ganges*, whose remarks could easily have been misconstrued—and were—in Alexander Starbuck, *The History of Nantucket: County, Island and Town* (Boston: C. E. Goodspeed and Company, 1924), and parenthetically misquoted by others.

6. Samuel L. Dana, Linus Child, Alex[ander] Wright, and James B. Francis, *A Report on Rosin Oil, by a Committee Appointed by the Agents of the Manufacturing Companies at Lowell, Massachusetts*, 2nd ed. (Lowell: S. J. Varney, 1852), 5.

7. United States, Commission of Fish and Fisheries, "Notes upon the History of the American Whale Fishery," by F. C. Sanford. In *Report of the Commissioner for 1882* (Washington, D.C.: Government Printing Office, 1884), 212. Winship reached that coast as early as 1806, and in that year met trader John D'Wolf at Sitka. D'Wolf had seen whales also—right whales—and soon afterward wrote: "This tract of ocean, from longitude 130° west, along the entire coast of Alashka and through the seas of Kamchatka and Ochotsk, was at that time the great place of resort of the right whale. Persecuted in all its other haunts, it had sought a refuge in this northern region, where as yet a single whale-ship had never made its appearance. We were frequently surrounded by [whales]."; John D'Wolf, *A Voyage to the North Pacific and a Journey Through Siberia More Than Half a Century Ago* (Bristol, R.I.: Rulon-Miller Books, 1983), 73–74. The presence of sperm whales in the northerly waters was much later corroborated by the narrative of Peter Corney's voyaging in the North Pacific between 1813 and 1818. Unpublished until the end of the century, Corney had written: "Between the latitudes of 30° and 46° north, and longitude of 180° and 123° west, we saw many shoals of sperm whale"; Peter Corney, *Voyages in the Northern Pacific: Narrative of Several Trading Voyages from 1813 to 1818, between the Northwest Coast of America, the Hawaiian Islands and*

    China, with a Description of the Russian Establishments on the Northwest Coast (Honolulu: Thomas G. Thrum, 1896), 42.

8. United States, Commission of Fish and Fisheries, "Notes," 212.

9. Frederick C. Sanford, [Account of Whale Ship Departures from Nantucket (1815–1868)]. Untitled ms (microfilm). ODHS.

10. United States Commission of Fish and Fisheries, "Notes," 212; and Sanford, [Whale Ship Departures], 42. "Pudder Bay" is a typographic error for Rudder Bay, near Cape San Lucas, Baja California, and is not to be mistaken for Pedder Bay, British Columbia.

11. The *Ganges*, built 1809 at Haverhill, Massachusetts, measured 91' x 23'10" x 12'11" and 265 24/95 tons at the registration of 1832; Certificates of Registry Issued at Nantucket, NHA (67), 1832, fo. 3.

12. Wood, 1: 202. The *Ganges* brought home 1,344 barrels of sperm oil and 340 barrels of whale oil (presumably all from the Northwest Coast). By the time Alexander Starbuck compiled his massive registry of American whaling voyages through the year 1876, the entire cargo—1,644 barrels—was credited as sperm oil.

13. A. G. E. Jones, *Ships Employed in the South Seas Trade 1775–1861 (Parts I and II) and Registrar General of Shipping and Seamen: Transcripts of Registers of Shipping 1787–1862 (Part III)*, Roebuck Society Publication 36 (Canberra: Roebuck Society, 1986), 234.

14. Frederick Debell Bennett, *Narrative of a Whaling Voyage Round the Globe, from the Year 1833 to 1836: Comprising Sketches of Polynesia, California, the Indian Archipelago, Etc.; with an Account of Southern Whales, the Sperm Whale Fishery, and the Natural History of the Climates Visited*, 2 vols. (London: Richard Bentley, 1840).

15. *Ibid.*, 1: 262. Bennett gave the ship's position as 50° North, 133° West, suggesting they were in fact south of the Charlottes.

16. *Ibid.*.

17. Ship *Canton* of New Bedford, 21 March 1836. KWM (252).

18. Ship *Endeavour* of New Bedford, 18 February 1836. KWM (81). Captain Stetson's visit aboard the French whaleship is corroborated in another journal kept aboard the *Endeavour*, under date of 19 February 1836. KWM (271).

19. Samuel S. Stafford, a "green hand" bound for the Northwest Coast in 1845–46 in the bark *Laurens* of Sag Harbor, New York later told an entirely different version of the discovery of the northern whaling grounds. It had been explained to him that a Russian bark bound from Petropavlovsk to San Francisco had been blown off course in inclement weather, and "being some hundred miles to the Southward of Bherings Straits, fell in with immense schools of Right whales, most of them of enormous size." On arrival at San Francisco, then a hide depot, the crew told the story, and "in a few months the news had reached New Bedford and other Whaling ports, when a perfect furore was created, exceeded only by the excitement occasioned by the discovery of gold in California a few years later"; [Samuel S. Stafford], "There She Blows: Journal of a Whaling Voyage Around the World in the Barque Laurens,' by 'Uncle Sam'." Unpublished typescript. KWM.

20. See Thierry Du Pasquier, *Les Baleiniers Français au XIXe Siecle (1814–1868)* (Grenoble: Terre et Mer, 1982), 163–64. Vessels employing a French captain and at least a two-thirds French crew earned 40 francs per ton; those with non-French captains, provided at least half the crew were French, earned 30 francs per ton.

21. Ship *Constance* of Havre, 15 June 1833. FNA 5JJ354 (microfilm). The *Constance* was not the first French whaling ship to reach California; the *Océan*, Thimothy Gardner, had sailed there during the voyage of 1827–29, and the *Cap Horn*, Thébaud, arrived during its voyage of 1830–32; see Du Pasquier, *Les Baleiniers Français*; 235–36; this whaling

ground below 36° North was also familiar to American whalemen moving northward from the coast of Mexico and Baja California.

22. Belford Hinton Wilson (British Consul General at Paita, Peru), "Return No. 1: Return of the French Whale Ships Which have Anchored in the Port of Payta, during the Year Ending the 31st of December 1835." PRO, BT 2/8, 308.

23. Gustave Duboc, *Les Nuées Magéllaniques: Voyage au Chili, au Pêrou et en Californie à la Pêche de la Baleine*, Part 1 (Paris: Amyot, 1853), 40. Duboc's fictionalized account appears to retain correct details of the voyage of the *Gange*.

24. Ship *Gange* of Havre, 15–16 June 1835. FNA 5JJ355 (microfilm).

25. *Ibid.*, 9 June 1835. Chaudière reported his longitudes west of Greenwich at a time when most French captains continued to use Paris as their 0° meridian—a difference of about five degrees. It may also be important to note that the *Gange* carried a chronometer, an expensive navigating instrument used for determining longitude, rarely found aboard whaleships prior to 1840; its presence implies that the officers were more than casually interested in their position.

26. Duboc claimed later that the ship reached 50° North and 160° West; Duboc, *Les Nuées Magéllaniques*, Part 1, 171.

27. Du Pasquier, *Les Baleiniers Français*, 170, 239; thirty whales were taken on the voyage, all told.

28. Duboc, *Les Nuées Magéllaniques*, Part 1, 173–74; Louis La Croix, *Les Derniers Baleiniers Français: Un Demi-Siècle d'Histoire de la Grande Pêche Baleinière en France de 1817 à 1867* (Nantes: Aux Portes du Large, 1947), 285. The date of the wreck is variously given, often 22 November. Lloyd's List reports the loss date as 12 November and names Mareschal as captain, though this seems to be in error. The crew were saved; Lloyd's List, 22 March 1836, cited in Jones, *Ships Employed in the South Seas Trade*, 110.

29. Humphrey W. Seabury, "Right Whaling." Unpublished typescript (1868), 4–5. KWM.

30. [Stafford], "There She Blows," 104. Other aspects of this account are suspect; this statistic may be at odds with truth.

31. Bark *Pantheon*, 16 June 1843. On 13 July the *Pantheon* spoke the *Alciope* of Sag Harbor, whose men reported that one of the eleven whales they took on the Northwest had made 253 barrels; *Pantheon*, 13 July 1843.

32. Duboc, *Les Nuées Magéllaniques*, Part 1, 172–73.

33. Ship *American* of Nantucket, 4 July 1838 - 25 July 1841, undated end matter. NHA. The attribution is in question, but probably the *Mary* of Edgartown, Fisher, at sea from 1838 to 1840 and returned with 2,200 barrels of [right] whale oil among its cargo. Otherwise perhaps the *Maria* of Nantucket, at sea during 1836–40 under the command of Elisha H. Fisher.

34. Sanford, quoted in Henry T. Cheever, *The Whale and His Captors: Or the Whaleman's Adventures and the Whale's Biography, as Gathered on the Homeward Cruise of the "Commodore Preble"* (New York: Harper and Brothers, 1850), 104.

35. Sanford, [Whale Ship Departures], 42.

36. The *Ville de Bordeaux* sailed 5 January 1837 and returned to Havre on 5 January 1841. Largeteau traversed the world, and reached 52° North, 155° East near the Kamchatka coast; see [Louis] Thiercelin, *Journal d'un Baleinier: Voyages en Océanie*, 2 vols. (Paris: L. Hachette et Cie., 1866). Largeteau's travels around Kamchatka, the Kuril Islands, and the Japan Sea are noted in "Extract uit een Verslag aan den Minister den Fransche Marine, door den Kapitein ter Zee Cecille, Commanderende de Korvet *l'Héroine*; ter Bescheriving der Walvischvangers in de *Indische* en *Australische* Zeeën, Gedurende de

Jaren 1837, 1838, en 1839," in C. Brandligt, *Geschiedkundige Beschouwing van de Walvisch-Visscherij* (Amsterdam: T. Nusteeg, Jr., 1843), 36.

37. Sandra Truxtun Smith, "A History of the Whaling Industry in Poughkeepsie, N.Y. (1830—1845)." Unpublished typescript (1956), 4, 11. FDR. The latter firm was named after Dutchess County, in which Poughkeepsie is situated.

38. Ship *Elbe* of Poughkeepsie, 14 December 1835. IMA (A105).

39. *Ibid.*, 27 December 1834. The first meeting took place on 27 November.

40. Smith, "Whaling Industry in Poughkeepsie," 19.

41. Poughkeepsie *Telegraph*, 31 May 1837; quoted in Smith, "Whaling Industry in Poughkeepsie," 20.

42. United States, Commission of Fish and Fisheries, "History of the American Whale Fishery: From Its Earliest Inception to the Year 1876," by Alexander Starbuck. In *Report of the Commissioner for 1875—1876*. Part 4, Appendix A (Washington, D.C.: Government Printing Office, 1876), 352—53. Hereafter cited as Starbuck. The *Elbe* sailed again in 1840 under Merrihew and was lost in Cook's Straits, New Zealand, in December 1841.

43. M. E. Bowles, "Some Account of the Whale-Fishery of the N. West Coast and Kamschatka," *Polynesian*, 4 October 1845, 82.

44. Bennett, *Narrative of a Whaling Voyage*, 261.

45. George Pelly to William Smith, Honolulu, 27 November 1837. HBC A.11/61, fo. 33d. It is possible that the term "N W Coast" was taken to mean Alta California, but Hudson's Bay Company terminology is usually clear.

46. United States, Works Projects Administration, *New Bedford Whalemen's Inward Manifests*, WPA Report 836. Typescript, 54. KWM.

47. Ship *Timoleon* of New Bedford, 3 May - 11 September 1838. RP, Wh/T585/ 18351. Two others whales were killed but sunk.

48. Ship *Ajax* of Havre, n.d.; ship *Gange* of Havre, n.d.; ship *Cachalot* of Havre, n.d. All FNA CC^{5}611.

49. Starbuck, 362—63.

50. See, for example, ship *Abigail* of New Bedford, 24 October 1835 - 22 October 1838, p. 191. MSM (96).

51. A circular tip of baleen at the distal end of the bayonet helped prevent injury during practice; such a bayonet is illustrated in Francis Bannerman Sons, *Illustrated and Descriptive Catalogue of Military Goods* (New York: Francis Bannerman Sons, 1929), 47.

52. Henry H. Crapo, *The New Bedford Directory*, 5th ed. (New Bedford: Benjamin Lindsey, 1845). Whale oil was also on the rise, but not so dramatically; it showed a steady increase from 31 3/4 cents per gallon in 1841 to 36 1/2 cents in 1844. Sperm oil fluctuated wildly, coming in at 94 cents per gallon in 1841, dipping to 73 cents and 63 cents in 1842 and 1843, respectively, and finishing 1845 at 90 1/2 cents.

53. Ship *Logan* of New Bedford. Maury 71, no. 1. See also Wood, 1: 287. The *Logan* returned to New Bedford 13 December 1841 and turned out 1,339 barrels of sperm oil and 946 barrels of whale oil, virtually all of the latter apparently taken on the Northwest Coast during the summer of 1840; it had not been on board when the ship left Paita, Peru, on 15 March 1840 but was there when the vessel returned to Paita the following November.

54. Ship *Phoenix* of New London. KWM (21).

55. [Whales seen and taken], abstract, ship *Superior* of New London. NLCHS Ms P.B495m 1914.04.78B.

56. Starbuck, 368—69.

57. N. and W. W. Billings to Captain William Fitch, New London, 4 October 1842. NLCHS Ms P.B495 m 1914.04.46A; fos. 1–2.

58. *Ibid.*, fo. 3. They would have had an argument with Sir George Simpson. At San Francisco in 1842 on behalf of the Hudson's Bay Company, Simpson thought Yerba Buena a better port than the Sandwich Islands, by way of access and departure and its superior sources of supply, but found it an indolent place, and strapped with fiscal regulations which prevented whaleships from using it freely. He complained that the whaleman "is by law forbidden to remain more than forty-eight hours, unless he has previously presented himself at Monterey and paid duty on the whole of his cargo. What wonder, then, is it that with such a government and such a people Whalers' Harbor is merely an empty name?" Sir George Simpson, *Narrative of a Voyage to California Ports in 1841–42, together with Voyages to Sitka, the Sandwich Islands & Okhotsk, to Which are Added Sketches of Journeys Across America, Asia, & Europe*, ed. Thomas C. Russell (San Francisco: Thomas C. Russell, 1930), 42–43. The whalemen who visited San Francisco in 1841 seem to have found supplies and provisions in short supply; see Nathan Spear to Thomas Oliver Larkin, in Thomas O. Larkin, *The Larkin Papers: Personal, Business, and Official Correspondence of Thomas Oliver Larkin, Merchant and United States Consul in California*, ed. George P. Hammond (Berkeley and Los Angeles: University of California Press, 1951–64), 1: 132.

59. Starbuck, 394–95.

60. [N. and W. W. Billings] to Alexander Hart, New London, 26 September 1842. MSM, VFM 1385 MC 81.69, fo. 3. Hart apparently found no reason to use his expanded permissions; the Bering Strait was not breached by a whaleship until 1848.

61. The first New London departures declaring for the Northwest Coast, as noted in Starbuck's 1876 tables, were the *Mogul* and the *Helvetia*, in 1842.

62. Ship *John Wells*, Wood; cited in Harry B. Weiss, Howard R. Kemble, and Millicent T. Carre, *Whaling in New Jersey* (Trenton: New Jersey Agricultural Society, 1974), 66. For other returns see Starbuck.

63. George Simpson, *An Overland Journey Round the World, During the Years 1841 and 1842*, Part 2 (Philadelphia: Lea and Blanchard, 1847), 93.

64. Du Pasquier, *Les Baleiniers Français*, 182.

65. [Eugène] Duflot de Mofras, *Duflot de Mofras' Travels on the Pacific Coast*, trans. and ed. Marguerite Eyer Wilbur (Santa Ana, Calif.: The Fine Arts Press, 1937), 1: 269. During the 1841 summer campaign the *Elisa* reported having sailed in 50–57° North, 145–159° West, and having struck twenty-seven whales, caught seven, for 800 barrels and 146 "paquets" of baleen; FNA CC⁵611.

66. Ship *Magnolia* of New Bedford, 7 August 1841. KWM (359). The other two whaleships spoken were the *South Carolina* and the *Milton*, both of which arrived at Lahaina in the fall.

67. Ship *Magnolia*.

68. Ship *Orozimbo* of New Bedford, 29 July 1841. Harvard F6870.65F.

69. *Ibid.*, 23 July 1841.

70. Joseph G. Clark, *Lights and Shadows of Sailor Life, as Exemplified in Fifteen Years' Experience, Including the More Thrilling Events of the U.S. Exploring Expedition* (Boston: John Putnam, 1847), 226–27.

71. Ship *Orozimbo*, 6 August 1841.

72. Clark, *Lights and Shadows*, 227; and Charles Wilkes, *Narrative of the United States Exploring Expedition During the Years 1838, 1839, 1840, 1841, 1842* (Philadelphia: Lea and Blanchard, 1845), 4: 489. In his published reports, Wilkes later discounted the effects

of diet as a cause of scurvy, attributing that "horrible disorder" to mental rather than physical deficiency, less to a shortage of fresh meat and vegetables than to "the long period passed at sea, aggravated by the despondency arising from want of success"; Wilkes, *Narrative*, 5: 501. In the case of the *Orozimbo*'s crew, at least, there had been no "want of success."

73. Briton Cooper Busch, *Alta California* 1840–1842: *The Journal and Observations of William Dane Phelps, Master of the Ship "Alert"* (Glendale, Calif.: Arthur H. Clark Co., 1983), 219. The *Alert* is best known for having brought home Richard Henry Dana from the California hide-droughing trade, a voyage made famous in Dana's *Two Years Before the Mast* (1840). Captain Bartlett was probably unaware that Phelps's vessel had been so memorialized.

74. Stephen Reynolds to Thomas Oliver Larkin, Oahu, 17 October 1841; in Larkin, *Larkin Papers*, 1: 128. Hoyer came to the Sandwich Islands in 1836 as master of the New Bedford whaling bark *Ospray*; he took the *Fama* from 1839 until it was sold out of whaling in 1841 and later returned to the Northwest Coast as master of the Bremen whaleship *Sophie*. Hoyer's consultation with Simpson is noted in Boyd Huff, *El Puerto de los Balleneros: Annals of the Sausalito Whaling Anchorage* (Los Angeles: Glen Dawson, 1957), 15.

75. The *Fama* was spoken in 53°30' North, 153° West by the whaleship *Sapphire*; ship *Sapphire* of Salem, 15 June 1841. Essex M.656–1839S2.

76. Abel Du Petit-Thouars, *Voyage Autour du Monde sur la Frégate la Vénus, Pendant les Années* 1836–1839 (Paris: Gide, 1840–46); see particularly 1: xiv-xvi. See also Léonce Jore, *L'Océan Pacifique au Temps de la Restauration et de la Monarchie de Juillet* (1815–1848), Vol. 1 (Paris: Éditions Besson et Chantemerle, 1959).

77. Duflot de Mofras, *Duflot de Mofras' Travels*, 1: 268–69. He recommended ten safe harbours along the Northwest Coast, from Unalaska and Unimak in the north to Nootka and "Puget's Bay in the Strait of Juan de Fuca" in the south, as well as San Francisco and Monterey in California.

78. Wilkes, *Narrative*, 5: 495–96.

79. Ship *Leonidas* of Bristol, Benjamin L. West, keeper, 31 January 1844. KWM (624).

80. Charles Wetherby Gelett, *A Life on the Ocean: Autobiography of Captain Charles Wetherby Gelett, a Retired Sea Captain, Whose Life Trail Crossed and Recrossed Hawaii Repeatedly* (Honolulu: Hawaiian Gazette Company, 1917), 32. Maury's abstract of the 1841 *India* voyage reveals that Gelett took eighteen whales between 51° and 55°48' North and returned to New Bedford 14 February 1843 with 2,541 barrels of whale oil, 679 barrels sperm oil, and 30,000 pounds of baleen. He promptly returned to the North Pacific in the *Uncas*; see ship *India* of New Bedford. Maury 71, no. 5; and Starbuck, 364–65.

81. Charles Haskins Townsend, "The Distribution of Certain Whales as Shown by Logbook Records of American Whaleships," *Zoologica*, 19 (3 April 1935); Tables A, C.

82. Charles M. Scammon, *The Marine Mammals of the North-Western Coast of North America, Together with an Account of the American Whale-Fishery* (San Francisco: John H. Carmany and Company, 1874), 66. Some captains further identified a smaller area within the gulf, lying between Pamplona Reef and the mainland abeam Mount Fairweather in 58°15' North, as the "Fairweather Ground"; this area is so listed yet for the benefit of fishermen.

83. "North West Coast Fishery," *Shipping & Commercial List*, 19 January 1848. The tabular material is furnished without a description of the grounds, but seems to omit vessels returning from Kamchatka:

Year	Ships	Barrels Each (avg.)	Total
1839	2	1,400	2,800
1840	3	587	1,760
1841	20	1,412	28,200
1842	29	1,627	47,200
1843	108	1,349	146,800
1844	170	1,528	259,570
1845	263	953	250,600
1846	292	869	253,800

84. "Sandwich Islands," *Sailor's Magazine*, 16 (May 1844): 290–91. Not all of these vessels sailed to the Northwest Coast; many were bound for the sperm-whaling grounds off Japan.

85. Chester S[mith] Lyman, *Around the Horn to the Sandwich Islands and California 1845–1850*, ed. Frederick J. Teggart (New Haven: Yale University Press, 1924), 179.

86. "Abstract of Hawaiian Laws, Respecting Vessels, Harbors and Customs" (broadside), n.p. [ca. 1850]. Lucas.

87. Ship *Julian* of New Bedford, F. Cady, keeper, 18 April 1848. KWM (404).

88. Giles G. Hempstead to Mary Hempstead, Mowee (Maui), 27 March 1844. NLCHS Ms. .H378g m1956.05.68. Hempstead had shipped in the New London whaleship *Mentor* the previous year, at the age of 17; crew list, ship *Mentor*. FRC, RG 36, Records of the Collector of Customs, District of New London.

89. "Maritime Resources of the United States," *Times* (London), 9 June 1846; reprinted in *Sailor's Magazine*, 19 (September 1846): 8–9. He wrote: "Perhaps on the coast of Vancouver's Island, or of the Oregon Territory, settlements might be made for the relief and improvement of the whale fisheries. . . . But nothing of this kind has been effected by British enterprise."

90. Quoted anonymously in *WSL*, 24 December 1844, 167.

91. William Armstrong Fairburn, *Merchant Sail* (Center Lovell, Maine: Fairburn Marine Educational Foundation, 1945–55), 2: 1009.

92. J. Ross Browne, *Etchings of a Whaling Cruise, with Notes of a Sojourn on the Island of Zanzibar: To Which is Appended a Brief History of the Whale Fishery, its Past and Present Condition* (New York: Harper and Brothers, 1846), 544; see also Du Pasquier, *Les Baleiniers Français*, 163–64; and "Bounties to Men and Ships Employed in the French Fishery," *Fisher's National Magazine and Industrial Record* (New York) 3 (October 1846): 410. After expiry, the new law of 22 July 1851 returned the bounties to the 1832 level but this encouragement was largely too late to again reinvigorate the declining French whale fishery.

93. Winslow, see Du Pasquier, *Les Baleiniers Français*, 102.

94. D. Baldwin, "Bibles, Tracts, and Papers Wanted at Lahaina," *Sailor's Magazine*, 14 (August 1842): 375.

95. Ship *Nancy* of Havre, Thomas Jay, master, took nine whales between 16 June and 23 July 1842 and reached 57° North, 149°30' West; ship *Faune* of Havre, O. de Grandsaigne, master, struck nine whales from 26 May to 26 August 1842 in 57° North, 158° West; ship *Nil* of Havre, Gilbert Smith, master, struck twelve whales between 16 May and 6 August in 50–58° North, 148–160° West; and ship *Adele* of Havre, Luhrs, master, struck ten from 1 June to 14 August in 52–57°30' North, 148–152° West; ship *Liancourt* in Du Pasquier, *Les Baleiniers Français*, 244; all others in FNA CC⁵611.

The *Adele* was spoken by the *Orozimbo* on the Northwest Coast on 1 August 1842, nine months out with 1,300 barrels; Ship *Orozimbo*, 1 August 1842.

96. Ship *Elisa* of Havre, Malherbe, master, 4 April-30 August 1842, in 34–56° North, 152–174° West; forty-four whales struck, fifteen caught, 2,000 barrels of oil, 281 "paquets" of baleen. Ship *Angelina* of Havre, Edouard Sebastian L'hynne, master, 23 May-26 August 1842, 39°-56°07' North, 150°47'-161°04' West; twenty whales struck, fourteen taken. The *Angelina* called in at San Francisco during this cruise. Both in FNA CC⁵611.

97. Ship *Orozimbo*, 15 June 1842. This whaleship cannot otherwise be identified.

98. Details of the French whaleship in Andre Manguin, *Trois Ans de Pêche de la Baleine: D'Après le Journal de Pêche du Capitaine Dufour (1843–1846)* (Paris: J. Peyronnet et Cie., 1938), 18, 24.

99. Du Pasquier, *Les Baleiniers Français*, 220–21, 230. Thayer was afterward captain of both the *Pallas* and the *Réunion*, in which he seems to have finished his whaling career on return to France in 1847.

100. Ship *Tiger*, 23 July 1846; the French captain is not named.

101. Noted in *The Friend*, 4 (2 March 1846), 37.

102. "Sandwich Islands," *Sheet Anchor* 3 (20 December 1845), 187.

103. Statistics published in *WSL*, 30 January 1844, 533; in the mid-1840s much of the right-whale oil collected by American crews found its way to markets in northern Europe.

104. Wanda Oesau, *Die Deutsche Südseefischerei auf Wale im 19. Jahrhundert* (Gluckstadt: J. J. Augustin, 1939), 32.

105. *Ibid.*. Some of the whale oil and baleen was collected on the Northwest Coast during the summer of 1843.

106. *WSL*, 10 September 1844, 107. A few of these probably went to Kamchatka instead of the American coast.

107. Ship *Montpelier* of New Bedford, 4 August 1846. Kelly; ship *Edward* of New Bedford, 18 August 1846. ODHS (551).

108. Ship *Virginia* of New Bedford, John Francis Akin, keeper, 12 January 1845. KWM (407).

109. Noted in *WSL*, 21 January 1845, 183; and in *Sheet Anchor*, 16 August 1845, 127. The *Clementine* was on its first voyage for J. F. Iken and Company, under Captain Hilken, and did not return home until July 1848; Oesau, *Die Deutsche Südseefischerei*, 69.

110. Ship *Columbus* of New Bedford, 25 July 1844. PMS 656/1844C. The deceased captain was probably Johan Diedrich Otten, the master who sailed from Germany for the owners, D. H. Wätjen and Company, since he was replaced before the end of the voyage by Samuel H. Austin. Cornelius Hoyer, master of record during 1843–44, was engaged at the Sandwich Islands, perhaps in response to Otten's death.

111. "Frederick C. Sanford's Letters," Nantucket, 22 March 1878. NHA (129), Box 9, Book 19, 107.

112. "Bark Rica, Darmer, of and from Wolcast [sic], Prussia, arrived at this port July 31st. Put in for officers and to complete fitting out for Indian Ocean and N.W. Coast, whaling"; *WSL*, 5 August 1845, 87.

113. Ship *William Hamilton* of New Bedford, 17 May 1846. KWM (548). The American bark *Favorite* reported a Dutch brig in 56° North, 146° West as late as 1859, but this may not have been a whaling vessel; bark *Favorite* of Fairhaven, 24 April 1859. KWM (512).

114. *The Friend*, 1 November 1844, 104; ship *Eugene* of Stonington, 30 July 1845. MSM (69).

115. "List of Whale Ships at Honolulu, Oahu, Sandwich Islands, from January 1, to November 25, 1842," *Sailor's Magazine*, 15 (August 1843): 374–75. Soldering data in Frederick P. Schmitt, *Mark Well the Whale!: Long Island Ships to Distant Seas* (Port Washington, N.Y.: Kennikat Press, 1971), 103.

116. There may have been two vessels of similar or identical name. O'May notes one under Captain Clements and another under William Lovett, Jr. at the same time. Tonnage reports vary; 237 tons reported by Lawson, 300 elsewhere. The 237-ton vessel was wrecked in Japan in 1851. See Harry O'May, *Wooden Hookers of Hobart Town* (Hobart Town: L. G. Shea, [n.d.]), 39–41; and Will Lawson, *Blue Gum Clippers and Whale Ships of Tasmania* (Melbourne: Georgian House, 1949), 62.

117. Spoken on the Northwest Coast by the ship *Montpelier* of New Bedford, 12 July 1846. Yale (HM 21, v. 2). *Maria Orr* also noted in Max Colwell, *Ships and Seafarers in Australian Waters* (Melbourne: Lansdowne Press, 1973), 42. The *Maria Orr* was the first square-rigged ship built in Van Dieman's Land; Captain Scott McArthur took the vessel to Kodiak; see Lawson, *Blue Gum Clippers*, 53.

118. The first Maritime whaling began in 1784, when Governor Parr of Nova Scotia induced some Yankee families to begin the business at Dartmouth; for a discussion of whaling from Maritime ports, see Frederick William Wallace, *In the Wake of the Wind-Ships: Notes, Records and Biographies Pertaining to the Square-Rigged Merchant Marine of British North America* (New York: George Sully and Company, 1927), 2–7.

119. Including the Bedford Commercial Insurance Company of New Bedford ($15,000), Mutual Marine Insurance Company of New Bedford ($11,000), Neptune Insurance Company of Boston ($5,000), and the Tremont Insurance Company of Boston ($5,000); see Wood, 1:insert at 608–9 (microfilm).

120. James Arnold, "Report on the Whale Fishery," in *Proceedings of the National Convention for the Protection of American Interests* [1841] (New York: Greeley and McElrath; Thaddeus B. Wakeman, 1842), 41. The British tariff, £26.12 plus "5% additional" on the Imperial ton of 302.4 gallons was, according to the National Convention, "meant to be prohibitory."

121. Edward T. Perkins, *Na Motu: Or Reef-Rovings in the South Seas* (New York: Pudney and Russell, 1854), 46–47.

122. Charles B. Reynolds, "Cure for Romance, or a Cruise on the 'North West'," *The Friend*, 1 October 1845, 150–51.

123. Sir George Simpson, in Thomas Lowe, "Journal of a Voyage from London to Sandwich Islands and Fort Tako, N.W. Coast of America, in the Barques 'Vancouver' & 'Cowlitz' 1841 & 2," 67. HBC E.25/1.

124. Ship *Robin Hood* of Mystic, 15 July 1846. MSM (48).

125. Ship *Java* of Fairhaven, 4 May 1855, in 51°54′ North; 7 May 1855, in 53°31′ North. Typescript. SAG.

126. Cushman noted in *WSL*, 20 January 1846, 182; ship *Gratitude* of New Bedford, 21 June 1844. KWM (433).

127. Ship *Edward*, 23 June 1848. [Stafford], "There She Blows," 102–4 recounted that the firing of a musket did not prevent two whaleboats' crews from being separated from the whaleship *Akbar* on the Alaskan coast during the summer of 1844. Allegedly, these lost mariners made their way to the coast, where they found fresh water, shot an elk and harpooned a "40-pound salmon" for provisions. They then sailed southward five days and arrived at the Columbia River in time to fall in with their own whaleship, which had called in for fresh water. This voyage does not appear in Starbuck's comprehensive 1876 tally and the events of the separation must be considered suspect.

128. "Epitaph," entered in a manuscript journal of three whaling voyages, including the

Northwest Coast voyage of the ship *Eliza Adams* of New Bedford (1852–54); the verse was probably written by the journal keeper, John Jones. KWM (319).

129. Ship *Callao* of New Bedford, 13 September 1843. KWM (41). The ship was in about 52°30′ North, 142°40′ West.

130. Ship *Gratitude*, 18 August 1844.

131. Ship *Columbus*, 25 July 1844.

132. Erastus Bill, *Citizen: An American Boy's Early Manhood Aboard a Sag Harbor Whale-Ship Chasing Delirium and Death Around the World, 1843–1849, Being the Story of Erastus Bill Who Lived to Tell It*, with notes by Robert Wesley Bills (Anchorage: O. W. Frost, 1978), 90.

133. Ship *Armata* of New London. KWM (21); Ship *Golconda* of New Bedford. KWM (244); Ship *Splendid* of Edgartown. KWM (187); Ship *Java* of Fairhaven.

134. Bark *Pantheon*, 24 July 1843, in 55° North, 148° West.

135. [Whales Seen and Taken], ship *Superior* of New London, 1841; Ship *Magnet* of Warren. KWM (294); Ship *Julian* of New Bedford.

136. Ship *South Carolina* of New Bedford. KWM (445); Ship *Cabinet* of Stonington. KWM (566); Ship *William Hamilton* of New Bedford. KWM (548).

137. Ship *Golconda*.

138. *WSL*, 27 January 1846, 186.

139. *WSL*, 3 March 1846, 206.

140. Ship *Gratitude*, 9 June 1844.

141. Cheever, *The Whale and His Captors*, 99.

142. Bowles, "Whale-Fishery of the N. West Coast," 83.

143. Ship *Java* of Fairhaven; 15 [June?] 1855. The cutting stage was a wooden platform suspended over the whale. The flensing crew stood on it and cut-in the blubber with long-handled knives and spades.

144. The *Phoenix* matter was apparently adjudicated, since the cruise continued with great success and nothing more was entered in the logbook; ship *Phoenix*, 10 July 1840; ship *Cachalot* of Havre, [n.d.].

145. Noted in *The Friend*, 15 August 1846, 124–25.

146. *Ibid.*, 126; and *WSL*, 12 January 1847, 175.

147. Both arrived at Honolulu on 6 June 1846. Captain Lester abandoned his pledge and reported the mutiny to the U.S. consul; more than a half dozen of his crew were confined in the fort, and five were eventually sent to the United States for trial, accompanied by Ira Horton—the first mate—and a seaman, Samuel L. Main, to stand as witnesses. Captain Lester shipped another crew and sailed again on 24 June; see *The Friend*, 15 August 1846, 126. A recent re-evaluation of the *Meteor* mutiny suggests that it may have resulted from more than customary hard service; in 1845 the crew had attempted to have the ship condemned at Honolulu as unseaorthy but failed, and there is a strong suspicion of inept seamanship and brutality on the part of the vessel's thirty-four-year-old captain. See Jane Litten, "Greenhand Hero or Mutineer?: Mutiny Aboard the Whaleship *Meteor*, 1846," *Log of Mystic Seaport*, 39 (Summer 1987): 54–64.

148. *The Friend*, 15 April 1846, 60–61.

149. Ship *Columbus*, 11 July 1844.

150. Ship *Gratitude*, 21 June 1844.

151. Thomas Joseph and Antonio Sylva, obituaries in *The Friend*, 2 November 1846, 166; William Maui, in *The Friend*, 15 October 1849, 64.

152. In *The Friend*, 2 November 1846, 166.

153. Weber, in *The Friend*, 1 February 1845, 24; Wate, in *WSL*, 9 May 1848, 40.

154. In *The Friend*, 24 September 1844, 88.

155. Seamen's Bethel, New Bedford, Massachusetts; also quoted in [M. V. Brewington], *Full Fadom Five* (Sharon, Mass.: Priceless Pearl Press, 1968), no. 4.

156. Ship *Tiger*, 24 June 1846.

157. Ship *Magnolia*, July 1841.

158. *The Friend*, 1 October 1846, 150.

159. Herman Melville, *Moby-Dick: Or, the Whale* (New York: Harper and Brothers, 1851), 628.

160. Ship *Golconda*, 26 April, 12 May, and 16 May 1852.

161. *WSL*, 19 November 1844, 147.

162. Ship *Magnet*, 12 August 1844.

163. S[amuel] C[henery] Damon, "Good Tidings from the Pacific," *Sailor's Magazine*, 18 (October 1845): 56. Damon was chaplain of the Seamen's Chapel, organized in Honolulu for the benefit of mariners by the American Seamen's Friend Society of New York. Sailors in port were encouraged to sign a written pledge, which was exhibited on board for the crew's edification: "We the undersigned, mariners of the [vessel's rig and name], believing that intemperance is the greatest bane to human happiness and usefulness, and that temperance eminently fits us for all the duties and enjoyments of life, do solemnly pledge one to the other, that we will in future abstain from all intoxicating drinks, except strictly for Medical use"; see *The Friend*, 2 December 1844, 117.

164. Bowles, "Whale-Fishery of the N. West Coast," 83.

165. Nathan B. Heath (master, ship *Hope* of Providence) to Pearce and Bullock (agents), Oahu, 21 September 1847; quoted in Brad Hathaway, "Letters from a Whaleman," New Bedford *Standard-Times*, 24 September 1978. Heath's decision was ill-starred; the *Hope* was lost two months later while whaling in Magdalena Bay, Lower California. Heath returned to the Pacific Northwest in 1848 as master of the *Edward*, replacing John S. Barker who went home ill.

166. *New London News*, quoted in *WSL*, 19 January 1847, 179.

167. Roys's vessel was spoken by the outward-bound whaleship *Majestic*; ship *Majestic* of New Bedford, 25 January 1849. PMS 656/1848M.

168. Roys gave credit to Captain Freeman H. Smith in the *Huntsville* for taking the "new-fashion" whale in the Sea of Okhotsk in 1848, but Charles M. Scammon believed that the first bowheads were taken by the *Hercules*, Ricketson, and the *Janus*, Turner, on the Kamchatka ground during 1843. The Danish whaleship *Neptune* seems to have taken one in the North Pacific in 1845, and it is likely that the 97-foot whale reported by the crew of the *Pantheon* on the Northwest Coast—reported earlier in this chapter—was not a right whale but one of the "Polar" whales, which had not been well documented and were unknown to most Pacific whalemen at that time. The "polear" species begins to appear in logbooks of Northwest Coast voyages during 1847; the *Edward* reported one in about 51° North on 11 August of that year. The *Julian* noted two in the following season, one in about 55°36′ North on 4 June 1848, and the other in 54°40′ on 1 August; ship *Edward*, 11 August 1847; ship *Julian*, 4 June 1848, 1 August 1848.

169. *WSL*, 24 December 1850, 170.

170. "A Polar Whale's Appeal," *The Friend*, 15 October 1850, 82–83.

171. Scammon, *Marine Mammals of the North-Western Coast of North America*, 69.

172. Ship *Abigail* of New Bedford, David Barnard, master and keeper, 26 July 1846. KWM (360).

NOTES TO CHAPTER 3

1. John McLoughlin to Governor, Deputy Governor, and Committee, Fort Vancouver, 18 November 1843, quoted in E. E. Rich, ed., *The Letters of John McLoughlin from Fort Vancouver to the Governor and Committee: Second Series,* 1839–44 (London: Hudson's Bay Record Society, 1943), 6: 178.

2. Caleb Cushing, *The Treaty of Washington: Its Negotiation, Execution, and the Discussions Relating Thereto* (New York: Harper and Brothers, 1873), 214.

3. William R. Sampson, ed., *John McLoughlin's Business Correspondence,* 1847–48 (Seattle: University of Washington Press, 1973), 146.

4. George Pelly to William Smith, Honolulu, Oahu, 21 March 1840. HBC A.11/61, fo. 55.

5. George Pelly and George T. Allan to Governor, Deputy Governor, and Committee, Honolulu, 21 May 1844. HBC A.11/62, fo. 18.

6. George Pelly and George T. Allan to the Governor and Council, [Honolulu], 28 April 1846. HBC D.5/15, fo. 486.

7. "Statement of Cash Received at the Sandwich Islands on account of the Hon'ble Hudsons Bay Company Outfit 1845." HBC B.191/d/4, fo. 2. Such retail outfitters included Sampson and Ford, Charles Brewer and Company, and Isaac Montgomery.

8. [Exchange note] payable to T. B. Marpillero, Honolulu, 16 January 1846. HBC A.11/62, fo. 117.

9. George Pelly and George T. Allan to Governor, Deputy Governor, and Committee, Honolulu, 10 January 1845. HBC A.11/62, fos. 32–32d.

10. Wilkes, *Narrative,* 5: 122.

11. Simpson wrote: "[whalemen] escape the exorbitant harbour dues exacted at Honolulu, get cheaper & perhaps better provisions, are at all times ready for sea, and run no risk of being detained, as we were, both in getting in and coming out of the intricate and dangerous harbour of Honolulu"; Simpson, in Lowe, "Journal of a Voyage from London to Sandwich Islands and Fort Tako, N.W. Coast of America," 66.

12. Sir George Simpson to Governor, Deputy Governor, and Committee, Honolulu, 1 March 1842, quoted in Joseph Schafer, "Letters of Sir George Simpson, 1841–1843," *American Historical Review* 14 (October 1908): 83.

13. Simpson to Governor, Deputy Governor and Committee, quoted in Schafer, "Letters of Sir George Simpson," 84.

14. Simpson, *An Overland Journey Round the World,* Part 2, 93.

15. Schafer, "Letters of Sir George Simpson," 84.

16. Sampson, *John McLoughlin's Business Correspondence,* xxxiii-xxxv.

17. Rich, *Letters of John McLoughlin,* 289. It seems unlikely that Simpson was properly informed about the right whales, at least in the Gulf of Georgia; perhaps his informants had seen humpback whales and reported them without exact reference to species.

18. Sir George Simpson to Charles Ross, Red River Settlement, 20 June 1844, quoted in W. Kaye Lamb, "Five Letters of Charles Ross, 1842–44," *British Columbia Historical Quarterly* 7 (1943): 117.

19. A. Barclay to [George] Pelly and [George] Allan, London, 18 September 1844. HBC B.191/c/1, fo. 1.

20. Charles Ross (chief trader at Fort Victoria) to Donald Ross, Fort Victoria, 10 January 1844; quoted in Lamb, "Five Letters of Charles Ross 1842–44," 111. Ross wrote: "Immense numbers of whalers have been upon the Coast this last Summer, some say as many as 300 sail! One or two of them called in at Neweté to look for Beaver, and it is said have done us a good deal of injury." The rumour eventually spread to Upper Cali-

fornia; in 1844 Thomas O. Larkin, American consul at Monterey, advised the American secretary of state that the whaling business at his place was constantly expanding: "it is said that one hundred sail will be on this coast next year, three hundred being this year on the N.W. Coast"; Thomas O. Larkin to John C. Calhoun, Monterey, 18 August 1844, in Larkin, *Larkin Papers*, Vol. 2 (1952), 206. He was particularly concerned that whaleships would call at California ports to discharge sick seamen and incorrigibles at the expense of the American consulate.

21. Simpson, *An Overland Journey Round the World*, Part 2, 92–93.

22. A few writers leave an impression that whalemen's contact with the natives of the Northwest Coast occurred as early as the 1830s. James Deans reported in 1899 that a sea captain named Jefferson left a whaleship while wintering over at Skidegate in the Queen Charlottes about 1832, and lived there among the Haida Indians until his death in the later 1830s. This is not reflected in the whaling record; the only known master named Jefferson sailed in the bark *Ann Maria* of Fall River, Massachusetts, on the voyage of 1845. The bark did not return, and was counted for lost; if Deans's informants were off by just one decade, it might have been reasonable to suppose that Jefferson finished up in the Queen Charlottes. Or perhaps Deans's Captain Jefferson signed off from a fur-trade vessel. See James Deans, *Tales from the Totems of the Hidery* (Chicago, 1899); quoted in Edward L. Keithahn, *Monuments in Cedar* (Ketchikan, Alaska: Roy Anderson, 1945), 34–35.

23. [James] Douglas to Sir George Simpson, Fort Vancouver, 23 October 1843. HBC D.5/9, fos. 117–118. The master was probably Alexander Duncan, later master of the Company barks *Vancouver* and *Columbia*; see Hartwell Bowsfield, ed., *Fort Victoria Letters, 1846–1851*. Publications of the Hudson's Bay Record Society 32 (Winnipeg: Hudson's Bay Record Society, 1979), xxxii.

24. Barton Ricketson, Benjamin S. Rotch, William J. Rotch, Ezekiel W. Davis and James Rider. All except Rider lived in New Bedford; Rider was from the nearby town of Dartmouth. United States, Works Projects Administration, Survey of Federal Records, *Ship Registers of New Bedford, Massachusetts*, Vol. 1, 1796–1850. Typescript (Boston: National Archives Project, 1940); no. 383. The *Canada* was described as a two-decked, three-masted vessel, square stern and billethead, 131'6" long, built New York in 1823, brought up from that state in 1841 to Boston, and reregistered in New Bedford in 1842 for the whale fishery.

25. United States, Works Progress Administration, Survey of Federal Archives, *New Bedford Crew Lists*, Report no. 1610, typescript, 166. KWM.

26. Samuel S. Munro, aboard the whaleship *Magnet* of Warren, Rhode Island, noted under date 18 June 1844 that the "Doctor from the Canada Came On Board to See the Sick," a visit perhaps prompted by the second mate of the *Magnet* having seriously cut his hand the previous afternoon when his whaleboat was stove by a Northwest Coast whale; ship *Magnet*, 18 June 1844.

27. Topham, quoted in Wood, 1: 112.

28. The *Maine* was 95'11" x 26'5" x 13'2 1/2", two decks, three masts, square stern and billethead. The owners of register for the voyage of 1842 were Ezekiel Sawin of Fairhaven, Massachusetts, and Bela Hunting of Boston; United States, Works Projects Administration, *Ship Registers of New Bedford*, 1, no. 1907.

29. "An Attempt to Murder the Master of the Whale Ship Maine, of Fairhaven," *WSL*, 11 July 1843, 311; ship *Europa*, 12 June 1843. Makee's long convalescence changed the course of his life. In November 1844, after the *Maine*'s return, the oil and bone were shipped home and the vessel fitted for sea under Edwards, the former mate. Makee remained at Oahu and was still "unable to rejoin his ship"; *WSL*, 8 April 1845, 19. In

1851 he outfitted one of the first whaleships belonging to the Sandwich Islands, the *Chariot*, Thomas Spencer, which returned in October 1852 with 39 barrels sperm oil, 1595 whale oil, and 22,688 "bone"; Reginald B. Hegarty, *Returns of Whaling Vessels Sailing from American Ports 1876–1928* (New Bedford: Old Dartmouth Historical Society and Whaling Museum, 1959), 48; also "List of Honolulu Whalers, with their Annual Catch," in *Hawaiian Annual*, date unknown [post-1880], 63.

30. A contemporary newspaper report indicates that a senior officer from the whaleship *Sabina*, then in port, was put in charge; but this is refuted by a legal document prepared later; see "Ship Maine—Wm. M. Smith, Abstract from Protest and other Adjustments," Boston, 4 August 1847. NHA (15), Box 8, Folder 114, fo. 1.

31. *Ibid.*.

32. Topham, quoted in Wood, 1: 112. The location of "Minetta Bay" is not clear; perhaps this was a corruption of "Nahwitti." One is tempted to place Topham at Minette Bay, up Kitimat Arm at the head of Douglas Channel, until one discovers that the name "Minette" was not adopted at that place until 1898, by Louis Coste, chief engineer of the Department of Public Works in Ottawa, who named it for his wife; see John T. Walbran, *British Columbia Coast Names 1592–1906, to Which are Added a Few Names in Adjacent United States Territory; Their Origin and History* (Ottawa: Government Printing Bureau, 1909; reprint ed., Vancouver: Library Press, 1971), 339.

33. Douglas to Simpson, 23 October 1843; fos. 117–118.

34. *Ibid.*.

35. John Kennedy to Sir George Simpson, Fort Simpson, 25 February 1844. HBC D.5/10, fos. 289d-290.

36. *Ibid.*. This whaleship may have been the *Canada* or the *Maine*, but more likely was another vessel entirely. The post at Stikeen, well into Russian-American territory, was maintained by the Hudson's Bay Company under the terms of an Anglo-Russian pact effected in 1840, by which the British company leased certain trade rights north of 54°40' in exchange for an annual payment of 2,000 otter skins.

37. Topham, quoted in Wood, 1: 112.

38. In 1909 John Walbran described both gravestones, the inscriptions yet legible, in his directory of British Columbia place names. He gave 4 May 1844 as the date inscribed on Thompson's stone, but an earlier and more plausible date of 6 January 1844 is recorded in *The Friend*, 9 October 1844, 96. Walbran learned from an Indian residing at Nahwitti village on Hope Island that Edward Rice was a young man who died from the breaking of a blood vessel. See Walbran, *British Columbia Coast Names 1592–1906*, 456.

39. Ship *Maine*, "Abstract from Protest." The "off-shore" ground as understood by whaling historians was a sperm-whaling ground off the west coast of South America; Captain Smith here referred to some place known to him, probably the usual right-whaling area offshore from the Northwest Coast.

40. Ship *Milo* of New Bedford, 6 July 1844. Huntington (HM 26538). The *Canada* apparently did well in the summer of 1844; when spoken by Captain Hamley of the ship *Stonington* in September of that year, in 26° N., Topham reported 3,000 barrels of oil aboard; ship *Stonington* of New London, 15 September 1844. MSM (335).

41. Bowles, "Whale-Fishery of the N. West Coast," 83.

42. "Port of Honolulu: Arrived," *Polynesian*, 4 October 1845, 93.

43. [James Ward], *Perils, Pastimes, and Pleasures of an Emigrant in Australia, Vancouver's Island and California* (London: Thomas Cautley Newby, 1849). The identity of this vessel is not conclusively determined, since the author may have exercised his literary preroga-

tive in changing of names. Two Australian whaling barks named *Jane* were active in the Pacific during the 1840s; one a 365-ton vessel which does not appear to have gone to the North Pacific, and the other a 250-ton hull about which only a scant itinerary survives; Honore Forster to Author, Canberra, 10 July 1985.

44. [Ward], *Perils, Pastimes, and Pleasures*, 185–89.

45. See "Comparative Statement of Tonnage of Vessels Employed in the Whale Fishery Jan'y 1, 1846, and Jan'y 1, 1847," *WSL*, 5 January 1847, 172.

46. Ship *Kutusoff* of New Bedford, William H. Cox, master. Maury 2, no. 1. The 1841–45 voyage of this vessel formed the basis of a giant whaling panorama—8.5 feet by 1,275 feet—painted and exhibited beginning in 1848 by Benjamin Russell and Caleb Purrington; Russell had been among the crew of the *Kutusoff*. The first narrator was Daniel McKenzie, former master of the *Caroline*. Although the panel includes a scene of right-whaling on the Northwest Coast, there is no depiction of West Coast natives. Part of the panorama is on display at the Old Dartmouth Historical Society Whaling Museum in New Bedford, Massachusetts; see "The Prodigious Panorama," *American Heritage*, 12 (December 1960): 55–62; and Old Dartmouth Historical Society, "A Lad Before the Wind," *Bulletin from Johnny Cake Hill* (Fall 1987): 2. Origin of the vessel is noted in "Frederick C. Sanford's Letters." NHA (129), Box 9, Book 19, 187.

47. United States, Works Progress Administration, *New Bedford Whalemen's Inward Manifests*. Report 836 (typescript), 69, KWM. Starbuck indicated a return of 1,746 barrels of sperm-whale oil and no [right-]whale oil; see Starbuck, 352–53. His statistics suggest that the vessel spent most or all of its time on the sperm-whaling grounds; the presence of beaver skins is therefore mysterious.

48. Aderial Smith was another of the French-American captains, born 25 May 1801 in Edgartown, Massachusetts. He became a whaling captain in 1836 and commanded two voyages in the *Oriental* before taking the *Réunion*. His 1843 visit in the Pacific Northwest was his last; he died on 31 July 1844, five days after returning the *Réunion* to Havre. He was the older brother of Gilbert Smith, master of the whaleship *Nil*; see Du Pasquier, *Les Baleiniers Français*, 229. The voyage of the *Caroline* was likewise the last for Daniel McKenzie; he had commanded eight whalers since 1818 and had already retired once from the sea, in 1840. In New Bedford in 1846 he became the weigher and gauger of the New Bedford customs house and assisted Matthew Fontaine Maury in abstracting logbook data for Maury's seminal study of wind and currents. He died in New Bedford in April 1854; see "A Lad Before the Wind," 2; and Raymond Calkins, *The Life and Times of Alexander McKenzie* (Cambridge: Harvard University Press, 1935), 5–6, 56; also personal communication, Richard C. Kugler, New Bedford, Massachusetts, 5 November 1987.

49. Appendix A includes a transcription of journal entries kept aboard the *Caroline* during September 1843 between Sitka and the Strait of Juan de Fuca.

50. Ship *Maine*, "Abstract from Protest," fo. 1.

51. William Hooper to Thomas Oliver Larkin, Sandwich Islands, 10 October 1844; quoted in Larkin, *Larkin Papers*, 2: 256. The sailing date of 1842 is erroneous.

52. Larkin to Hooper, Monterey, 4 November 1844, in Larkin, *Larkin Papers*, 2: 274.

53. Ship *Maine*, "Abstract from Protest," fo. 2. The survey was conducted on 11 November by Hiram Weeks, George Drew, and Cornelius "Heger"—probably the ubiquitous Cornelius Hoyer. The owners paid the surveyors thirty dollars.

54. Ship *Caroline* of New Bedford, George W. R. Bailey, keeper, 11 July 1843. KWM (596). The ship was then in 54°30′ North, 147°22′ West.

55. *Ibid.*, 14 August 1843.
56. *Ibid.*, 15 August 1843; the "Aconitum" being the plant reputedly used by native whalemen to poison their prey.
57. *Ibid.*, 5 September 1843. Doubtless the *Columbia* belonging to the Hudson's Bay Company, which serviced Sitka under private arrangements negotiated as part of a long-term agreement to provision Russian outposts in America; for discussion see James R. Gibson, *Imperial Russia in Frontier America: The Changing Geography of Supply of Russian America 1784–1867* (New York: Oxford University Press, 1976), 199–208.
58. Donald C. Davidson, "Relations of the Hudson's Bay Company with the Russian American Company on the Northwest Coast, 1829–1867," *British Columbia Historical Quarterly* 5 (1941): 34. Bancroft, citing statistics that the quantities of furs taken from Alaskan islands fell off during 1821–42, notes that the Russians blamed this on the "encroachments of foreign traders, and especially of American whaling vessels, whose masters touched at various points in the Russian possessions... and paid much higher prices for furs than those fixed by the company tariff"; Hubert Howe Bancroft, *History of Alaska 1730–1885*, The Works of Hubert Howe Bancroft, Vol. 33 (San Francisco: Bancroft and Company, 1886), 582. Though whalemen were charged, there is as yet no supporting evidence to acknowledge that these early American fur traders simultaneously engaged in whaling.
59. "Ukase of September 4, 1821," text published in "Appendix to Case of Her Majesty's Government, Vol. 1," in United States, Senate, *Proceedings of the Tribunal of Arbitration, Convened at Paris Under the Treaty Between the United States of America and Great Britain Concluded at Washington February 29, 1892, for the Determination of Questions Between the Two Governments Concerning the Jurisdictional Rights of the United States in the Waters of Bering Sea*, Senate 53:2 [177] (Washington, D.C.: Government Printing Office, 1895), 4: 213–14. Gardner's return on 15 April 1823 was recorded by a New Bedford diarist, but in insufficient detail to reveal whether the captain had himself been on the Northwest Coast. If so, he was thirteen years ahead of Narcisse Chaudière in the *Gange*; Joseph R. Anthony, *Life in New Bedford A Hundred Years Ago: A Chronicle of the Social, Religious and Commercial History of the Period as Recorded in a Diary Kept by Joseph R. Anthony*, ed. Zephaniah W. Pease (New Bedford: Old Dartmouth Historical Society, 1925), 38.
60. P. Tikhmenieff, "Historical Review of the Formulation of the Russian-American Company," translated in United States, Senate, *Tribunal of Arbitration* 4: 265. Neither captain nor ship was named by Tikhmenieff, but may have been Captain Wilcox in the *Parachute*, whose meeting with the Russians was noted by Sir George Simpson.
61. United States, House of Representatives, "Resources of Alaska," 350.
62. A. Etholeuff to Lewis L. Bennett, New Archangel, 30 September 1843; quoted in *WSL*, 14 May 1844, 3. The *Ann Mary Ann* of Sag Harbor sailed 25 November 1842 under Captain Winters and returned 27 May 1845 with 2,600 barrels whale oil, 75 barrels sperm, and 23,000 pounds of baleen—a very fine voyage; see Starbuck, 396–97.
63. P. A. Tikhmenev, *A History of the Russian-American Company*, trans. and ed. Richard A. Pierce and Alton S. Donnelly (Seattle: University of Washington Press, 1978), 318–19.
64. Tikhmenieff, "Formation of the Russian-American Company," 4: 266.
65. Bill, *Citizen*, 57. Surviving documents suggest that the whaleship spent the summer of 1845 on the Kamchatka coast and around the Bonin Islands, but visited the Northwest Coast in later years; ship *Citizen* of Sag Harbor. Maury 78, no. 25.
66. Ship *Stonington*, 7 June, 1 July 1846.
67. United States, House of Representatives, "Resources of Alaska," 389, noted that "The

dried fish is generally stored in the dwellings, being piled up along the walls; but if the supply is great it frequently happens that the floor is covered with them several feet high, and the family live on the top of their food until they gradually eat their way to the floor." Radou likely did not find this much food. Ship *Narwal* of Havre, 5 July 1846. KWM (559).

68. Ship *Narwal*, 7 July 1846.

69. *Ibid.*, 6 September 1846; "their harmony," he added, "is far poorer than the soft and melancholy song of the Tahitians or even the Sandwich Islanders."

70. *Ibid.*, 5 September 1846.

71. *Ibid.*, 7 September 1846.

72. Daily journal, Fort Victoria, 1846–50. HBC B.226/a/1, fos. 60–60d.

73. *Ibid.*, fos. 60–61.

74. A veteran of the French whaling fleet, he had been second-captain of the *Nil* in 1836, and later officer on the *Guillaume Tell* (1837–38), the *Elisabeth* (1841–42), and the *Nérée* of Nantes (1843–46); see DuPasquier, *Les Baleiniers Français*, 227.

75. See *The Friend*, 1 May 1847, 71; 1 April 1848, 32; 1 May 1848, 40; and 1 September 1849, 44–45.

76. Douglas, quoted in Bowsfield, *Fort Victoria Letters*, 13.

77. Bowsfield, *Fort Victoria Letters*, 13.

78. Daily journal, Fort Victoria, 31 January 1848, fo. 76d.

79. *Ibid.*, fo. 70. Morin's official report makes no mention of his fruitless cruise in the Strait of Georgia; he does give details of his other North Pacific cruises: from 22 May to 31 July 1847 in 48–55°49' North, 139–164°28' West, his men struck twelve right whales, brought four alongside, and recovered four other dead carcasses; they also obtained one sperm whale. In the summer of 1848 he made two cruises between 40°30' and 57°17' North along the Asian coast, where the crew struck twenty-eight whales and recovered twenty-two—a fine showing—as well as one carcass. FNA CC⁵611.

80. As of the first week of January, 1846 the American whaling fleet numbered 735 hulls, 692 of which were then at sea. Almost 250 had been recently reported at the Sandwich Islands or elsewhere in the North Pacific. The British, by comparison, possessed only 43 South-Seas whaleships; see Robert Lloyd Webb, "The American Whaleman as a Deterrent to Pacific War, 1845–47," *Proceedings* of the International Commission for Maritime History/North American Society for Oceanic History joint meeting, Charleston, South Carolina, 1987 (in press).

81. *Correspondence Politique Etats Unis*, quoted in George Vern Blue, "France and the Oregon Question," *Oregon Historical Quarterly* 34 (March 1933): 46.

82. "P.," "South Sea Whalers," *Times* (London), 18 June 1846, 8.

83. Quoted in *Sheet Anchor*, 5 July 1845, 103.

84. The American whaling fleet played an auxiliary rôle in the conquest of California during 1846–47; at least one whaleship was employed as a troop transport, and the officers and men of two others—the *Edward* and the *Magnolia* of New Bedford—stayed on shore a week at San José del Cabo, Lower California, during a siege of that town by Mexican patriots in the autumn of 1847; see Webb, "American Whaleman as a Deterrent"; ship *Edward*.

85. Douglas to Sir George Simpson, 23 October 1843, 117–18.

86. Roderick Finlayson to John McLoughlin, Fort Victoria, 10 October 1844. HBC B.226/b/1, fo. 6. Finlayson replaced Charles Ross, who had died; see Bowsfield, *Fort Victoria Letters*, xx.

87. Roderick Finlayson to John McLoughlin, Fort Victoria, 10 June 1845. HBC B.226/b/1, fo. 27d.

88. Roderick Finlayson to John McLoughlin, Fort Victoria, 5 August 1845 (postscript 6 August). HBC B.226/b/1, fo. 33d.
89. Daily journal, Fort Victoria, fo. 2d.
90. *Ibid.*, fo. 4.
91. *Ibid.*, fo. 29d.
92. *Ibid.*, fo. 59.
93. In August 1845, for example, Finlayson notified McLoughlin that 400 gallons of whale oil had been loaded in the schooner *Cadboro*, while a further 600 gallons were being retained at Fort Victoria for shipment to England in the autumn; Finlayson to McLoughlin, 5 August 1845, fo. 32d.
94. An account of the *Morrison*'s voyage from New London to the Sandwich Islands kept by a passenger, the Rev. Thomas Douglass, survives in the collections of Mystic Seaport Museum in Mystic, Connecticut; ship *Morrison* of New London, Thomas Douglass, keeper. MSM (343).
95. He was mate of the ship *Flora* of New London during 1835–36 and assumed command from Captain Richard S. Smith during that voyage. On his return home he went two voyages in the *Neptune*, the first to the South Pacific and the second only as far as the South Atlantic. Returning in May 1844, he almost immediately went to sea again in the *Morrison*.
96. The gam among the *Morrison*, the *Levant* of Sag Harbor and the *George* is documented in the Rev. Douglass's journal, 23 October 1844. The *George* was not the *George* of New Bedford, as Douglass supposed, but the *George* belonging to Fairhaven, just across New Bedford harbor. It had departed 16 September, bound for the Northwest Coast. The *Levant* was also bound out to the same grounds; Starbuck, 412–13.
97. Ship *Morrison*, 23 October 1844.
98. For a more detailed account see Robert Lloyd Webb, "Connecticut Yankees in Queen Victoria's Fort: New London Whalemen on the Northwest Coast," *Log of Mystic Seaport* 39 (Summer 1987): 43–53.
99. James M. Green to Lyman Allyn, aboard ship *Louvre*, 15 August 1845. ODHS MSS 76, S-g 1, Series E, S-s 1, Folder 1, fo. 1.
100. Green to Allyn, 15 August 1845, fo. 2.
101. Wood, 2: 410.
102. Ship *Luminary* of Warren, 16 August 1846. KWM (463). The following day the *Luminary* took on board two Kanakas from the *Jonas*, as "supernumaries... to work their passage to Oahu"—apparently the French captain did not intend to return by that route.
103. James Douglas to Governor, Deputy Governor, and Committee, Fort Victoria, 16 December 1845. HBC A.11/72, fo. 3.
104. Clark, *Lights and Shadows*, 225. Wilkes's own account notes that as many as 40 canoes came alongside his ships, the occupants clamoring to offer fish, venison, and other articles for sale, "without much regard to the priority of rank, station, or any thing else"; Wilkes, *Narrative*, 4: 488.
105. Douglas to Governor, Deputy Governor, and Committee, 16 December 1845.
106. "Ship Morrison Book of Accounts," Samuel Green, Jr., keeper, 1845–47, at "Port Victory," 1 October 1845. MSM Misc. Vol. 318, [fo. 2].
107. Bark *Cowlitz*, 16 October 1845. HBC C.1/259, fo. 232d.
108. *Ibid.*, 19 October 1845.
109. Roderick Finlayson to John McLoughlin, Fort Victoria, 23 October 1845. HBC B.226/b/1, fo. 39.
110. [untitled receipt], ODHS M-SS 76, S-g 1, Series E, S-s 1, Folder 43, fo. 1.

111. "Ship Morrison Book of Accounts," 20 October 1845, [fo. 2].

112. Douglas to Governor, Deputy Governor, and Committee, 16 December 1845, fos. 3–3d.

113. *Ibid.*, 16 December 1845, fo. 3d.

114. "Ship Morrison Book of Accounts," 20 October 1845, [fo. 2].

115. H.M.S. *America*, scouting the Strait of Georgia, observed "several American whalers at anchor" there on 1 October 1845, but did not think it important to investigate. H.M.S. *America*, Lieutenant Davies, keeper, 30 September - 1 October 1845. NMM JOD 42.

116. Douglas to Governor, *et al.*, 16 December 1845, fo. 4.

117. "Three Young Men Drowned in Gray's Harbor, North West Coast," *The Friend*, 14 March 1846, 42. The three survivors are not named. Royce was an eighteen-year-old native of Tolland, Connecticut, Kirby an Englishman from Birmingham. They joined the ship in New London. Church, supposed to be a native of Massachusetts, joined ship sometime after the crew list was prepared; "Ship Morrison Book of Accounts," [fos. 46, 69]; crew list, ship *Morrison*. FRC, RG 36, Records of the Collector of Customs, District of New London.

118. "Ship Louvre & Owners To The Vice consulate San Francisco," 17 December 1845. ODHS MSS 76, S-g 1, Series E, S-s 1, Folder 18, fo. 1. The certificate was filed in the office of Vice-Consul Leidesdorff; the use of the plural in James Green's account suggests that his *Louvre*, too, lost more than one man by desertion at Cape Flattery. Sam Green indicated that Church, Kirby, and Royce had jumped ship in the strait; seven more deserted upon arrival at San Francisco. Consular certificate, William. A. Leidesdorff, San Francisco, 17 December 1845. FRC, RG 36, Records of the Collector of Customs, District of New London. The three ships had arrived in San Francisco together, the *Montezuma* and *Morrison* coming in on 24 November, and the *Louvre* the following day; noted in ship *Charles Phelps* of Stonington, 24–25 November 1845. KWM (368).

119. James Douglas to Captain Heath [bark *Cowlitz*], Fort Victoria, 16 December 1845. HBC A.11/72, fos. 11–11d.

120. Alex G. Abel [consul] to [James] Green, Honolulu, 27 February 1846. ODHS MSS 76, S-g 1, Series E, S-s 1, Folder 2, fo. 1.

121. "Wh Oil, Sperm, Bone Purchased of Louvre Crew Apl 14 1847." ODHS MSS 76, S-g 1, Series E, S-s 1, Folder 5, fo. 1; "Division of Cargo to Crew of Ship Louvre, Apl 15 1847." ODHS MSS 76, S-g 1, Series E, S-s 1, Folder 5, fo. 1.

122. "Thomas P. Davis in a/c with Lyman Allyn." ODHS MSS 76, S-g 1, Series E, S-s 1, Folder 12, fo. 1. The *Louvre* "turned out" 2,860 barrels of whale oil, 133 barrels of sperm oil, and 8,500 pounds of baleen on return 6 April 1847. A further cargo of 20,191 pounds of baleen had been trans-shipped in 1846 in the bark *Angola* from the Sandwich Islands. The *Montezuma* returned 24 May 1847 with 3,177 whale, 56 sperm, 31,900 pounds of baleen; Wood, 2: 410, 475. The *Morrison* arrived, belatedly, on 5 May 1848, with 3,982 whale, 18 sperm, and 15,000 pounds of baleen; a further 23,712 pounds of baleen had been trans-shipped previously; Starbuck, 418–19.

123. Daily journal, Fort Victoria, 17 July 1846, fo. 11.

124. *Ibid.*, fo. 37.

125. *Ibid.*.

126. Sampson, *John McLoughlin's Business Correspondence*, n. 15.

127. John S. Galbraith, "James Edward Fitzgerald Versus the Hudson's Bay Company: The Founding of Vancouver Island," *British Columbia Historical Quarterly* 16 (1952): 203.

128. See Enderby, *Proposal for Re-Establishing the British Southern Whale Fishery*.

129. From 1849 to 1858 Vancouver Island was administered as a colonial protectorate under the stewardship of the Hudson's Bay Company; James Douglas became governor in 1851 and ruled with the assistance of a legislative council until 1856 when a House of Assembly was created. In November 1858 mainland British Columbia became a Crown Colony and Douglas severed his ties with the company to assume the post of governor. The Fraser River gold rush brought such an influx of population that the Crown determined to bring the island under its control; the domination of the Hudson's Bay Company was terminated in 1859; see W. Kaye Lamb, "British Columbia Official Records: The Crown Colony Period," *Pacific Northwest Quarterly* 29 (January 1938): 17–25.

NOTES TO CHAPTER 4

1. Arne Odd Johnsen, "Granatharpunen: En Kort Utredning om Hvordan Svend Foyn Løste Projektil-problemet," *NHT* 29 (September 1940): 222.
2. Raymond M. Gilmore, "Rare Right Whale Visits California," *Pacific Discovery* 9 (July-August 1956): 24.
3. Charles H. Stevenson, "Whale Oil," *Scientific American Supplement* 57 (5 March 1904): 23550.
4. Ship *Emily Morgan* of New Bedford. NBFPL. The presence of the *William Gifford* entering "the north strait of Kadiak" is further documented in United States, Coast Survey, *Coast Pilot of Alaska, First Part, from Southern Boundary to Cook's Inlet*, by George Davidson (Washington, D.C.: Government Printing Office, 1869), 49.
5. J[oshua] F[illebrown] Beane, *From Forecastle to Cabin: The Story of a Cruise in Many Seas, Taken from a Journal Kept Each Day, Wherein Was Recorded the Happenings of a Voyage Around the World in Pursuit of Whales* (New York: Editor Publishing Company, 1905), 291–92.
6. United States, House of Representatives, *Letter from the Secretary of War, in Relation to the Territory of Alaska*, by William W. Belknap. Congress 42:1 [5] (Washington, D.C.: Government Printing Office, 1871).
7. Bark *Emma F. Herriman* of San Francisco, 6–7 June 1891, "cruising off Kodiak." KWM (603).
8. Beane, *From Forecastle to Cabin*, 292–93.
9. Ship *Java*, 31 July 1855; bark *Coral* of San Francisco, 21 May 1889. KWM (470).
10. Bark *Josephine* of San Francisco, Mrs. John McInnes, keeper, n.d., 22. KWM (253).
11. John A. Cook of the bark *Jessie H. Freeman*, noted in bark *Josephine*, 16 June 1892. Beginning in 1879 a number of steam-auxiliary barks were built for the western Arctic; their engines were intended to assist the whalemen in getting through pack ice.
12. United States, House of Representatives, *Letter from the Acting Secretary of the Treasury, Transmitting the Papers in the Claim of the North American Commercial Company for Transportation and Subsistence Afforded the Surviving Officers and Crew of the Whaling Bark Sea Ranger, Wrecked at Kayak Island May 26, 1893*, Congress 53:2 [186] (Washington, D.C.: Government Printing Office, 1894), 2–3. The bark was lost 24 May 1893 on a reef four miles north of Cape St. Elias, adjacent to Kayak Island, after 1943 named "Sea Ranger Reef"; U.S. Coast and Geodetic Survey Chart no. 8513 (1943). Subsequent to rescuing the whalemen, the North American Commercial Company petitioned the U.S. government for repayment of $949.55 in transportation and supplies costs.
13. See United States, Senate, *Message from the President of the United States in Response to Senate Resolution of January 8, 1895, Transmitting Information Relating to the Enforcement of the*

Regulations Respecting Fur Seals, Congress 53:3 [67] (Washington, D.C.: Government Printing Office, 1895), 304–05.

14. Matthew Howland to G. P. Pomeroy, Matthew Howland Letter Book, 1858–79, 15 December 1858; quoted in David Moment, "The Business of Whaling in America in the 1850's," *Business History Review* 31 (Autumn 1957): 286. American whaleships cruised in the Okhotsk Sea as early as 1845, and French vessels at least a year or two earlier. After 1850, whaleships more often frequented the Okhotsk as other grounds were "fished-out."

15. *Colonist*, 9 March 1869, 2.

16. R. W. Stevens, *Stowage of Cargo* (London, 1869); quoted in [James H. Hamilton], *Western Shores: Narratives of the Pacific Coast, by "Capt. Kettle"* (Vancouver: Progress Publishing Company, 1933), 184.

17. *Colonist*, 9 March 1869, 2.

18. Victoria *Gazette*, 9 September 1858; quoted in R.L. Reid, "The Whatcom Trails to the Fraser River Mines in 1858, Part 2," *Washington Historical Quarterly* 17 (October 1926): 275.

19. Confederate Secretary of State Judah P. Benjamin is said to have received a letter to this effect from a "southern association" operating in British Columbia and on Vancouver Island; it does not appear that the request was granted; see United States, Navy Department, Naval History Division, *Civil War Naval Chronology* 1861–1865 (Washington, D.C.: Government Printing Office, 1971), 1: III-148.

20. Benjamin F. Gilbert, "Rumours of Confederate Privateers Operating in Victoria, Vancouver Island," *British Columbia Historical Quarterly* 18 (July-October 1954): 241, 245.

21. Viator (pseud.), *Colonist*, 31 October 1859, 3.

22. Matthew Macfie, *Vancouver Island and British Columbia: Their History, Resources, and Prospects* (London: Longman, Green, Longman, Roberts and Green, 1865), 168–69.

23. Swan's work in Washington Territory is considered in Doig, *Winter Brothers*.

24. James G. Swan, *The Northwest Coast: Or, Three Years' Residence in Washington Territory* (New York: Harper and Brothers, 1857), 402.

25. Scammon, *The Marine Mammals of the North-Western Coast of North America*; 50. The whale was a proposed subspecies of the minke, or piked whale, *Balaenoptera acutorostrata* (Lacépède 1804); see Lyall Watson, *Sea Guide to Whales of the World* (New York: E. P. Dutton, 1981), 88. The whale was hauled into Port Townsend by fishermen.

26. Bark *Pantheon*, 22 July 1843, in 55° North, 148° West.

27. Bark *Pantheon*, 22 July 1843. "Imposebileity"; ship *Benjamin Tucker* of New Bedford, Sina Stevens, keeper, 14 April 1857. MSM MR 90.

28. Ship *Europa* of Bremen, 13 July 1843. KWM (333); ship *Julian* of New Bedford, 20 May, 4 July 1848.

29. Ship *William Hamilton*, 27 June 1846.

30. Cheever, *The Whale and His Captors*, 98.

31. Unidentified encyclopedia entry, ca. 1802, 163. KWM.

32. For a discussion of the development of harpoons and shoulder-guns see Thomas G. Lytle, *Harpoons and Other Whalecraft* (New Bedford: Old Dartmouth Historical Society Whaling Museum, 1984).

33. Ship *Navy* of New Bedford, 23 June 1871. de Lucia.

34. Ship *Navy* of New Bedford, 20 June 1871. KWM (156).

35. Ship *Navy*, 13 June 1871. de Lucia.

36. Bark *Sea Breeze* of New Bedford, 22 June 1854. KWM (472).

37. Ship *Emily Morgan*, 7–8 May 1867.

38. See Frederick P. Schmitt, Cornelis de Jong, and Frank H. Winter, *Thomas Welcome Roys: America's Pioneer of Modern Whaling* (Charlottesville: University Press of Virginia, 1980).

39. Thomas Roys to Obed N. Swift, typescript letter, 26 December 1860. KWM.

40. Bark *Pantheon*, 19 July 1843.

41. Schmitt, de Jong, and Winter, *Thomas Welcome Roys*, 169.

42. Shipping Register, Victoria. PABC RG 12, A1, Vol. 211, fo. 7.

43. See Richard A. Pierce, *Alaskan Shipping, 1867–1878: Arrivals and Departures at the Port of Sitka* (Kingston, Ontario: Limestone Press, 1972), 60. The *Emma* was carvel-built, round-bottomed, with two masts and two steam engines. Built at Port Madison, Washington Territory, probably in 1867, Spratt is listed as owner, and Holmes master, in December of that year. The *Emma* seems to have been owned by Spratt until late in 1872, all through the whaling period; then ownership transferred to Thomas John Burnes. Shipping Register, Victoria, fo. 7.

44. "On a Whaling Cruise," *Colonist*, 1 July 1868, 3.

45. *Colonist*, 2 November 1866; see also Bill Merilees, "Humpbacks in Our Strait," *Waters* 8 (1985): 12.

46. E. W. Wright, ed., *Lewis & Dryden's Marine History of the Pacific Northwest* (New York: Antiquarian Press, 1961), n. 441.

47. "Cargo of the Schooner Kate by Auction," *Colonist*, 31 July 1868, 3; and *Colonist*, 20 August 1868, 3.

48. "Whaling," *Colonist*, 2 September 1868, 3; and "Whaling," *Colonist*, 30 October 1868, 3. The casks were made of locally dried timber, important since whalemen were forced to pay a high duty on whaling gear imported into Victoria from San Francisco. "Our waters will soon be thronged with whalers," the local press lamented, "and it is well known that there is not a whale boat, a swivel gun—nay, even a whale line in the city, and why? Because by the time importers paid the duty chargeable here, they would be unable to sell the articles at a price at which these adventurers can buy them over the Sound when brought from San Francisco"; *Colonist*, 31 March 1869, 2.

49. Maud Emery, "Victoria's First Whaler Went to Sea," *Colonist*, 13 January 1963.

50. Oleaginous [pseud.], "The Importance of Our Whale Fisheries to Victoria," *Colonist*, 29 March 1869, 3.

51. "Whaling," *Colonist*, 26 August 1868, 3; "The Whaling Business," *Colonist*, 28 August 1868, 3; and "Whaling," *Colonist*, 30 October 1868, 3.

52. "The Whaling Movement," *Colonist*, 23 October 1868, 3. The managing group included J. R. Stewart, Edgar Marvin, T. Stahlechmidt, Captain Stamp, Captain Raymer, and John Kriemler; Kriemler was named secretary of the new firm.

53. *Colonist*, 7 January 1869, 3.

54. Moritz Lindeman, *Die Arktische Fischerei der Deutschen Seestädte 1620–1868* (Gotha: Justus Perthes, 1869), n. 99.

55. *Colonist*, 31 March 1869, 3.

56. "The Whaling Expedition," *Colonist*, 3 May 1869, 3.

57. "The Roys Whaling Expedition," *Colonist*, 21 May 1869, 3. Mr. Moore, sailing master of *Sparrowhawk* provided the charts.

58. The site, or near it, is known to this day as Whaletown, and a post office in that name was later established on the west side of the island; Canada, Permanent Committee on Geographical Names, *Gazetteer of Canada: British Columbia*, 2nd ed. (Ottawa, 1966). The *Gazetteer* lists thirteen place-names beginning with "whale" in British Columbia, including Whale Channel, Whale Passage, four different examples of Whale Rock,

Whaleback Mountain, Whaleboat Passage, Whaler Bay, Whaler Islets, Whaletown, Whaletown Bay, and Whaling Station Bay. There are also related names, such as Blubber Bay, and the Ballenas (Whales) Islands.

59. "Big Catch," *Colonist*, 27 July 1869, 3.

60. A. Douglass to Editor, in *Colonist*, 15 September 1869, 3. See also *Colonist*, 17 September 1869, 3.

61. *Colonist*, 31 March 1869, 3.

62. *Ibid.*; also *Colonist*, 13 March 1869, 3.

63. Notes of conversations between Major James Matthews and Mrs. James Walker, 23 September 1943 and 10 December 1943. VCA Add MSS 54. Harry Trim came out from New York in 1858—probably for the Fraser River gold rush—and was in the Cariboo in 1862, where he prospected for about six years. In 1868 he became involved in whaling, and later entered the steamboat business at Moodyville; Major Matthews, personal notes, 21 September 1943. Peter Smith was remembered as a resident of Brockton Point in Vancouver, a man who made his living "spearing" whales; conversation notes, Matthews with August Jack Khahtsahlano, 23 November 1934. Smith "used to catch [whales] off Bowen Island, and take them to [Worlcombe Island] and cut them up and make oil to sell to the Hastings Sawmill and in Westminster and Victoria"; August Jack Khahtsahlano, 23 November 1934. Smith is probably the enigmatic "Peter the Whaler" who has been mentioned in some undocumented accounts of early whaling in British Columbia; he may have been affiliated with Dawson and Douglass, or possibly with Lipsett's whaling operation in Howe Sound in 1870–71.

64. Pierce, *Alaskan Shipping*, 18–19.

65. *Ibid.*, 60.

66. *Colonist*, 14 June 1869, 3.

67. "The Whalers—Three More Whales Caught," *Colonist*, 18 January 1870, 3; also *Colonist*, 3 February 1870, 3.

68. *Colonist*, 27 July 1870, 3. The partners of record at this time included Thomas and James Lowe, Robert Wallace, and James Hutcheson; see Merilees, "Humpbacks," 12.

69. *Colonist*, 4 June 1870, 3.

70. Advertisement in *Colonist*, 20 February 1870, 2.

71. *Colonist*, 12 June 1870, 3, and 26 June 1870, 3.

72. *Blue Book* of 1870, quoted in H. L. Langevin, *Report of the Hon. H. L. Langevin, C.B., Minister of Public Works* (Ottawa: I. B. Taylor, 1872), 15.

73. Shipping Register, Victoria, fo. 20; and *Colonist*, 25 October 1871, 3. *Lloyd's Register* for 1871–72 lists the place of construction as Yarmouth; measurements 179 tons, 85.7' x 20.4' x 14.5'.

74. On one occasion he brought sixty-five Chinese passengers from Tahiti to Honolulu.

75. *Colonist*, 11 May 1871, 3.

76. Langevin, *Report*, 15. His description of Roys's rocket is grossly in error.

77. See Shipping Register, Victoria, fo. 34; and Pierce, *Alaskan Shipping*, 22–23.

78. Dawson may have withdrawn from the company at this time; see Merilees, "Humpbacks," 13.

79. Shipping Register, Victoria, fo. 31.

80. Langevin, *Report*, 16.

81. *Ibid.*.

82. Probably at the site now known as Whaling Station Bay, on the northeast side of the island; see Canada, Permanent Committee on Geographical Names, *Gazetteer of Canada*.

83. *Colonist*, 13 July 1873.

84. Shipping Register, Victoria, fo. 31. The same vessel was likely used later in the seal-

hunting business.
85. *Colonist*, 8 August 1871, 3 and 30 September 1871, 3.
86. Shipping Register, Victoria.
87. Thomas Welcome Roys, quoted in "Byzantium," *Underwater Archaeological Society of B.C. Newsletter* (January 1986), 3.
88. "Wreck of the Whaling Brig Byzantium: All Hands Saved!" *Colonist*, 25 October 1871, 3.
89. One of the crew later died in Victoria as a result of exposure.
90. "Wreck," 3; also "From the North—Omineca," *Daily Standard*, 25 October 1871.
91. "Ominica," *Colonist*, 22 November 1871, 3.
92. "The Byzantium," *Daily Standard*, 6 November 1871, 3.
93. On 27 January 1877; Frank H. Winter and Frederick P. Schmitt, "Captain Thomas Welcome Roys: America's First Scientific Whaler," *Oceans* 8 (May 1975): 39; see also Frank H. Winter and Mitchell R. Sharpe, "The California Whaling Rocket and the Man Behind It," *California Historical Quarterly* 50 (December 1971): 351.
94. Emery, "Victoria's First Whaler," 13.
95. See Wright, *Lewis & Dryden's Marine History*, n. 441. In 1893, Douglass was master of the sealing schooner *Arietes*.
96. For a discussion of these transactions see Winter and Sharpe, "California Whaling Rocket."
97. Lael Morgan, "Modern Shore-Based Whaling," *Alaska Geographic* 5 (1978): 37–38.
98. *Report of the British Columbia Fisheries Commission* 1905–07, 62; quoted in W. A. Carrothers, *The British Columbia Fisheries* (Toronto: University of Toronto Press, 1941), 125
99. "Whale Fishery," *Colonist*, 19 September 1886, 3.
100. Notes of telephone conversation, Major James Matthews with W. A. Grafton, 5 April 1940. VCA Add MSS 54.
101. Cadieux Coll., VCA Add MSS 782, Vol. 240.
102. "Capt. Whitelaw's Application," *Colonist*, 15 August 1890, 5.
103. "Another Whaler," *Colonist*, 12 August 1890, 5.
104. "A New Whaling Schooner," *Colonist*, 16 August 1890, 5.
105. "Neah Bay Siwashes Kill Whale off Flattery; 'Lorne' Crew See Hunt," *Colonist*, 15 June 1905.
106. Densmore, *Nootka and Quileute Music*, 52.
107. United States, Commission of Fish and Fisheries, "The Fisheries of the Pacific Coast," by William A. Wilcox. In *Report of the Commissioner for the Year Ending June 30, 1893*, Part 19 (Washington, D.C.: Government Printing Office, 1895), 256, 289.
108. Basil Lubbock, *The Arctic Whalers* (Glasgow: Brown, Son & Ferguson, 1955), 372. See also John R. Bockstoce, *Steam Whaling in the Western Arctic* (New Bedford: Old Dartmouth Historical Society, 1977), 7; and Robert Lloyd Webb, "Invented Too Late: The Introduction of Steam to the Arctic Whaling Fleet," *American Neptune* 44 (Winter 1984): 11–21.
109. The vessel was reportedly 81'9" x 15'9" x 8'6"; "Nylands Verksted," *NHT* 26 (June 1937): 188. A second set of measurements, 94'10" x 14'11" x 7'9" (length between perpendiculars, 87 feet) has been provided also; see Paul Budker, *Whales and Whaling* (London: George C. Harrap & Co., 1958), 104. See also "Nylands Verksted, 1854–1954," *NHT* 43 (August 1954): 296.
110. Johnsen, "Granatharpunen," 231.
111. See Sigurd Risting, "Den Norske Hvalfangst: Gjennem 50 Aar," *NHT* Supplement (1914): 63.

112. The harpoon line ran through a pulley near the masthead, which hung from a frame filled with sixteen rubber strings of just over one-inch circumference, plus other tackles and ropes; see Schmitt, de Jong, and Winter, *Thomas Welcome Roys*, 147.
113. Johnsen, "Granatharpunen," 232.
114. *Ibid.*, 233; Risting, "Den Norske Hvalfangst," 64.
115. Alfred Heneage Cocks, "The Finwhale Fishery on the Coast of Finmarken," *Zoologist* 9 (1884): 4–5. The harpoon itself had reached a weight of 56 kilograms—more than 123 pounds— without the shell or whale line; Alfred Heneage Cocks, "The Finwhale Fishery of 1886 on the Lapland Coast," *Zoologist* 12 (June 1887): 14.
116. *The Great Fisheries of the World* (London: T. Nelson and Sons, [ca. 1878]), 402–3.
117. Statistics from F. V. Morley and J. S. Hodgson, *Whaling North and South* (London: Methuen and Company, 1927), 67.
118. Arne Odd Johnsen, *Norwegian Patents Relating to Whaling and the Whaling Industry: A Statistical and Historical Analysis* (Oslo: A.W. Brøggers, 1947); 16–17. Norwegian nationals received 406; Americans contributed 21 patents, and British, 11.
119. Risting, "Den Norske Hvalfangst"; 64.
120. Arne Odd Johnsen, "Causation Problems of Modern Whaling," *NHT* 36 (August 1947): 21.
121. H. J. Bull, *The Cruise of the "Antarctic" to the South Polar Regions* (London: Edward Arnold, 1896), 17–18.
122. "Svend Foyn's Whaling Establishment," *Deutsche Fischerei-Zeitung* 6 (1883), translated in United States, Commission of Fish and Fisheries, *Report of the Commissioner for 1883*, Part 11 (Washington, D.C.: Government Printing Office, 1885), 338.
123. United States, Commission of Fish and Fisheries, "Report Submitted to the Department of the Interior on the Practical and Scientific Investigations of the Finmark Capelan-Fisheries, Made During the Spring of the Year 1879," by G. O. Sars. Trans. Herman Jacobson. In *Report of the Commissioner for 1880*, Part 8 (Washington, D.C.: Government Printing Office, 1883), 176, 178–79.
124. Sigurd Risting, *Av Hvalfangstens Historie* (Christiania: J. W. Cappelens, 1922), 158.
125. *Ibid.*. Most of those incarcerated received minimum sentences of ten to twenty days on charges of property destruction.
126. See Morley and Hodgson, *Whaling North and South*, 69, and Charles Rabot, "The Whale Fisheries of the World," *La Nature* (14 September 1912), trans. in Smithsonian Institution, *Annual Report of the Board of Regents of the Smithsonian Institution . . . for the Year Ending June 30, 1913* (Washington, D.C.: Government Printing Office, 1914), 43. The affected districts in Norway included Nordland, Tromsö and Finmarken.
127. "Whaling in Prospect," *Colonist*, 29 November 1898, 6.
128. *Ibid.*.
129. "The Work of the Whalers," *Colonist*, 15 November 1899, 6.
130. Canada, Department of Marine and Fisheries, "Annual Report of the Fisheries of British Columbia for the Year 1888," by Thomas Mowat. In *Fifth Annual Report of the Deputy Minister of Fisheries for the Year 1888*, Appendix 8 (Ottawa, [ca. 1888]), 241.

NOTES TO CHAPTER 5

1. Rabot, "The Whale Fisheries of the World," 483.
2. Jack Shickell, quoted in "The Last Gam: The Old Whale Hunter Passes," *National Fisherman Yearbook* 57 (1977): 116.
3. *Ibid.*; also Bruce D. Berman, *Encyclopedia of American Shipwrecks* (Boston: Mariners

4. United States, Commission of Fish and Fisheries, "Aquatic Products in Arts and Industries," by Charles H. Stevenson. In *Report of the Commissioner for the Year Ending June 30, 1902*, Part 28 (Washington, D.C.: Government Printing Office, 1904), 184. The business was kept alive almost exclusively for sperm oil. Baleen, which had become so rare as to command seven dollars per pound by 1907, was supplanted by spring steel and other manufactured materials. A few sailing whaleships continued in the Atlantic Ocean sperm-whale fishery until 1925.

5. Tønnessen and Johnsen, *History of Modern Whaling*, 51.

6. For discussion of the role of oil, see *ibid.*, 228—33; also Karl Brandt, *Whale Oil: An Economic Analysis*, Fats and Oils Study 7 (Palo Alto, Calif.: Food Research Institute, Stanford University, 1940), 136—38.

7. See, for example, "Whale in Seine Net," *Victoria Daily Times*, 20 November 1907, 9.

8. What little enforcement there was resulted in the licensing of some of the last American whaleships to visit Hudson's Bay, in the eastern Canadian Arctic. Much of this work fell to Captain Joseph Bernier, whose voyages may be noted in J[oseph] E[lzear] Bernier, *Report on the Dominion of Canada Government Expedition to the Arctic Islands and Hudson Strait on Board the D.G.S. "Arctic"* (Ottawa: Government Printing Bureau, 1910); and in Yolande Dorion-Robitaille, *Captain J. E. Bernier's Contribution to Canadian Sovereignty in the Arctic* (Ottawa: Indian and Northern Affairs Canada, 1978).

9. "Hvalfangst I Rusland," *NHT* 3 (30 September 1914): 135.

10. *Ibid.*; see also Japan, Ministry of Agriculture and Forestry, Fisheries Agency, *Japanese Whaling Industry* (Tokyo: Japan Whaling Association, 1954), 43.

11. "The Sealing Industry," in Corporation of the City of Victoria, *Victoria Illustrated: Containing a General Description of the Province of British Columbia, and a Review of the Resources, Terminal Advantages, General Industries, and Climate of Victoria* (Victoria: Ellis and Company, 1891), 74. One of these, named the *Kate*, may have been the same vessel used for coastal whaling by Dawson and Douglass a few years earlier.

12. Briton Cooper Busch to Author, Hamilton, New York, 27 May 1986. "The Sealing Industry," 74, reported forty hulls employing 660 whites and 368 natives. For discussion see Briton Cooper Busch, *The War Against the Seals: A History of the North American Seal Fishery* (Kingston, Ontario: McGill-Queen's University Press, 1985), 137—47.

13. A total of 97,000 skins was reported in British Columbia, Provincial Secretary, *Return to a Respectful Address Presented to His Honour the Lieutenant-Governor . . . Relative to the Grievances of the Sealers Referred to in the Answer of the Honourable the Attorney-General on the 12th Day of February Last*, 60 Vict. 965a (Victoria: Richard Wolfenden, 1897), 95a.

14. Master's Certificate no. 1279, Halifax, 13 September 1884. Balcom.

15. Balcom.

16. See "Pioneer Sealer Passes in City: Capt. G. W. S. Balcom Active in Whaling and Sealing Industries for Many Years," *Colonist*, 22 December 1925, 5. Balcom's descendants believe that he came west in the schooner *Zillah Mae*.

17. Family lore holds that this experience provided writer-poet Rudyard Kipling with the background for his story "The Devil and the Deep Sea." Kipling apparently met Balcom while visiting Victoria. Personal communication, Graeme Balcom, Vancouver, 24 September 1984.

18. Shipping Master's Office certificate, George Kirkandale, shipping master, Victoria. Balcom. In 1894, for example, he was master of the 76-ton *Walter L. Rich*.

19. Joan Goddard, "Reminiscences of Life on the Shore Whaling Stations of British Columbia 1905—1918" (typescript of paper presented at the Society for Historical Archaeology / Conference on Underwater Archaeology annual meeting, Sacramento, California, 9 January 1986), 4.

20. J. G. Millais, *Newfoundland and Its Untrodden Ways* (London: Longmans, Green and Company, 1907), 168. The 1904 catch included 1,274 rorquals and one solitary sperm whale.

21. P. T. McGrath, "Wonderful Whale-Hunting by Steam," *Cosmopolitan* 37 (May 1904): 56.

22. Cocks, "The Finwhale Fishery of 1886 on the Lapland Coast," 13.

23. Tønnessen and Johnsen, *History of Modern Whaling*, 105.

24. F. Gourdeau (Deputy Minister of Marine and Fisheries) to Charles Patton, Ottawa, 12 June [1902]. PAC, RG 23, Vol. 242, File 1536 [1], 58.

25. J[ohn] Leckie (John Leckie Ltd., Toronto) to R[aymond] Préfontaine, Toronto, 11 July [1903]. PAC, RG 23, Vol. 242, File 1536 [1], 61.

26. W[illiam] Templeman to [Raymond Préfontaine], Ottawa, 20 July 1903. PAC, RG 23, Vol. 242, File 1536 [1], 63.

27. The applications, each dated 1 May 1904, defined sites from Sooke Inlet in the Strait of Juan de Fuca north to a location five miles from Skidegate Inlet on the east coast of the Queen Charlotte Islands. The eight original applicants were Birrell, Bryman, Ewen, Forrester, Hays, Holcross, Livingston, and Mackie; see Raymond Préfontaine to Aulay Morrison, 9 May 1904. PAC, RG 23, Vol. 242, File 1536 [1], 102.

28. *Ibid.*.

29. R. E. Finn to Raymond Préfontaine, Halifax, 11 July 1904. PAC, RG 23, Vol. 242, File 1536 [1], 112.

30. *Ibid.*.

31. R. E. Finn to Raymond Préfontaine, telegram, Halifax, 22 July [1904]. PAC, RG 23, Vol. 242, File 1536 [1], 114; and R[aymond] Préfontaine to R. E. Finn, telegram, Ottawa, 22 July 1904. PAC, RG 23, Vol. 242, File 1536 [1], 115.

32. R. E. Finn to Raymond Préfontaine, Ottawa, 29 July 1904. PAC, RG 23, Vol. 242, File 1536 [1], 120. The precise wording was: "On either side of two small islands in close proximity to each other known as Hunter's Island and Calvert Island, or in the alternative on a point in Nahmu Harbour on the east side of Fitz Hugh Sound"; "At Ucluelet Arm on the west coast of Vancouver Island, on the entrance of Barclay Sound"; and "Rose Harbour in Esperanza Inlet, north-east coast of Catala Island"; and "From a point five miles north of Nanaimo on the east coast of Vancouver Island or in the alternative on the main land at a point opposite Nanaimo."

33. *Ibid.*.

34. Corbett was added at Morrison's request; see PAC, RG 23, Vol. 242, File 1536 [1], 123 (6 August 1904).

35. M. Foley to Raymond Prafontaine [sic], Brooklyn, N.Y., 15 November 1904. PAC, RG 23, Vol. 242, File 1536 [1], 161.

36. Chapter 45 of the Revised Statutes of Canada, 4 Edw. 7, ch. 13.

37. 4 Edw. 7, ch. 13; see, for example, "Whale Factory License," PAC, RG 23, Vol. 242, File 1536 [3], 508d.

38. The minister had received at least one inquiry, from T. LeMessurier of Vancouver, regarding the capture of whales in nets. Deputy Minister Gourdeau answered that any such device would be considered illegal; see F. Gourdeau to T. LeMessurier, Ottawa, 3 February 1906. PAC, RG 23, Vol. 242, File 1536 [1], 268.

39. Plant and machinery specifications, Sechart, B.C.. PAC, RG 23, Vol. 242, File 1536 [1], 129.

40. Sprott Balcom to Raymond Préfontaine, Victoria, 14 August 1904. PAC, RG 23, Vol. 242, File 1536 [1], 130.

41. E. E. Prince (Dominion Commissioner of Fisheries) to E. G. Taylor (Inspector of

Fisheries, Nanaimo), [Ottawa], 5 May 1905. PAC, RG 23, Vol. 242, File 1536 [1], 217.

42. W. H. Riddall to Raymond Préfontaine, Ottawa, 23 June 1905. PAC, RG 23, Vol. 242, File 1536 [1], 224; and John O. Townsend to Raymond Préfontaine, Victoria, [June] 1905. PAC, RG 23, Vol. 242, File 1536 [1], 226.

43. "Memorandum," prepared by R. N. Venning, Ottawa, 12 October 1905. PAC, RG 23, Vol. 242, File 1536 [1], 244.

44. F. Gourdeau to William Sloan, [Ottawa], 8 December 1905. PAC, RG 23, Vol. 242, File 1536 [1], 258.

45. Charles H. Lugrin to L. P. Brodeur, Victoria, 13 July 1906. PAC, RG 23, Vol. 242, File 1536 [1], 304.

46. Charles H. Lugrin and John A. Watson to [L. P. Brodeur], Victoria, 31 July 1906. PAC, RG 23, Vol. 242, File 1536 [1], 316.

47. Charles H. Lugrin to [F. Gourdeau], Victoria, 31 July 1906. PAC, RG 23, Vol. 242, File 1536 [1], 318.

48. D. G. Macdonell (barrister and solicitor) to [L. P. Brodeur], Vancouver, 15 August 1906. PAC, RG 23, Vol. 242, File 1536 [1], 320.

49. John I. Bostock to [L. P. Brodeur], Vancouver, 18 August 1906. PAC, RG 23, Vol. 242, File 1536 [1], 323.

50. Preference shares amounted to $104,000, common shares $96,000; "Victoria's Garnering the Leviathan Crop; Details of Whaling Industry Published for First Time, Show Enormous Extent of Operations," *Victoria Daily Times*, 28 July 1906, 1. Financial data were later denied by Balcom; see *Victoria Daily Times*, 30 July 1906, 4.

51. R. E. Finn to W. S. Fielding (Minister of Finance), Halifax, 15 March 1905. PAC, RG 23, Vol. 242, File 1535 [1], 212.

52. Oslo after 1 January 1925.

53. Personal communication, Jimmy Wakelen, Victoria, 25 August 1983.

54. Lewis L. Robbins, "Letters Written of Whaling Expeditions 1936, 1937, & 1938," typescript letter book, Rose Harbour, B.C., 11 July 1936, 1. Robbins.

55. John H. Harland to Author, Kelowna, B.C., 13 October 1983.

56. Bill of Sale, J. Lyman Allyn to N. and W. W. Billings, one-fourth of ship *Louvre*, New London, 5 July 1844. NLCHS Ms P .B495 m 1914.04.74.

57. Finn to Fielding, 15 March 1905.

58. Roy Chapman Andrews, *All About Whales* (New York: Random House, 1954), 3.

59. Personal communication, Herb Smith, Victoria, 25 August 1983.

60. Finn to Fielding, 15 March 1905.

61. See Raymond Préfontaine to William Sloan, Ottawa, 14 February 1905. PAC, RG 23, Vol. 242, File 1536 [1], 198.

62. "Only One Original Left," *The Fisherman*, 18 December 1961, 12. Nils Nilsen, who came out as steersman under Reuben Balcom, stayed on as gunner. Otto Gaustad came as engineer and stayed. Olsen is said to have come in the *Orion*. He worked in British Columbia awhile, and then took another whaling job at a station at Port Armstrong, Alaska until his wife, "being pregnant for the third time, put her foot down and took them all home to Norway"; Joan Goddard to Author, Portland, Oregon, undated (November 1985). Other Norwegians would come later: Karl Larsen, the blacksmith at Rose Harbour in 1937, had come out from Norway in 1910 or 1911 aboard one of the chaser boats delivered to B.C. from Christiania; see Robbins, "Letters Written of Whaling Expeditions," Rose Harbour, 30 July 1937, 5. The original crew of the *Orion*, including Balcom, Nilsen, Gaustad, second-steersman Carl Sorensen and engineer P. Rasmussen are mentioned in J. N. Tønnessen and Arne Odd Johnsen, *Den Moderne*

Hvalfangsts Historie: Opprinnelse og Utvikling (Sandefjord: Norges Hvalfangstforbund, 1967), 2: 140.

63. James Clark to the Department of Labour, New Alberni, B.C., 27 April 1905. PAC, RG 23, Vol. 242, File 1536 [1], 219.

64. F. Gourdeau to W. L. Mackenzie King, Ottawa, 12 May 1905. PAC, RG 23, Vol. 242, File 1536 [1], 221.

65. Tønnessen and Johnsen, *Den Moderne Hvalfangsts Historie*, 2: 139; and "Famous Sealer is Back in Halifax," *Halifax Herald*, 5 June 1913.

66. Much later, Balcom wrote that his Pacific Whaling Company had purchased the rights to Rissmüller's patent processes and installed them in a "previously built" station at Sechart; "Affidavit of Sprott Balcom," Dominion of Canada, Province of British Columbia, January 1921; unsigned copy in Balcom.

67. "Agreement: Ludwig Rismuller and The Pacific Whaling Company Limited," prepared by R. T. Elliott, barrister and solicitor, Victoria, copy dated 25 January 1907. Balcom.

68. "Agreement," fos. 3–4. The company also issued 2,018 shares of "7% preferred cumulative" stock.

69. Millais, *Newfoundland*, 184.

70. "Conditions of the Whaling Industry: Capt. Balcom and Dr. Rissmuller Return from Visit to the West Coast Stations," *Colonist*, 16 March 1906, 6.

71. Ben McCourt, *The Story of the Whale* (Pacific Whaling Company, n.d.), 8.

72. Rissmüller serviced one such contract with Anders Ellefsen, who in 1903 received $24 for a blue whale carcass, $14 for a finback, and $10 for a humpback; Tønnessen and Johnsen, *Den Moderne Hvalfangsts Historie*, 2: 117–18. This seems a high price, considering that carcasses were traded fifteen years later in the Antarctic whale fishery for £1 each regardless of species.

73. *Ibid.*, 2: 117.

74. Millais, *Newfoundland*, 184.

75. This was their hull no. 144; an early photograph shows the vessel's full name, *Saint Lawrence*, painted on its side; soon thereafter ship-painters shortened it to *St. Lawrence*.

76. Tønnessen and Johnsen, *Den Moderne Hvalfangsts Historie*, 2: 139.

77. *Ibid.*, 2: 117.

78. Balcom.

79. "Apparatus for Drying, Grinding and Screening," Dominion of Canada Patent Papers No. 88875, dated 23 August 1904; copy, dated 13 June 1908. Balcom.

80. "Process of Extracting Fatty Substances from Meat," Dominion of Canada Patent Papers No. 88351, dated 19 July 1904; copy, dated 13 June 1908. Balcom.

81. *Ibid.*

82. "Apparatus for Drying, Grinding and Screening," fo. 2.

83. Alfred Hustwick, "The War on the Whale, Part 2," *British Columbia Magazine* 8 (January 1912): 9.

84. "Acid Proof and Oil Proof Tanks," Dominion of Canada Patent Papers Number 94083, 11 July 1905; copy, dated 18 July 1910. Balcom.

85. "Particulars of the Rissmüller Patented Whale Reduction Process." PAC, RG 23, Vol. 242, File 1536 [1], 179–78.

86. R. V. Sinclair to [Raymond Préfontaine], Ottawa, 6 January 1905. PAC, RG 23, Vol. 242, File 1536 [1], 181.

87. Millais, *Newfoundland*, 193. Apparently, the author thought that a man who properly utilized the corpses of whales was more to be praised than, say, a successful real estate agent.

88. "Auditor's Report, Balance Sheet and Manufacturing & Profit & Loss Account to 31st

December 1906," prepared by W. Curtis Sampson, Victoria, 18 January 1907, fo. 2. Balcom. See also "Year Was Good for Whalers, Interesting Report Made to Shareholders," *Victoria Daily Times*, 21 January 1907, 5. Rissmüller was by then an officer in the company.

89. See "Year Was Good for Whalers," 5; and "Victoria's Garnering the Leviathan Crop," 1.

90. C[ereno] J[ones] Kelley to L. P. Brodeur, Victoria, 21 August 1906. PAC, RG 23, Vol. 242, File 1536 [1], 326.

91. John O. Townsend to L. P. Brodeur, Victoria, 7 September 1906. PAC, RG 23, Vol. 242, File 1536 [1], 328.

92. L. P. Brodeur to William Templeman, Ottawa, 22 September 1906. PAC, RG 23, Vol. 242, File 1536 [1], 331.

93. "Hvalfangst ved Vestkysten av Amerika," *NHT* 6 (March 1917): 71.

94. "Memorandum—RE Applications for Whale Factory Licenses on the Pacific Coast," prepared by R. N. Venning, Ottawa, 21 September 1906. PAC, RG 23, Vol. 242, File 1536 [1], 336–35.

95. [F. Gourdeau] to unidentified recipient, draft copy, Ottawa, 1 October 1906. PAC, RG 23, Vol. 242, File 1536 [1], 339–38.

96. Sprott Balcom to E. G. Taylor, Victoria, 26 September 1906. PAC, RG 23, Vol. 242J File 1536 [1], 349.

97. Edward G. Taylor to R. N. Venning, Nanaimo, B.C., 18 October 1906. PAC, RG 23, Vol. 242, File 1536 [1], 356.

98. W[illiam] Templeman to L. P. Brodeur, Ottawa, 11 October 1906. PAC, RG 23, Vol 242, File 1536 [1], 353.

99. *Ibid.*, 353–52.

100. Charles H. Lugrin to William Templeman, copy, Victoria, 19 October 1906. PAC, RG 23, Vol. 242, File 1536 [1], 358–57.

101. W[illiam] Sloan to L. P. Brodeur, Nanaimo, 20 October 1906. PAC, RG 23, Vol. 242, File 1536 [1], 369.

102. C[ereno] J[ones] Kelley to L. P. Brodeur, Victoria, 22 October 1906. PAC, RG 23, Vol. 242, File 1536 [1], 367.

103. Sprott Balcom to L. P. Brodeur, two letters, both Victoria, 20 October 1906. PAC, RG 23, Vol. 242, File 1536 [1], (letter 32223); 377, and (letter 32224); 379.

104. L[udwig] Rissmüller to L. P. Brodeur, Victoria, 20 October 1906. PAC, RG 23, Vol. 242, File 1536 [1], n.p.

105. F. Gourdeau to Charles Lugrin, Ottawa, 14 November 1906. PAC, RG 23, Vol. 242, File 1536 [1], 388.

106. Tønnessen and Johnsen, *History of Modern Whaling*, 116.

107. "Strike at Whaling Station," *Nanaimo Free Press*, 27 July 1907, 1; and John Cass, "Nanaimo's Whaling Station," unpublished typescript, n.d. VMM.

108. Pacific Whaling Company [Sprott Balcom?] to L. P. Brodeur, Victoria, 10 January 1907. PAC, RG 23, Vol. 242, File 1536 [1], 403 3/4–403 1/2.

109. The pilot house was almost immediately replaced by an open bridge similar to that on the *Orion*.

110. Jimmy Wakelen, 25 August 1983.

111. "Pacific Whaling Company's Good Season," *Nanaimo Free Press*, 5 October 1907, 1. A large quantity of bones was also brought down in the *Amur*, to be ground for bone meal. Several mills were to be tested, the most satisfactory to receive a contract. Clearly, the two stations had not yet received bone-grinding apparatus, and this is corroborated by the *Nanaimo Free Press* reporter who described the skeletons of "innumerable whales"

piled in "four mountains of bones, two at each station."

112. "Whale Sailed Away with Stuart City Wharf: Tragic End of Great Fish Drowned in His Native Element," *Province*, 29 April 1907, 1.

113. "Thought Beelzebub Was In Towing Business: Mystery of Floating Wharf Driven Through Water by Northern Whale," *Province*, 16 May 1907, 11.

114. "To Catch Nanaimo Whales Next Week," *Nanaimo Free Press*, 25 October 1907, 1.

115. Sprott Balcom to L. P. Brodeur, Victoria, 22 November 1907. PAC, RG 23, Vol. 242, File 1536 [1], 563–62.

116. *Ibid.*, 562.

117. "Whaling, Latest Nanaimo Sensation," *Nanaimo Free Press*, 18 November 1907, 1.

118. "Local Company's Big Success," *Nanaimo Free Press*, 27 November 1907, 1–2. The reference made is to Upton Sinclair's book *The Jungle*, published to wide international acclaim the preceding year (1906), which exposed the plight of immigrants in the factories of Chicago; therein a new arrival brags that the modern meat-packing plants "use everything about the hog except the squeal."

119. *Ibid.*, 2; and "Some More Whale Talk," *Nanaimo Free Press*, 13 December 1907, 7.

120. "Many Whales Were Caught This Week," *Nanaimo Free Press*, 7 December 1907, 1; see also 11 December 1907, 1.

121. "Year Was Good for Whalers," 5.

122. J. A. Cates to W. H. Macpherson, Vancouver, 22 November 1907. PAC, RG 23, Vol. 242, File 1536 [1], 574.

123. L. P. Brodeur to R. G. Macpherson, 28 December 1907; cited in R. G. Macpherson to L. P. Brodeur, Ottawa, 17 January 1908. PAC, RG 23, Vol. 242, File 1536 [1], 588.

124. Macpherson to Brodeur, 17 January 1908, 588.

125. Edward G. Taylor to unidentified recipient, Nanaimo, 20 March 1908. PAC, RG 23, Vol. 242, File 1536 [2], 38–37.

126. "Page's Lagoon No. 3," Pacific Whaling Company factory report for weeks ending 16 November 1907 to 25 January 1908. PABC, Add MSS 21, Vol. 1.

127. Manuscript notebook, probably kept by George Wilbur Piper, ship *Europa* of Edgartown, 1866–72, 131. KWM (A-194).

128. Robbins, "Letters Written of Whaling Expeditions," Rose Harbour, 30 July 1937, 5–6. Robbins attributed the remark to Larsen, the Rose Harbour blacksmith.

129. "Whaling Co. Pays Big Dividend: Annual Meeting of the Pacific Whaling Company Held Yesterday," *Colonist*, 25 January 1908, 10.

130. "Memorandum—RE Whale Factory Licenses on the Pacific Coast," prepared by R. N. V[enning], Ottawa, 20 June 1907. PAC, RG 23, Vol. 242, File 1536 [1], 501.

131. Sprott Balcom to F. Gourdeau, Victoria, 13 June 1907; quoted in V[enning], "Memorandum—RE Whale Factory Licenses," 501.

132. *Ibid.*.

133. Richard Hall to William Templeman, Victoria, 27 June 1907. PAC, RG 23, Vol. 242, File 1536 [1], 514.

134. F. Gourdeau to William Templeman, Ottawa, 12 September 1907. PAC, RG 23, Vol. 242, File 1536 [1], 538.

135. The licence was issued on 23 July 1907; C[ereno] J[ones] Kelley to L. P. Brodeur, Victoria, 15 July 1907. PAC, RG 23, Vol. 242, File 1536 [1], 520. Issuance of licence noted in R. N. Venning to John T. Williams (Inspector of Fisheries, Port Essington, B.C.), [Ottawa], 23 July 1907. PAC, RG 23, Vol. 242, File 1536 [1], 521.

136. Sprott Balcom to R. P. [sic] Venning, Victoria, 2 September 1907. PAC, RG 23, Vol. 242, File 1536 [1], 528–27.

137. D. Stevens (British Columbia Information Agency) to W[illiam] Templeman, Victoria,

24 February [1908]. PAC, RG 23, Vol. 242, File 1536 [2], 9–8.

138. L. P. Brodeur to W[illiam] Templeman, [Ottawa], 8 March 1908. PAC, RG 23, Vol. 242, File 1536 [2], 17.

139. "The Prince Rupert Whaling Company, Ltd." (stock prospectus), n.p., [ca. 1908], 2. PAC, RG 23, Vol. 242. File 1536 [2], microfilm, UBC AW1R 5474, Vol. 41.

140. "Prince Rupert Securities Ltd." (pamphlet), 8 October 1908, 13. PABC NWp971.1PV; Prince Rupert Regional Archives 982–11, MS #53.

141. "Big Dividend Will Be Paid: Pacific Whaling Company Authorizes Division of $69,513 Among Shareholders Next June," *Colonist*, 28 February 1910, 14.

142. George A. Huff to L. P. Brodeur, application for whaling licence, Victoria, 10 September 1907. PAC, RG 23, Vol. 242, File 1536 [1], 535.

143. F. Gourdeau to George A. Huff, Ottawa, 24 September 1907. PAC, RG 23, Vol. 242, File 1536 [1], 544; and Sprott Balcom to F. Gourdeau, Victoria, 24 October 1907. PAC, RG 23, Vol. 242, File 1536 [1], 549.

144. George A. Huff to L. P. Brodeur, [February 1908]. PAC, RG 23, Vol. 242, File 1536 [2], [1]; also L. P. Brodeur to William Templeman, Ottawa, 7 March 1908, File 1536 [2], 13.

145. "Whaling Station at Rose Harbour," *Colonist*, 10 September 1908, 10.

146. Certificate of Incorporation no. 2734; see G. A. Huff to Ralph Smith (M.P., Ottawa), Alberni, B.C., 30 December 1909. PAC, RG 23, Vol. 242, File 1536 [3], 278.

147. Sprott Balcom to Ralph Smith, Victoria, 30 December 1909. PAC, RG 23, Vol. 242, File 1536 [3], 279.

148. "Directors' Report, Balance Sheet etc. to 31 December 1908," Pacific Whaling Company, prepared by W. Curtis Sampson, 20 January 1909, fo. 1. Balcom.

149. The remaining whales processed at Sechart and Kyuquot during the 1908 season included 99 sulphur-bottom (blue) whales and 29 finbacks; Sprott Balcom to [L. P. Brodeur], Victoria, 24 February 1909. PAC, RG 23, Vol. 242, File 1536 [2], 100–99. The two whales unaccounted for were apparently sperms. A second set of statistics record that Sechart processed 201 humpback, 32 blue, 16 finback, 1 sperm; Kyuquot 242 humpback, 66 blue, 10 finback, 1 sperm. This equals 569 whales, with oil production reaching 11,583 barrels at Kyuquot, 8,084 barrels at Sechart; fertilizer manufacture totalling 1,220 tons at Kyuquot, 770 tons at Sechart; [Consolidated Whaling Corporation], whaling station statistics, 1908–23. Unpublished typescript. Parks Canada Anthropological Research Project, Victoria. Early catch statistics are equivocal.

150. Common shares earned but 7 per cent; "Whaling Company Pays Dividend: Fourteen and Seven Per Cent on Stock Will Be Given to Shareholders," *Colonist*, 23 January 1909, 10.

151. Sprott Balcom to G. J. Desbarats (Acting Deputy Minister of Marine and Fisheries), Victoria, 24 March 1909. PAC, RG 23, Vol. 242, File 1536 [2], 134. Bone meal fertilizer appears to have been marketed domestically.

152. Memorandum of agreement, between Ludwig Rissmüller of St. John's, Newfoundland, and the Tyee Company of San Francisco, California, dated 10 July 1908. Balcom.

153. "Agreement: Ludwig Rismuller and The Pacific Whaling Company Limited," Victoria, 25 January 1907, fo. 4. Balcom. Rissmüller's involvement was clearly spelled out in the Tyee contract; the doctor agreed to provide plans and specifications "for building and equipping steam vessels"; memorandum of agreement, fo. 1.

154. Risting, *Av Hvalfangstens Historie*, 573.

155. "Whaler Tyee Jr., Here, Calls for Her Crew," *Colonist*, 24 October 1907, 10.

156. "Twenty-Five Years Ago Today," *Victoria Daily Times*, 11 April 1932.

157. The whaling at that time was primarily conducted in the vicinity of Cape Fanshaw; Andrews also mentioned a group of rocks called "The Five Fingers" and also Storm Island; see Roy Chapman Andrews, "Whale-Hunting As It Is Now Done," *World's Work* (December 1908), 11031–33, 11043.

158. Edward G. Taylor to R. N. Venning (Superintendent of Fisheries), Nanaimo, 30 March 1909. PAC, RG 23, Vol. 242, File 1536 [2], 148. The United States Whaling Company, a Norwegian concern, was formed in 1907 and was reportedly based in Port Angeles; G. B. Sword (Inspector of Fisheries, New Westminster, B.C.) to [L. P. Brodeur], New Westminster, 22 June 1909. PAC, RG 23, Vol. 242, File 1536 [2], 195–94. The firm did not begin whaling at that time, perhaps because of the effect of the 1907–08 economic slump on whale products, but later established a whaling station in southeastern Alaska; see Chapter 6.

159. Balcom to [Brodeur], 24 February 1909, 101.

160. The cited figure was $49,325; Balcom to Desbarats, 24 March 1909, 133.

161. "Memorandum: RE Whale Fishing On Pacific Coast," prepared by R. N. Venning, dated Ottawa, 22 May 1909. PAC, RG 23, Vol. 242, File 1536 [2], 184.

162. Section 1 of Chapter 20 of the Statutes of 1910 formed the replacement. L[udwig] Rissmüller to [A. Johnston (Deputy Minister of Marine and Fisheries)], Victoria, 9 January 1912. PAC, RG 23, Vol. 242, File 1536 [4], 3.

163. T. D. Regehr, *The Canadian Northern Railway: Pioneer Road of the Northern Prairies 1895–1918* (Toronto: Macmillan of Canada, 1976), 219–20. Mackenzie, Mann and Company incorporated as a joint-stock company in 1902.

164. Petition to establish whaling factory, signed by John M. Macmillan, [1909]. PAC, RG 23, Vol. 242, File 1536 [2], 216.

165. Harvey Fitzsimons (Canadian Northern Railway Company, Land Immigration and Industrial Departments) to Secretary, Department of Marine and Fisheries, Toronto, 26 November 1913. PAC, RG 23, Vol. 242, File 1536 [4], 105. See also Regehr, *Canadian Northern Railway*, 227.

166. R. N. Venning to John T. Williams, [Ottawa], 25 November 1909. PAC, RG 23, Vol. 242, File 1536 [2], 248.

167. John M. Macmillan to [L. P. Brodeur], [Vancouver], 15 April 1910. PAC, RG 23, Vol. 242, File 1536 [3], 388.

168. "Big Dividend Will Be Paid," 14.

169. "Agreement for Sale. Queen Charlotte Whaling Company Limited and Pacific Whaling Company Limited," Elliott and Shandley, Barristers and Solicitors, 1 August 1910, unsigned copy. Balcom. This copy specifies the sale of the Queen Charlotte Whaling Company to the Pacific Whaling Company for $212,937.50, on or before 1 December 1910. By its terms, the unnamed new firm created by the sale would hand over five shares of its stock, at par value of £1 each, to the Queen Charlotte Whaling Company for each share outstanding from the old firm, up to 3,407 shares, presumably the total then outstanding.

170. "Crowding Out the Americans: Canada Will Protect Her Whaling Industry," *Prince Rupert Optimist*, 14 September 1910, 1.

171. "Memorandum: RE Whale Factory licenses on the Pacific Coast, B.C.," prepared by William A. Found, Ottawa, 29 November 1910. PAC, RG 23, Vol. 242, File 1536 [3], 520. The mortgage was registered in Victoria and secured by trust deed on 25 November 1910; CNPF debenture share certificate. KWM.

172. Herschel Island, note particularly John R. Bockstoce, *Whales, Ice, and Men: The History of Whaling in the Western Arctic* (Seattle: University of Washington Press, 1986).

173. "Crowding Out The Americans," 1.

174. "Memorandum: RE Application for Whale Factory licenses, Arctic Ocean," prepared by William A. Found, Ottawa, 17 August 1910. PAC, RG 23, Vol. 242, File 1536 [3], 497.

175. *Ibid.*.

176. "Whaler Sebastian Enters Pacific: New Vessel for Pacific Whaling Company Reported from Puntas Arenas on Her Way Here," *Colonist*, 13 February 1910, 14.

177. John H. Harland to Author, Kelowna, B.C., 13 October 1983.

178. Akers no. 222; the *Germania* did not begin whaling until 1903, together with sistership *Island* (no. 223) at Faskrudfjord; see Klaus Barthelmess, "Julius Tadsen—Manager des Walfanges: Die 'Germania Walfang-und-Fischindustrie Aktiengesellschaft' zu Hamburg," *Nordfriesland Tageblatt*, 5 October 1981; and Klaus Barthelmess, "Deutsche Walfanggesellschaften in Wilhelminischer Zeit: Germania AG und Sturmvogel GmbH," *Deutsches Schiffahrtsarchiv* 9 (1986); 232–38. Also Harland to Author, 13 October 1983.

179. "New Whaler Starts Today: Steamer William Grant Shipped Crew Yesterday and Had a Satisfactory Test Trial Trip During the Afternoon," *Colonist*, 22 June 1910, 14. Also Klaus Barthelmess to Author, Cologne, 12 November 1985.

180. "General Arrangement of Whaling Steamer N^r 306–309–310–312–313," Akers Mek. Verksted Tegning N^r 4694; copy provided by John Harland, 25 September 1985, courtesy of Tjerand Skåré, Aker Engineering A/S, Oslo. In actuality their tonnage varied slightly, from 105.74 to 106.13; Harland to Author, 25 October 1983.

181. W. A. Hagelund, *Flying The Chase Flag: The Last Cruise of the West Coast Whalers* (Toronto: Ryerson Press, 1961), 141; also Harland to Author, 25 October 1983.

182. "General Arrangement of Whaling Steamer."

183. Probably John Macmillan, though the name of Rissmüller's adversary is not mentioned in most accounts.

184. Balcom's role was recalled by a grandson, Lawrence Balcom; Graeme Balcom, 24 September 1984.

185. "Whaling Season Will Be Busy," *Evening Empire* (Prince Rupert), 4 February 1911, 4.

186. "General Arrangement of Whaling Steamer."

187. Peter Madison, "British Columbia Whaling Venture," *Canadian Fisherman* 35 (September 1948): 22.

188. Alfred Hustwick, "The War on the Whale, Part 1," *British Columbia Magazine* 7 (December 1911): 1249.

189. "S.S. 'Gray': Built in Workington, England; Official No. 124395," and "Docking Data—Steamer 'Gray'," typescripts prepared by the Victoria Whaling Company or North Pacific Fisheries Limited, ca. 1914–15. VMM. The *Petriana* was powered by a triple-expansion steam engine built by Ross and Duncan, which produced 585 IHP. The 706.24-ton ship (1140 tons displacement, loaded), measured 27.9 feet across the beam; holds were 12.3 feet in depth. Accommodation for ten passengers was provided, and a refrigeration plant for cold-storage was installed for the cartage of perishable products. The purchase is noted in the *Victoria Daily Times*, 20 December 1910, 11.

190. "The Queen Charlotte Whaling Co.," *Colonist*, 24 February 1910, 15.

191. Cass, "Nanaimo's Whaling Station." Page's—now Piper's—Lagoon, is held as a preserve by the city of Nanaimo, which has expanded to its very edge.

192. Douglas Raine to Author, Victoria, 19 September 1985.

193. "Plenty of Whales," *Prince Rupert Optimist*, 13 September 1910, 2.

194. See PAC, RG 23, Vol. 242, File 1536 [3], 524. The Naden Harbour station was built on the Kelley-Hall licence originally issued for Big Dundas Island. The CNPF had asked to abandon "Port Stanley" since "no suitable site for the erection of a station is

available on Dundas Island"; see E. V. Bodwell (barrister and solicitor for the CNPF) to L. P. Brodeur, Victoria, 29 November 1910. PAC, RG 23, Vol. 242, File 1536 [3], 516. Also, one of the licences was shifted to Nasoga Gulf to provide for a station—never built—at that place. And the "Rose Harbour," (Esperanza Inlet) licence was eventually redrawn to read "Kyuquot Sound, Vancouver Island"; PAC, RG 23, Vol. 242, File 1536 [3], 525.

195. "Largest Whaling Station Yet," *Evening Empire*, 13 December 1910, 1. On one of these trips, the *Henriette* came to grief while departing the Michigan-Puget Sound Lumber Company wharf in Victoria, taking out thirty feet of dock belonging to the British Columbia Soap Company at Laurel Point. The vessel and its cargo of 250,000 board-feet of lumber were not seriously damaged so the captain proceeded to Naden Harbour; see "Henriette Had Smash in Victoria Harbor," *Evening Empire*, 15 December 1910, 1.

196. "Affidavit of Sprott Balcom," January 1921. Perhaps this is the arrangement that Balcom had prevously worked out with George A. Huff at Rose Harbour back in 1908.

197. "Memorandum of Agreement Between: Ludwig Rissmuller and Sprott Balcom and Canadian North Pacific Fisheries, Limited," dated 25 August 1911, unsigned copy. Balcom.

198. Lucile McDonald, "Whaling on the Washington Coast," *Sea Chest* 6 (September 1972): 3.

199. "Whaling Vessel," *Prince Rupert Journal*, 16 June 1911, 4. The *Paterson*, launched on 15 June 1911, was named for J. V. Paterson, at that time president of Moran. Each vessel measured 87'3" x 18'0" x 11'4", 120 tons gross, 77 tons net.

200. The *Aberdeen* and the *Westport* were nine inches longer, one foot broader, and drew one-tenth of a foot more water than the the *Moran* or the *Paterson*, but measured four tons less, gross.

201. Whaling station statistics, 1908–23. Rose Harbour processed 310 whales in 1911.

202. "Oil Production—All Stations, 1908–1923," in whaling station statistics; tonnage reported in F. H. Cunningham (Chief Inspector of Fisheries) to Department of Marine and Fisheries, New Westminster, B.C., 9 March 1912. PAC, RG 23, Vol. 242, File 1536 [4], 39.

203. Whaling station statistics, 1908–23. Chief Inspector of Fisheries Cunningham reported a total catch (excluding Bay City, Washington) of 1,623 whales in 1911, including 1,023 humpback, 361 finbacks, 205 blue, and 34 sperm whales; Cunningham to Department of Marine and Fisheries, 39. To this would be added the Bay City catch, which totalled 173 humpbacks, 5 finbacks, 2 blues, and 2 sperm whales.

204. Based on oil at $75 per ton; body bone at $50 per ton; fertilizer worth $30 per ton, baleen at $16 per hundredweight; Adam Shortt and Arthur G. Doughty, eds., *Canada and its Provinces: A History of the Canadian People and Their Institutions by One Hundred Associates* (Toronto: Publishers' Association of Canada, 1914), 22: 474–75.

205. Hustwick, "The War on the Whale, Part 2," 12.

206. Whaling station statistics, 1908–23.

207. "Whale Oil and Fertilizer Marketed—British Columbia, 1905–37," Table 32 in Carrothers, *British Columbia Fisheries*, 126.

208. "We like to make the Government the following proposition... ," typescript prepared by the Canadian North Pacific Fisheries Limited, or by the Victoria Whaling Company, undated, fos. 1, 3. VMM. The *Germania* operated under charter as a Fisheries inspection/patrol vessel during 1911; Klaus Barthelmess to Author, 12 November 1985.

209. "Famous Sealer is Back in Halifax."

210. British Columbia, Royal Commission, *Claims by Pelagic Sealers Arising out of the Wash-*

ington Treaty, 7th July 1911 *and the Regulations Made Under the Paris Award which Came into Force* 1894, White Claims, Vol. 13 (Victoria, 1914). PAC, RG 33/107, Vol. 3, 21.

211. "Pioneer Sealer Passes in City," 5. Reuben Balcom died in Victoria on 15 February 1929, aged 74 years; see "Capt. R. Balcom Passes to Rest: Well-Known Local Mariner Had Interesting Career at Sea from Age of Twelve," *Colonist*, 16 February 1929, 5.

212. Balcom.

213. "Canadian North Pacific Fisheries Ltd: Weekly Report, Department 2—Whaling," prepared by G[eorge] LeMarquand, manager, Naden Harbour, 28 June-4 July 1914. VMM.

214. Weekly report, prepared by Alfred Gosney, manager, Sechart, 12—18 April 1914.

215. Tønnessen and Johnsen, *History of Modern Whaling*, 117.

216. Harvey Fitzsimons to G. J. Desbarats, Ottawa, 2 July 1915. PAC, RG 23, Vol. 764, File 716—14—3 [1], 11—10.

217. See "To Engage in Fisheries," *Prince Rupert Journal*, 12 May 1911, 6; and "Fish Proposition," *Prince Rupert Journal*, 21 March 1912, 1.

218. "Trawlers on Way," *Prince Rupert Journal*, 4 May 1912, 2; see also J. D. Hazen to Sir George Doughty, [Ottawa], 22 January 1912. PAC, RG 23, Vol. 242, File 1536 [4], 11; and "To Engage in Fisheries," 6.

219. Hazen to Doughty, 22 January 1912, 10.

220. "Application for a Fishery License in British Columbia, Season of 1912," signed by Wilfred Doughty, Victoria, 9 September 1912. PAC, RG 23, Vol. 242, File 1536 [4], 56.

221. C. Stanton (Assistant Deputy Minister of Marine and Fisheries) to Wilfred Doughty, [Ottawa], 20 September 1912. PAC, RG 23, Vol. 242, File 1536 [4], n.p.

222. Bark *Andrew Hicks* of New Bedford, Charlotte E. Church, keeper, [8 May 1906]. KWM (607).

223. *Ibid.*, [17 May 1906].

NOTES TO CHAPTER 6

1. Knut Birkeland, *The Whalers of Akutan: An Account of Modern Whaling in the Aleutian Islands* (New Haven: Yale University Press, 1926), 30.

2. Reported in "Hvalfangst ved Vestkysten av Amerika," 71.

3. "Steam Freighter for the Whaling Company," *Colonist*, 11 December 1907, 10. It is not clear which two sites were in question, but one may have been John O. Townsend's.

4. Risting, *Av Hvalfangstens Historie*, 573.

5. "Steam Freighter," 10.

6. "Alaskan Whalers," *Prince Rupert Journal*, 27 October 1911, 5. The stock was reportedly sold to European and American subscribers. The company must have organized in 1907, since reports of that year said the firm was based in Port Angeles, Washington; see G. B. Sword to [L. P. Brodeur], 22 June 1909, 195—94.

7. Susan Mathews to Author, Sitka, Alaska, 23 August 1984.

8. A. J. Villiers, *Whaling in the Frozen South: Being the Story of the* 1923—24 *Norwegian Expedition to the Antarctic* (Indianapolis: Bobbs-Merrill Company, 1925), 191—93.

9. Risting, *Av Hvalfangstens Historie*, 575.

10. *Ibid.*.

11. The vessels measured 100' x 19.2' x 12.4'; American Bureau of Shipping, *Record of*

American and Foreign Shipping 1917 (New York: American Bureau of Shipping, 1917), 437, 785.

12. *Admiralen* noted in Brandt, *Whale Oil*, 60; and Hans S.I. Bogen, *70 år Lars Christensen og hans samtid* (Oslo: Johan Grundt Tanum, 1955), 211–12. *Coronado* in *Victoria Daily Times*, 25 July 1912; cited in Frank Clapp, "Odd Notes," ms. VMM.

13. Birkeland, *Whalers of Akutan*, 3.

14. "Consul Lars Christensen, in Memoriam," *NHT* 55 (January 1966): 4.

15. Tønnessen and Johnsen, *History of Modern Whaling*, 121.

16. Noted in Clapp, "Odd Notes"; see also Berman, *Encyclopedia of American Shipwrecks*, 206.

17. Risting, *Av Hvalfangstens Historie*, 580.

18. Finn Koren, cited in "Walfangstsæsongen paa Vancouvers Vestkyst Avsluttet," *NHT* 2 (15 January 1913): 14.

19. Birkeland, *Whalers of Akutan*, 2. Minnesota registry is noted in L. W. Stanley (North Pacific Sea Products Company) to Victoria Whaling Company, Seattle, 9 March 1916, fo. 1. VMM. Bonuses noted in "Alaska Whaling Co. Rekonstrueres," *NHT* 3 (31 March 1914): 42. A brief summary of the firm's activities and the reconstruction is found in Bogen, *70 år Lars Christensen*, 211–12.

20. " 'North Pacific Sea Products Company'," *NHT* 6 (February 1917): 50, 53; see also "Aktieselskapet 'United States' Wh. Co.," *NHT* 3 (31 March 1914): 41.

21. "United States Whaling Company," *NHT* 4 (30 April 1915): 46. United States Whaling Company production totalled 8,101 barrels of oil, 7,536 sacks of fertilizer, all sold in the United States; profits reached 17,000 kroner; see Risting, *Av Hvalfangstens Historie*, 575. The North Pacific Sea Products' station at Akutan returned 9,000 barrels of oil.

22. In 1910–11, for example, the company operated the *Tyee Junior*, the *Fearless*, and the *Resolute*; see "Whaling in North," *Prince Rupert Journal*, 17 October 1911, 5. The barge *Diamondhead* was apparently towed to Kodiak Island and returned to Admiralty Island at the close of the season in the fall; see Tønnessen and Johnsen, *History of Modern Whaling*, 118.

23. E. W. Outerbridge (Harvey and Outerbridge, Shipping and Commission Merchants) to Ludwig Rissmüller, New York, 3 May 1910, fo. 1. Balcom.

24. "Whaler Sunk By a Whale: Lizzie Sorensen of Tyee Company Foundered After Being Struck by Tail of Enraged Mammal off Alaskan Coast," *Colonist*, 18 May 1910; details of the *Sorensen* may be found in "Steam Whaler for Port Angeles Company," *Colonist*, 25 March 1909; loss accounts in Tønnessen and Johnsen, *History of Modern Whaling*, 118; see also Berman, *Encyclopedia of American Shipwrecks*, 215, and Wilson Fiske Erskine, *White Water: An Alaskan Adventure* (London: Abelard-Schuman, 1960), 11.

25. Tønnessen and Johnsen, *History of Modern Whaling*, 118. A powered schooner belonging to the North Pacific Sea Products Company, the 61-ton *Halcyon*, was also lost, in November 1918, but not to a whale. The schooner, which had been used during 1917–18 as a "whaling vessel," slipped its mooring at Akutan and was blown to sea. No one was aboard at the time; see United States, Department of Commerce, Bureau of Fisheries, "Alaska Fisheries and Fur Industries in 1917," by Ward T. Bower and Henry D. Aller. Appendix 2 in *The Report of the United States Commissioner of Fisheries for* 1917, Document 847 (Washington, D.C.: Government Printing Office, 1918), 51. See also United States, Department of Commerce, Bureau of Fisheries, "Alaska Fisheries and Fur Industries in 1918," by Ward T. Bower. Appendix 7 in *The Report of the United States Commissioner of Fisheries for* 1918, Document 872 (Washington, D.C.: Government Printing Office, 1919), 64.

26. Thomas H. Jenkins, *Bark Kathleen Sunk By a Whale, as Related by the Captain, Thomas H. Jenkins, to Which is Added an Account of Two Like Occurrences, the Loss of Ships Ann Alexander and Essex* (New Bedford: H. S. Hutchinson and Company, 1902), 12, 16. The *Kathleen*, like the *Lizzie S. Sorensen*, settled quickly and sank.

27. "Whaler Sunk By a Whale," 14.

28. Built as a bark in 1874 by the W. Rogers yard in Bath, Maine, and later cut down for a barge, the *Fresno* survived until 4 April 1923, when it burned at Bellevue, Washington. It was 197.2' x 38.6' x 23.1' and 1,245 tons gross.

29. "Whaling Review," *Pacific Fisherman Year Book*, January 1916, 139.

30. Tønnessen and Johnsen, *History of Modern Whaling*, 125.

31. Roy Chapman Andrews, *Ends of the Earth* (New York: G. P. Putnam's Sons, 1929), 27. He clearly did not understand the derogatory connotation in the term "Siwash."

32. Birkeland, *Whalers of Akutan*, 125.

33. Hustwick, "The War on the Whale, Part 2," 8, 10.

34. Personal communication, Sam Kosaka, Victoria, 30 November 1984.

35. White flensers at Akutan are noted in Birkeland, *Whalers of Akutan*, 125.

36. Thirty-two "natives" were also employed, but these were probably counted from the beluga-hunting operations at Nome run by Ben Nygren and S. Torkensen. See United States, Department of Commerce, Bureau of Fisheries, "Alaska Fisheries and Fur Industries in 1916," by Ward T. Bower and Henry D. Aller. Appendix 2 in *The Report of the United States Commissioner of Fisheries for 1916*, Document 838 (Washington, D.C.: Government Printing Office, 1917), 73–74.

37. [Simeon Oliver], *Son of the Smoky Sea, by "Nutchuk"*. With Alden Hatch (New York: Julian Messner, 1941), 108.

38. Robbins, "Letters Written of Whaling Expeditions," Rose Harbour, 12 July 1936, 2.

39. The author's informants who worked the stations in the 1920s and 1930s do not recall Indians working there, although the local Haida sold fresh fish to the whalemen at Rose Harbour in the 1920s; Sam Kosaka, 30 November 1984. By the 1930s the fish was purchased at Prince Rupert, but an Indian butcher in Masset provided meat for the stations; personal communication, Harry Osselton, Victoria, 30 November 1984.

40. Goddard, "Reminiscences," 13. The burning may have resulted from an outbreak of infectious disease.

41. *Ibid.*, 12–13.

42. Andrews, *Ends of the Earth*, 27.

43. Andrews described the flensing process in "Shore-Whaling: A World Industry," *National Geographic* 22 (May 1911), 414–15.

44. Herb Smith, 25 August 1983.

45. Personal communication, Jim O'Shea, Vancouver, 26 September 1984.

46. Andrews, *Ends of the Earth*, 28–29.

47. Goddard, "Reminiscences," 15–16.

48. See Goddard, "Reminiscences"; and Donald B. Macpherson, "Discovery: 1917," *British Columbia Historical News* 16, no. 4; also J[ames] F[arrell] (Assistant Treasurer, Victoria Whaling Company) to E. E. Blackwood (agent, Northern Pacific Railway Company), [Victoria], 9 March 1916; fo. 1. VMM.

49. Gosney arrived in 1909 to manage the station at Sechart; Rolls was bookkeeper at Sechart in that year but soon moved to manage Rose Harbour. LeMarquand left the sealing industry in the east to come to Rose Harbour in 1912 and was afterwards general manager of the whaling company. Their families came out and spent summers at the stations. See Joan Goddard, "North to the Station," in Regional Council, Senior Citizens of the Queen Charlotte Islands, *Tales From the Queen Charlotte Islands, Book 2*

(Masset, B.C.: Regional Council, 1982), 8–9; and Macpherson, "Discovery: 1917," 15.

50. "Grub was always an item of the utmost importance to the crew," another wrote, "as the class of cooks met with in those days always followed the line of least resistance. They never could seem to get the best of tough beef and salt pork; so at every opportunity at least half the crew went off hunting or fishing, and most anything brought back was tried out at least once"; 733 [pseud.], "Whaling—and Some Whalers," *Canadian Merchant Service Guild Yearbook* (1926), 6.

51. Ship *Eliza Adams*, John Jones (probable keeper), 4 January 1852.

52. Andrew R. Boone, "Wings for Whale Shooters," *Popular Mechanics Magazine* 52 (September 1929): 353.

53. Frances Diane Robotti, *Whaling and Old Salem: A Chronicle of the Sea*, 1st ed. (Salem, Mass.: Newcomb & Gauss, 1950), 147.

54. See, for instance, Birkeland, *Whalers of Akutan*, 94–95.

55. *Ibid.*, 30.

56. The author's informants do not recall it in use at that time; see Hustwick, "War on the Whale, Part 1," 1253. The shooting lance was described on the reverse of a photograph of Harry Balcom whaling from Naden Harbour in 1916. Balcom.

57. 733 [pseud.], "Whaling—and Some Whalers," 6.

58. *Ibid.*.

59. Hustwick, "War on the Whale, Part 1," 1520.

60. Ship *Milo*, Charles Goodall, keeper, 17 July 1844.

61. John Jewitt, *Narrative of the Adventures and Sufferings of John R. Jewitt; Only Survivor of the Crew of the Ship Boston, During a Captivity of Nearly Three Years Among the Savages of Nootka Sound: With an Account of the Manners, Mode of Living, and Religious Opinions of the Natives* (New York: printed for the publisher, ca. 1815), 108.

62. Harry Osselton, 30 November 1984.

63. Almost certainly John Fransen, who worked the British Columbia stations during the 1930s and early 1940s.

64. [Hamilton], *Western Shores*, 92.

65. "Issue Warrant for A Rest," *Victoria Daily Times*, 10 September 1910, 2.

66. Jimmy Wakelen, 25 August 1983.

67. "Capt. Harry Balcom Succumbed Yesterday," *Colonist*, 30 August 1918, 7.

68. Jimmy Wakelen, 25 August 1983.

69. Photo album, Lewis L. Robbins, photographs taken at Rose Harbour 1936–37. KWM.

70. Jimmy Wakelen, 25 August 1983.

71. Kean made at least two trips aboard the whaling steamers. In 1916 he filmed a whale hunt aboard the *Black* with Willis Balcom, and returned the following summer to go with Heater to the seal rookeries. His whaling film was apparently commissioned by the Department of Publicity and Industries of the City of Vancouver, and screened at the Dominion Theatre there in late November 1916. The following month it played at the Dominion Theatre in Victoria. Kean's footage of the seal rookeries is apparently lost. A print of the whaling film, sanitized and subtitled for theatre use, is held by the National Film Archives in Ottawa; see A. D. Kean, "Whale Hunting off the Coast of British Columbia," *Industrial Progress and Commercial Record* 4 (November 1916): 125–28; also correspondence from Kean's Canada Films to the Victoria Whaling Company, including Kean to [Sidney C.] Ruck, Vancouver, 16 September 1916; and H. W. Davison to S. C. Ruck, Vancouver, 29 November 1916. VMM.

72. A compound powder and wadding came into use after 1926 in the cannons built in Nor-

way by Kongsberg Vabenfabrikk and in Sweden by Bofors; see Johnsen, *Norwegian Patents Relating to Whaling and the Whaling Industry*, 74. In these weapons a "shock absorber" piston filled with a glycerine-base fluid replaced the gutta-percha recoil pads. None of the Pacific Northwestern companies operating before World War II upgraded to breech-loading harpoon-cannons, which became available during 1926–27.

73. William Heater, "Victoria's Whaling Fleet," *Canadian Fisherman* 22 (November 1935): 22.

74. *Ibid*. In 1925 E. Keble Chatterton described 60 fathoms of four-inch Italian hemp foregoer attached to 120 fathoms of five-inch whaling line, although this was a general specification probably obtained from the Antarctic fishery; E. Keble Chatterton, *Whalers and Whaling: The Story of Whaling Ships up to the Present Day* (London: T. Fisher Unwin, 1925), 215.

75. Personal Communication, Lorne Hume, Delta, B.C., 4 August 1983.

76. Hagelund, *Flying the Chase Flag*, 128–29. Hagelund also recalled an instance aboard the "Ebony" (*Black*) when a harpoon recoiled off the tail of a whale and tore down the foremast before exploding. The lookout in the barrel was knocked into the sea but—as always—escaped injury.

77. Henry Deacon, a seaman in the S.S. *White* out of Rose Harbour about 1929; quoted in Joseph E. Forester and Anne D. Forester, *Fishing: British Columbia's Commercial Fishing Industry* (Saanichton, B.C.: Hancock House Publishers, 1975), 194.

78. S.S. *Green*, Chief Officer's Pilot House Log Book, 4 June-8 October [n.d.]; various dates. PABC, Add MSS 21, Vol. 16.

79. S.S. *Green*, Chief Officer's Pilot House Log Book, 17 March-15 October [n.d.]; 20 August [n.d.]. PABC Add MSS 21, Vol. 16.

80. S.S. *W. Grant*, Chief Officer's Pilot House Log Book, 18 April-17 September 1937; 2 July 1937. PABC Add MSS 21, Vol. 18.

81. S.S. *Green*, Engineer's Log Book, 23 April-16 September 1936; 15 May 1936. PABC Add MSS Vol. 18.

NOTES TO CHAPTER 7

1. Hjort, "The Story of Whaling: A Parable of Sociology," 32.

2. "Proposed Gunners Now at Home and Vacant," list prepared by Isak Kobro, [Christiania], 25 January 1915; and Isak Kobro to the Canadian North Pacific Fisheries Limited, Christiania, 6 February 1915, fo. 1. VMM. Hansen was gunner aboard the chaser boat *Juarez* attached to the factory ship *Capella*. Peder Bogen, who organized the United States Whaling Company in Alaska, sent the *Capella* into Mexican waters in November 1913; in May 1914 the ship put into San Diego for repairs and then sailed to the South American coast, where whaling continued until September. World War I and the "unsettled state of political affairs in Mexico" prompted Bogen to withdraw; see Isak Kobro to C. Rogers Brown, Christiania, 20 August 1915. VMM; also "Hvalfangeraktieselskapet 'Capella,' *NHT* 4 (31 January 1915): 11; and "A/S Capella: Nyt fangstfelt paa Kysten av Mexico," *NHT* 2 (August 1913): 90–91.

3. "Copy of Telegrams Exchanged with Messrs. The Canadian North Pacific Fisheries Ltd.," manuscript sheet kept by Isak Kobro, Befragtnings Assurance and Commissions Agent, Christiania, 6 February 1915; telegram of 30 January 1915. VMM.

4. Isak Kobro to Canadian North Pacific Fisheries Limited, Christiania, 19 March 1915. VMM.

5. C. R[ogers] B[rown] to Harvey Fitzsimons, [Victoria], 6 July 1915, fo. 2. VMM. See

also C. R. B. to G. J. Desbarats, Victoria, 25 August 1915. PAC, RG 23, Vol. 716–14–3 [1]; 31.

6. Personal communication, William S. Lagen, Bellevue, Washington, 16 May 1987; see also McDonald, "Whaling on the Washington Coast," 3. The railroad connection is mentioned—erroneously as the Canadian Pacific—in an obituary (unidentified news clipping) in the W.P. Schupp File, UW.

7. W[illiam] A[lexander] Lawson (general manager, Victoria Whaling Company) to Bodwell and Lawson, Victoria, 22 October 1915. VMM.

8. Personal communication, William S. Lagen, Bellevue, Washington, 17 August 1983.

9. Personal communication, William Hagelund, Vancouver, 2 October 1984.

10. See [C. Rogers Brown] to Hon. John McDougall (Commissioner, Customs Department, Ottawa), Ottawa, 9 February 1915, fos. 1–2. VMM.

11. [Brown] to McDougall, 9 February 1915, fos. 1–2.

12. See Victoria Whaling Company Limited to Harvey Fitzsimons, telegram, Victoria, 2 July 1915, fos. 1–2; C. Rogers Brown to Harvey Fitzsimons, telegram, Victoria, 5 July 1915; and John McDougall to Collector of Customs, Victoria, telegram, [Ottawa], 15 July 1915. VMM.

13. Victoria Whaling Company to Harvey Fitzsimons, 2 July 1915, fos. 1–2. Other purchasers included National Red Oil and Soap Company, Harrison, N.J.; Atlas Refinery, Newark, N.J.; Frank L. Young and Company, Boston; Virginia Products Company, Baltimore; W. A. Robinson and Company, New Bedford; and Atkins, Kroll and Company, San Francisco.

14. Harvey Fitzsimons to William Schupp, Ottawa, 3 July 1915, fo. 2. VMM.

15. Balfour, Guthrie and Company to Victoria Whaling Company Limited, Seattle, 30 October 1916, fo. 1. VMM.

16. S[idney] C. R[uck] to Balfour, Guthrie and Company, [Victoria], 17 November 1916, fo. 2–3. VMM.

17. W[illiam] K[elly] (superintendent engineer, Victoria Whaling Company) to Canadian Northern Railway Company, [Victoria], 6 November 1916. VMM.

18. "Memorandum: RE Fitting Boat for Halibut Fishing," [n.d.]. VMM. This may not have been the first such effort; in 1910 it had been proposed to outfit the chaser boats for halibut fishing in the winter as a means of keeping the crews employed year-round. A freezing plant was to be installed at Rose Harbour. This plan does not seem to have been carried out. See "Will Engage in Halibut Industry: Pacific Whaling Company's Effort to Break Monopoly of American Fishermen—Five Vessels to Be Operated," *Colonist*, 6 July 1910, 1.

19. "Victoria Whaling Company Limited: Memo RE Halibut Fishing," [n.d]. VMM. Wages ranged from $125 per month for captain to $50 per month plus bonuses for firemen and the cook. The mate and fishermen worked on a "lay" system similar to that found in whaling, and received 1.25 cents for every pound of fish landed.

20. Halibut fishing, see S[idney] C. R[uck] to George Heater, [Victoria], 29 February 1916, fos. 1, 3; George Heater to Victoria Whaling Company, telegram, Prince Rupert, 4 June 1916; and [Sidney C. Ruck] to George Heater, [Victoria], 31 May 1916. All VMM.

21. E. Read to S[idney] C. Ruck, Vancouver, 11 March 1916, fo. 1. VMM

22. W[illiam] Kelly to E. T. McLennan (Canadian Tug Boat Company), Rose Harbour, 15 August 1916, fo. 2. VMM.

23. Kelly to McLennan, 15 August 1916, fo. 1.

24. William N. Kelly to S. C. Ruck, Rose Harbour, 3 August 1916, fo. 1. VMM.

25. Kelly to Ruck, 3 August 1916, fo. 1. Kelly wrote: "On June 22nd the 'Orion' while

towing a sperm whale alongside, position 25 miles E.S.E. Cape St. James began to leak. On examination it was found that the point of an unexploded bomb had pierced the hull on the Starboard side at the line pen, second plate down. A temporary repair was made by the blacksmith, the vessel being listed to port while coaling to permit of this. No entry was made, in the log of the 'Orion' in fact the keeping of the daily log has not been observed at all by the late master J. H. McGregor." The unexploded bomb was not an issue, nor was the situation unique; the *St. Lawrence* had already received damage on 16 June "by the head of a bomb which had not exploded, projecting out of the whale and bumping against the side of the vessel." See [William A. Lawson] to Aetna Insurance Company, Victoria, 30 June [1916]. VMM.

26. William N. Kelly to Victoria Whaling Company Limited, Naden Harbour, 26 August 1916, fo. 1. VMM.

27. Probably Joe Yamasaka; Kelly to Ruck, 3 August 1916; and [William Kelly] to the Steamboat Inspector, Department of Marine, Rose Harbour, 4 August 1916, fo. 1. VMM.

28. T. H. Rankin to W. N. Kelly, aboard S.S. *Grant*, 15 August 1916, fos. 1, 1d. VMM.

29. William N. Kelly to S. C. Ruck, Naden Harbour, 26 August 1916, fos. 1–2. VMM.

30. [Kelly] to the Steamboat Inspector, 4 August 1916, fos. 1–2.

31. William N. Kelly to S. C. Ruck, Kyuquot, 2 October 1916, fo. 1. VMM.

32. C. Harrison (provincial coroner) to Manager, Naden Harbour Whaling Station, Massett, B.C., 26 February 1916. VMM.

33. Price noted in E. W. Igard (Yarrows Limited) to the Victoria Whaling Company, Victoria, 28 November 1916. VMM; return of broken harpoons in William Kelly to C. Nickerson, Rose Harbour, 2 August 1916. VMM. See also Kelly to Ruck, 26 August 1916, fos. 2–3; preferences noted in Kelly to Ruck, 26 August 1916, fo. 3.

34. See [Isak] Kobro to Victoria Whaling Company Limited, Christiania, 17 June 1916, fos. 1, 1d.

35. Birkeland, *Whalers of Akutan*, 101–02.

36. H. Browning (secretary-treasurer, Wellington-Comox Agency Limited) to Victoria Whaling Company, Victoria, 4 October 1916, fo. 1. VMM.

37. R. W. Large (medical superintendent, Port Simpson General Hospital) to C. Nickerson, Port Simpson, 18 October 1915. VMM; and "Whaling Company to Extend Operations," 11.

38. S. C. R[uck] to G. J. Hawes (master, S.S. *Gray*), [Victoria], 26 July 1916, fo. 2. VMM. Correspondence pertaining to the San Pedro labour dispute may be found in the Victoria Whaling Company papers held by the VMM, particularly [G. J. Hawes] to Victoria Whaling Company Limited, San Pedro, 11 July 1916; Hawes to Company, San Pedro, 15 July 1916; and Hawes to S. C. Ruck, S.S. *Gray*, 25 July 1916.

39. G. J. Hawes to Victoria Whaling Company Limited, Prince Rupert, 13 November 1916, fo. 1. VMM.

40. Ludwig Rissmüller to [Sprott] Balcom, San Diego, 13 June 1915. Balcom.

41. California State Board of Health, Bureau of Vital Statistics, Certificate of Death no. 650 535, filed 19 April 1916. See also [Sprott Balcom] to Manager, First National Bank of San Diego, Victoria, 10 October 1916, fo. 1. Balcom. The obituary published in the 19 April edition of the San Diego *Union* erroneously backdates the death to 14 April; his name there appears as Louis R. Rissmüller, perhaps an Anglicized version of Ludwig adopted at a time of strong anti-German sentiment; San Diego *Union*, 19 April 1916; Rhoda Kruse to Author, San Diego, 31 January 1986.

42. "Balcoms and Rissmullers Holdings," ms., 26 March 1915. Balcom.

43. For a discussion of markets, see T. H. Brown, "Pacific Coast Whaling," *G.A.*, 1921–22; 7. Typescript copy, VCA, Cadieux Coll., Vol. 240.
44. R. Bruce Scott, *Barkley Sound: A History of the Pacific Rim National Park Area* (n.p.: by the author, 1972), 185.
45. Lorne Hume, 4 August 1983; also personal communication from unidentified museum visitor, Kendall Whaling Museum, Sharon, Massachusetts, 1984.
46. The young son of the captain of the sailing whaleship *Charles W. Morgan* recalled having eaten "tenderloin of whale" just after 1900, but did not specify whether it was porpoise, baleen whale, or sperm whale; see James Earle, quoted in Patricia Biggins, "Doughnuts in the Tryworks: A Child's Life Aboard the *Charles W. Morgan*," *Log of Mystic Seaport* 27 (May 1975): 10. Henry Deacon, who sailed in the *White* in 1929 said that the boat crews ate mostly salt-brine pork, as in sailing-ship days. "The only time we got any fresh meat was when we got into the station and saw a whale on the slip stripped of its blubber. Some of us crew would go over and cut off a few steaks." Deacon, quoted in Forester and Forester, *Fishing*, 194.
47. Andrews, "Shore-Whaling: A World Industry," 411–42.
48. "Amerikansk Hvalfangst," *NHT* 8 (February 1919): 25. Henry Schupp, general sales manager and brother of William Schupp, was placed in charge of the cold-storage operation.
49. "Prepare Storage for Whale Meat Business," *Colonist*, 22 July 1917, 19.
50. "Amerikansk Hvalfangst," 25.
51. *Colonist*, 1 October 1917 and 19 November 1918; cited in "Notes on Whalers out of Victoria," mss. notes from newspapers prepared by Frank Clapp, Victoria, September 1984.
52. See Frances Diane Robotti, *Whaling and Old Salem: A Chronicle of the Sea*, 2nd ed. (New York: Fountainhead Publishers, 1962), 52; and "Hvalkjøtt," *NHT* 7 (August 1918): 171.
53. Robotti, *Whaling and Old Salem*, 1st ed., 152.
54. "Packing Sea Beef," *Colonist*, 27 February 1919, 12.
55. David Wales, "Whale Meat and Hooverism," *Illustrated World* (June 1918); cited in Robotti, *Whaling and Old Salem*, 1st ed., 151.
56. [Hamilton], *Western Shores*, 98; see also Birkeland, *Whalers of Akutan*, 60.
57. "The Introduction of Whale Meat," *Pacific Fisherman Year Book* (1919), 117a.
58. Carrothers, *British Columbia Fisheries*, 127.
59. "Remarkable Catch of Whales," *Pacific Fisherman* 21 (August 1923): 56; see also "Naden Plant Will Ship Whale Meat," *Pacific Fisherman* 21 (April 1923): 56.
60. [Hamilton], *Western Shores*, 98.
61. "Consolidated Whaling Corporation Limited: Memorandum as to Reduction of Share Capital," [n.d.], 1. UW, Lagen Papers.
62. Schupp sold 4,000 shares of American Pacific Whaling Company (Bay City); 2,902 shares of North Pacific Sea Products preference (Akutan), and all the property and assets of the Victoria Whaling Company—all for $2.5 million in cash. He then purchased 24,986 shares of Consolidated common stock and 14 previously subscribed shares owned among the seven-man board of directors for the same amount, and walked away from the transaction with "the whole of the stock of Consolidated Whaling Corporation Limited"; Agreement to transfer, 21 May 1918; copy dated 24 February 1922. PABC Add MSS 21, Vol. 19. The directors included David Blythe Hanna, railway official; Alfred James Mitchell, comptroller; George Osborne Thomson, accountant; George Robinson Donovan, insurance broker; Reginald Herbert Montague Temple, barrister-

at-law; George Norman Limpricht, draughtsman; and William Bowler; secretary, all of the city of Toronto; see Letters Patent Incorporating Consolidated Whaling Corporation, Limited, 26 April 1918 (recorded 1 May 1918). PABC Add MSS 21.

63. W. P. Schupp and Geo[rge] H. Donovan to Department of Naval Service, Toronto, 22 November 1918. PAC, RG 23, Vol. 764, File 716–14–3 [2], fo. 1.

64. [G. J. Desbarats] (Deputy Minister of the Naval Service) to General Manager, Victoria Whaling Company Limited, [Ottawa], 5 March 1918, PAC, RG 23, Vol. 764, File 716–14–3 [1], fo. 1.

65. "and other places"; S. C. Ruck to the Hon. C. C. Ballantyne (Minister of Marine and Fisheries), telegram, Victoria, 7 February 1920. PAC, RG 23, Vol. 764, File 716–14–3 [2], fo. 1.

66. William S. Lagen, 16 May 1987.

67. The Norwegian oil cartel arranged to sell the output of Norwegian whaling companies, paying members an average price based on total sales per ton. After a few years, lots were sold in advance of a season at a contracted price, and by this means the oil market was soon reduced to "relatively few transactions between the highly concentrated demand and equally few corporations representing the supply side"; see Brandt, *Whale Oil: An Economic Analysis*, 145–47. The few purchasers included Lever Brothers and Jergens. The entrenchment toward a smaller number of large purchases created a dependency among whaling firms on the soap and margarine producers. The "pool" grew in importance following World War I, and after 1928 other countries' whaling firms were invited to join.

68. Specifically, 5,392 barrels of sperm oil and 2,339 barrels of whale oil; "United States Whaling Co.," *NHT* 9 (April 1920): 61.

69. S. C. Ruck, W. N. Kelly, and C. E. Ruck to E. G. Taylor, Victoria, 10 January 1920. PAC, RG 23, Vol. 764, File 716–14–3 [2], fos. 1–2.

70. In 1920 Ruck was mentioned as the former managing director of Consolidated, but "now of the Vancouver Island Whaling Company"; see "Notes on the Occurrence of a Humpbacked Whale Having Hind Legs," in Province of British Columbia, *Report of the Provincial Museum of Natural History for the Year* 1920. 11 Geo. 5. (Victoria: William H. Cullen, 1921), 11. See also S. C. Ruck to Chief Inspector of Fisheries, Vancouver; Victoria, 18 April 1921. PAC, RG 23, Vol. 764, File 716–14–3 [3], fo. 1.

71. *Ibid.*.

72. *Colonist*, 9 March 1920; cited in Clapp, "Notes on Whalers Out of Victoria."

73. Tønnessen and Johnsen, *Den Moderne Hvalfangsts Historie*, 2: 163; also "Fint Utstyret Hvalfanger," *NHT* 10 (April 1921): 48.

74. See Erskine, *White Water*, 82–87. The accident occurred in Prince William Sound.

75. "Consolidated Whaling Corporation Limited Papers" Finding Aid. PABC.

76. Pilchard fishing began in British Columbia about 1917 and the catch was canned or salted until 1924. Then, a sudden increase in their numbers—often attributed to a warming trend in the sea—led to the production of meal and fish oils; see P. A. Larkin and W. E. Ricker, *Canada's Pacific Marine Fisheries: Past Performance and Future Prospects*, Fifteenth British Columbia Natural Resources Conference (Vancouver, 1962), 55. The catch remained high until 1945, then dropped off to virtually nothing; see Canada, Fisheries Research Board, *Marine Oils: With Particular Reference to Those of Canada*, Bulletin no. 89, ed. B. E. Bailey, N. M. Carter, and L. A. Swain, Bulletin No. 89 (Ottawa, 1952), 317.

77. Statistics in "Whaling, 1925," *Canadian Merchant Service Guild Yearbook* (1926), 35; quotation from George Donovan to William A. Found (Director of Fisheries, Department of Marine and Fisheries), Toronto, 1 May 1926. PAC, RG 23, Vol. 764, File

716–14–3 [4], fo. 1. Donovan said that Kyuquot was "shut down permanently" as of 12 September 1925.

78. *Colonist*, 16 April 1926; cited in Clapp, "Notes on Whalers."

79. See "Hvalfangsten paa Pacifickysten," *NHT* 15 (December 1926): 12; also "New Developments in Pacific Whaling Industry," *NHT* 16 (April 1927): 58. Henry Schupp employment noted by William S. Lagen, 16 May 1987.

80. On 26 March a corporate bylaw decreased the Consolidated stock to $250,000 by reducing the par value of common shares from $50 to $10 each, and by redemption of the 1,355 shares of preference stock at $83.46 per share. The remainder of unissued stock was cancelled. Schupp, of course, owned all these shares personally. By a second bylaw, given the same day, the shares were further devalued from $10 each to nothing — "without nominal or par value." The 25,000 shares of worthless stock were next increased to 100,000 shares of worthless stock, at which point the board of directors voted to issue this new stock from time to time "for such consideration not exceeding $10.00 per share as may be fixed by the directors." On this same day — a busy one — 25,000 of these recoined shares were subscribed to, at $5 per share — presumably by William Schupp. See "Consolidated Whaling Corporation Limited: Memorandum As To Reduction of Share Capital," n.d., 1. UW, Lagen Papers.

81. "Calif. Whaler for Alaska," *NHT* 16 (June 1927): 90; also "California Whalers in Alaska," *NHT* 16 (August 1927): 135; and "Output of Pacific Whale Products Shows Substantial Gain," *NHT* 17 (March 1928): 51f.

82. The increase was 12.5 per cent on importation from Commonwealth countries, 30 per cent imported from elsewhere; see Hugh [?] (Canadian Manufacturers Association) to William Schupp, Vancouver, 27 June 1934. UW, Lagen Papers.

83. William Hagelund, 2 October 1984. Harry Osselton, former engineer on the S.S. *Gray* began his service with the company in 1933; he, too, remembered signing on for "half pay"; Harry Osselton, 30 November 1984. No insurance was carried on the chaser boats during 1933; Schupp advised his captains to take every precaution against accident. As regards wages, he seems to have cut them almost literally in half. Captain Andy Anderson went back to work in the *Green* that year for $75 per month, plus an end-of-season bonus varying from $5.25 to $10.50 per whale depending upon species; J. M. Lawson to Andy Anderson, Victoria, 30 May 1933. See PABC Add MSS 21, Vol. 4. By 1936 the base rate for captains had risen to $150 per month plus bonuses; A[lf] Garcin to P[eter] Bramsen, Victoria, 14 April 1936. PABC Add MSS 21, Vol. 41. In 1941 Captain Bramsen earned $165 per month; bonuses reached $14–16 depending upon species taken; A[lf] Garcin to Peter Bramsen, Victoria, 22 May 1941. PABC Add MSS 21, Vol. 4.

84. A. Mackie (Department of Fisheries) to William Found, Vancouver, 31 May 1933. PAC, RG 23, Vol. 764, File 716–14–3 [4], fo. 1.

85. "Extract from 1933 Annual Report of Inspector E. S. Richardson, Queen Charlotte Island Area, British Columbia." PAC, RG 23, Vol. 1082, File 721–19–5 [7], fo. 1. The catch included 190 sperm whales, 17 finbacks, 1 sei whale, and 1 blue whale.

86. William S. Lagen, 17 August 1983; and William S. Lagen, 16 May 1987.

87. Brandt, *Whale Oil*, 89.

88. See "Canada and the International Whaling Treaty," *Western Fisheries*, 10 (August 1935): 1.

89. *British Columbia: Official Centennial Record* (Vancouver: Evergreen Press, 1957), 127, in a caption accompanying a photograph of a right whale on a flensing slip.

90. Ralph W. Andrews and A. K. Larssen, *Fish and Ships* (New York: Bonanza Books, 1959), 162. California shore stations in operation during the 1920s and 1930s killed

only one right whale. Although these stations were located south of the right whale's usual range, this statistic may signify reduced population levels to the north; see Gilmore, "Rare Right Whale Visits California," 20–25; also Dale W. Rice and Clifford H. Fiscus, "Right Whales in the Southeastern North Pacific," *NHT* 57 (October 1968): 105–7.

91. E. S. Richardson to J. Boyd, Port Clements, B.C., 31 December 1934. PAC, RG 23, Vol. 764, File 716–14–3 [5], fo. 1. The gunner was not identified.

92. James E. Scarff, "Historic and Present Distribution of the Right Whale (*Eubalaena glacialis*) in the Eastern North Pacific South of 50° N and East of 180° W," International Whaling Commission *Reports* (1986), galley proofs. The minimum right whale catch in the northeastern Pacific, as noted in Scarff, table 5, must be: Alaska: 1916 (1), 1917 (1), 1923 (2), 1924 (1), 1925 (1), 1926 (2), 1927 (1), 1928 (6), 1932 (2), 1933 (1), 1935 (2); British Columbia: 1924 (3), 1926 (1), 1951 (1) (accidental); Japanese researchers took 3 additional right whales from the Gulf of Alaska in 1961 on a permit. To these statistics must be added a single right whale processed at Naden Harbour in 1914; a few others may have gone unreported.

93. Twenty-six nations signed; the eighteen that ratified included the United States, Canada, Denmark, Great Britain, Northern Ireland, Finland, France, Italy, Mexico, the Netherlands, New Zealand, Norway, Poland, Spain, Switzerland, the Union of South Africa, Turkey, and Yugoslavia. Countries which signed but did not ratify were Albania, Australia, Belgium, Colombia, Germany, Greece, India, and Rumania; see "Canada and the International Whaling Treaty," 1, 6; L. Larry Leonard, *International Regulation of Fisheries*, Carnegie Endowment for International Peace Monograph 7 (Washington, D.C.: Carnegie Endowment for International Peace, 1944), n. 102; also William Found to Consolidated Whaling Corporation, telegram, 27 May 1933. PAC, RG 23, Vol. 764, File 716–14–3 [4], fo. 1 (101). Canada did not ratify officially until the last Parliamentary session of 1935.

94. See Found to Consolidated Whaling Corporation, 27 May 1933.

95. J. M. Lawson (assistant treasurer, Consolidated Whaling Corporation) to Captain Andy Anderson (S.S. *Green*), Victoria, 30 May 1933. PABC Add MSS 21, Vol. 4, fo. 1.

96. Deacon, quoted in Forester and Forester, *Fishing*, 194.

97. William Schupp to A[lf] Garcin, [Bellevue, Washington?], 3 August 1934. UW, Lagen Papers.

98. By 1950 the Pacific North American coastal contribution had fallen to less than 0.5 per cent; see Robert T. Orr, "The Distribution of the More Important Marine Mammals of the Pacific Ocean, as it Affects Their Conservation," *Sixth Pacific Science Congress* [1939], 217; and Gordon C. Pike, "Whaling on the Coast British Columbia," *NHT* 43 (March 1954): 70.

99. Brandt, *Whale Oil*, 4. By 1935 Germany used whale oil as the raw material for 54 per cent of its lard and margarine production. In Great Britain, whale oil was utilized in 41 per cent of margarine, 28 per cent of lard compounds, and 16 per cent of soaps by 1937; see Great Britain, Imperial Economic Committee, *Vegetable Oils and Oilseeds*, Appendix 1 (London, 1938), quoted in Brandt, *Whale Oil*, 4–5. British manufacturers produced nearly 90,000 tons of margarine in Great Britain in 1937, and in Germany about 180,000 tons. "From the middle of the nineteen thirties," Arne Odd Johnsen wrote, "a comprehensive state-supported German whaling was started, one object of which was undoubtedly to procure a reserve of fat for the country in readiness for the war that was being prepared." Whale oil was second among commodities purchased by Great Britain's Food Department at that time, and the Japanese also began pelagic whaling to secure a material and financial supply "with a view to war"; see Johnsen, "Causation

Problems of Modern Whaling," 281. The annual consumption in the United States, lower but significant, was about 40,000 tons.

100. The Lagen Papers at the University of Washington include undated blueprint plans for converting the T.S.S. *Achilles* to a factory ship.

101. Henrietta Nesbitt, *White House Diary* (Garden City, N.Y.: Doubleday and Company, 1948), 8.

102. "Henry Nesbitt, 72, White House Aide," *New York Times*, 7 January 1938.

103. Harry Nesbitt to Henry Morgenthau, the White House, 14 April 1934. FDR, Morgenthau Papers, Box 205.

104. Harry [Nesbitt] to William Schupp, the White House, 26 May 1934. UW, Lagen Papers.

105. See "Agency Examiner's Report on Application for Direct Loans to Industry," Seattle Loan Agency, American Pacific Whaling Company, 28 November 1934, fos. 3—4. NARS, RG 234, Records of the Reconstruction Finance Corporation, Examining Division, Declined and Cancelled Loans 1939—1941, American Pacific Whaling Co., Loan #1, 1934, Box 420, Docket No. D11—1232.

106. *Ibid.*.

107. RFC Card Index to Business Loans, "American Pacific Whaling Company —Bellevue, Washington." NARS, RG 234, Records of the Reconstruction Finance Corporation, Examining Division, Records Relating to Declined and Cancelled Loans.

108. "Secretary to Mrs. Roosevelt" to Jessie Jones, the White House, 21 February 1935. FDR, Eleanor Roosevelt Papers, File 70, Box 655.

109. Many RFC documents have been destroyed by government order; see Card Index to Business Loans. The first loan was repaid in full in 1939; in that year, a $25,000 loan application—Schupp's fifth—was approved but immediately rescinded in favor of a larger amount of $98,645.45. This seems to have been the only other loan approved by the RFC. Extant records indicate that the second debt was repaid in 1944, but there may have been some moneys still owing after World War II.

110. "Memorandum to Dr. Ezekiel," prepared by L. R. Edminster (Chief, Import-Export Section, U.S. Department of Agriculture), [Washington, D.C.], 15 May 1935, fo. 1. FDR, O.F. Box 9, OF 2371; and "Memorandum: American Whaling Fleet," attached to Cordell Hull to Franklin D. Roosevelt, Washington, D.C., 9 May 1935. FDR, O.F. Box 9, OF 2371.

111. In 1931 lard represented 11 per cent of American hog products; "Iowa's Farmers Have a Whale of a Competitor," *Business Week* (11 March 1931); cited in Robotti, *Whaling and Old Salem*, 1st ed., 149.

112. "Memorandum: Norwegian Whale Oil—A Summary," attached to Hull to Roosevelt, 9 May 1935.

113. Hull to Roosevelt, 9 May 1935, fos. 1—2. Hull advised that "there is no immediate or close competition [between U.S. and Norwegian whalemen] on account of the different marketing localities."

114. See "Memorandum to Dr. Ezekiel."

115. Robotti, *Whaling and Old Salem*, 1st ed., 149. The companies then disposed of their oil in the foreign market, selling particularly to Germany, where demand had driven whale prices to $160 a ton by 1939. Norwegian entrepreneurs held much of the stock in these companies.

116. Only the United States, Great Britain, Northern Ireland, the Irish Free State, British colonies, New Zealand, Australia, Norway, and Argentina participated; Canada, Portugal, and South Africa sent observers. Japan, whose infant pelagic fleet was just then under construction, declined to take part, although Japan did participate with

twelve other countries at the subsequent International Whaling Conference held in June 1938; see Orr, "The Distribution of the More Important Marine Mammals of the Pacific Ocean," 219; and Brandt, *Whale Oil*, 95.

117. See Leonard, *International Regulation of Fisheries*, 100.

118. The taking of baleen whales south of 40° South was prohibited except during designated seasons, and closed areas were established in the Pacific east of 150° West, between 40° South and 35° North; and west of 150° West between 40° South and 20° North; Michael Douglas Bradley, "The International Whaling Commission: Allocating an International Pelagic Ocean Resource" (Ph.D. dissertation, University of Michigan, 1971. University Microfilms 71—23,707.) All Arctic seas except the Bering Strait were closed.

119. Blue whales must exceed seventy feet, finbacks fifty-five feet, humpbacks thirty-five feet, and sperm whales—included for the first time—thirty-five feet; see *International Agreement for the Regulation of Whaling* (unratified) (London: H. M. Stationers Office, 1937). Copy at PAC, RG 23, Vol. 1082, File 721—19—5 [10].

120. [A. Van De Venter], "Kodiak Whaling Station, Whaling Inspector's Report," Port Hobron, Alaska, 29 June 1937. KWM (466). This single animal was the only sei whale taken in the Pacific Northwest that year by the shore stations; see Table M in Committee for Whaling Statistics, *International Whaling Statistics* (Oslo: Grøndahl & Son, 1950), 23: 21. Few seis were taken before World War II, though hunting for the species began as early as the teens in Alaska; four were processed in 1918, for example, by the United States Whaling Company; "United States Whaling Co.," *NHT* 8 (May 1919): 98. In British Columbia the species was added to the bonus list after July 1937, but the award was only one-half that paid for a sperm whale, and few were taken by Consolidated's crews; see T. C. Hubard (station manager) to Captain P. Bramsen, Naden Harbour, 22 July 1937. PABC Add MSS 21, V.4.

121. Professor Harold Heath of Stanford University visited Akutan about 1918, for example; see Morgan, "Modern Shore-based Whaling," 41. A Dr. Cornish, from the University of California at Los Angeles, is thought to have visited British Columbia to study the concentration of nutritional value in whale livers; Harry Osselton, 30 November 1984.

122. The results of his research were published in medical journals; see, for example, E. M. K. Geiling and G. B. Wislocki, "The Anatomy of the Hypophysics of Whales," *Anatomical Records* 66 (1936): 17—41.

123. Among them Dr. Robert Walmsley of the Edinburgh University Medical School in Scotland, who studied the deep-diving capabilities of whales in an effort to solve the problem of the "bends" experienced by human divers. Another was Miss Frances Oldham, a Ph.D. candidate in pharmacology at the University of Chicago. She, too, studied the function of the pituitary and eventually received her doctoral degree on the basis of her investigations. A daughter of Colonel Oldham of "Balgonie," Cobble Hill, Vancouver Island, Frances (Oldham) Kelsey became one of the youngest Ph.D. candidates ever matriculated from the Pharmacology Department of the University of Chicago, and eventually earned a high post in the U.S. Food and Drug Administration; see Robbins, "Letters Written of Whaling Expeditions," Vancouver, 8 September 1937, 1.

124. *Ibid.*, 11 July 1936, 2.

125. *Ibid.*, 12 August 1937, 2.

126. *Ibid.*, 10 July 1937, 2. Iwabuchi was one of the better scrimshanders at the station, and Robbins watched in fascination one afternoon as he created a small sperm whale from a tooth: "He filed a little away here and chiseled a little away there. The whole seemed to grow in his hands and when he was finished a rough tooth had been transformed into a

small ivory replica of its original home. From a sperm whale into a sperm whale!"; Robbins, "Letters Written of Whaling Expeditions," Rose Harbour, 30 July 1937, 1.

127. Rose Harbour blacksmith Jimmy Wakelen claimed to have made at least two miniature harpoons from skeletal bones of whales, as well as others from iron; Jimmy Wakelen, 25 August 1983.

128. Not identified, but probably Louis Klum, a sixty-one-year-old fireman aboard the *Blue*, whom Robbins photographed and included in his scrapbook of snapshots from the whaling station. Robbins Photo Coll. KWM.

129. Robbins, "Letters Written of Whaling Expeditions," Rose Harbour, 4 August 1937, 9.

130. Shortly after Robbins returned in 1937, these scrimshandered items were displayed at the University of Chicago bookstore; a news reporter wrote that the majority of the sperm whale teeth collected in British Columbia were sold to Japan "to make trinkets for American five and dime stores." Unidentified typescript news clipping. KWM.

131. Robbins, "Letters Written of Whaling Expeditions," Rose Harbour, 26 August 1937, 1.

132. Personal communication, Harry Osselton to Helen Richmond, Victoria, 3 October 1984.

133. Harry Osselton, 30 November 1984.

134. S.S. *White*, Gunner's Report, 30 May 1942. PABC Add MSS 21, Vol. 11.

135. The *Brown* developed a leak while towing as early as 1916. Pumps had to be used, and the vessel was sent to Vancouver for repairs; W[illiam] A. L[awson] to Aetna Insurance Company, [Victoria?], 30 June 1916. VMM.

136. S.S. *Blue*, Chief Officer's Log Book, 30 May 1935. PABC Add Mss 21, Vol. 18.

137. None among the whalemen interviewed by the author ever saw this done, and did not think it was done; that it was employed to effect minor repairs even as late as 1930 is corroborated in the log book of the S.S. *Brown*, which notes, under date of 5 July: "Arrived at Station & put Brown on the Beach on account of rudder being slack." The chaser remained "beached" for eight hours; see S.S. *Brown*, Chief Engineer's Log Book, 5 July 1930. PABC Add MSS 21, Vol. 14.

138. A[lf] Garcin to William Schupp, Victoria, 15 March 1934. UW, Lagen Papers.

139. S.S. *Brown*, Chief Engineer's Log Book, H. D. Hornibrook, keeper; 15 June 1934. PABC Add MSS 21, Vol. 14. "Ruddered it in"; Jimmy Wakelen, 25 August 1983.

140. S.S. *Brown*, Chief Engineer's Log Book, 16 June 1934.

141. "Whaling" to "Whaling, Victoria BC," telegram, 18 June [1934]. UW, Lagen Papers.

142. S.S. *Brown*, Log Book, 23 June 1941. PABC Add MSS 21.

143. The method was described in William Hagelund's account of whaling in the *Brown* in 1941: "As the tide dropped the upper row of damaged plates were exposed . . . [they were] bolted up using cotton string for grummets. The sheared rivets had to be knocked out along with those that showed their damaged condition by loss of head or looseness. It was knuckle-scraping, tiring work, and required every loose bolt we could find on the ship, even stealing them from other less important jobs.
    "When this was completed, boards were cut and wedged between the frames . . . to form a crude form for the cement. This thick waterproof cement was made from the sand . . . from the river's bed nearly a mile up the river away from salt water. . . . Odd bits and pieces of wire and steel rod were shoved into the cement to add to its strength"; Hagelund, *Flying the Chase Flag*, 157.

144. S.S. *Green*, Pilot House Log Book, 25–29 June 1941. PABC Add Mss 21, Vol. 16.

145. S.S. *Brown*, Chief Engineer's Log Book, H. D. Hornibrook, keeper; 29 May 1935. PABC Add MSS 21, Vol. 14.

146. S.S. *Green*, Engineer's Log Book, 17 May 1936. PABC Add MSS 21, Vol. 16.

147. A[lf] Garcin to William Schupp, Victoria, 11 December 1934. UW, Lagen Papers.
148. W. M. S. to Mrs. H. B. Kristensen, [Bellevue, Washington], 26 June 1934. UW, Lagen Papers.
149. Alfred M. Peterson (master, S.S. *Unimak*) and A. Arnesen (mate, *Unimak*) to United States Shipping Commissioner, [?], 7 May 1934; and American Pacific Whaling Company to U.S. Steamboat Inspectors, [Seattle], 27 August 1934. UW, Lagen Papers.
150. U.S. Coast Guard Wreck Report, quoted in Betty L. Santos to Author, Valdez, Alaska, 20 January 1985.
151. See, for example, L. Clarke (Seafarers Industrial Union of Victoria) to William Schupp, Victoria, 15 March 1935; and [William Schupp] to L. Clark[e], [Seattle?], 16 March 1935. UW, Lagen Papers.
152. One such dispute apparently led to an informal dockside strike by the regular crews. This incident is recalled variously. The "strike" was called by boat crews before leaving Victoria for the whaling stations; Manager Garcin is said to have told the gunners to pick up Japanese crews at the stations. One version tells how William Schupp arrived in a chauffeured sedan, and, after ten minutes of private discussion with Garcin, announced a wage increase for the boat crews.
153. Reported 18 August 1939; see "Whaling," *Pacific Fisherman Year Book* (1940), 303.
154. J. A. Motherwell (Chief Supervisor of Fisheries) to J. J. Cowie (Acting Deputy Minister, Department of Fisheries), Vancouver, 16 March 1940. PAC, RG 23, Vol. 764, File 716–14–3 [6], fo. 1.
155. J. A. Motherwell to D. B. Finn (Deputy Minister, Department of Fisheries), Vancouver, 31 March 1941. PAC, RG 23, File 716–14–3 [6], fo. 1.
156. In 1940, cooks, deckhands and firemen earned $55 per month, plus a $3 bounty per whale; that was raised at the bargaining table to $70 monthly plus a $4 bounty; "Whaling," *Pacific Fisherman Year Book* (1942), 281. Wages may have been even higher; a company crew list from 1942 suggests that deckhands, firemen and cooks working for the Consolidated Whaling Corporation that year earned $95 monthly, second engineers $190, chief engineers $215, mates $125 and captains $200, with bonuses ranging from $5.20 to $6.20 depending on species and size; captains negotiated their own bonuses, up to $25 for sperm and $20 for others; see Consolidated Whaling Corporation Limited, "Whalers' Crews —Season 1942." UW, Lagen Papers.
157. "The Summer Season 1943," *NHT* 36 (May 1947): 174. "Whaling," *Pacific Fisherman* 42 (25 January 1944): 357 reported that two Canadian vessels had been made available to the California shore station, but evidence suggests that only the *White* made the trip; see, for example, J. A. Motherwell to D. B. Finn, Vancouver, 19 June 1943. PAC, RG 23, Vol. 764, File 716–14–3 [6], fo. 1.
158. Willi Frischauer, *Onassis* (London: Book of the Month Club, 1968), 85. The plant operated in 1940–44 and again in 1947–48. The catch was 29 (1940), 24 (1941), 26 (1942), 29 (1943), 5 (1944), 0 (1945–1946), 38 (1947) and 67 (1948); see Robotti, *Whaling and Old Salem*, 2nd ed., 54. Onassis would again impinge on the British Columbia whaling business in the mid-1950s; see Chapter 8.

NOTES TO CHAPTER 8

1. Paul Budker, *Baleines et Baleiniers* (Paris: Horizons de France, 1957), 126. "We are now very far from Melville and wooden ships, and each year sees the appearance of some new device, some murderous improvement, or some fiendish use for a part of this enormous carcass"; Budker, *Whales and Whaling* (English ed.), 121.

2. Attributed to Harvey Beck; Jimmy Wakelen, 25 August 1983.

3. "Certain other creditors" noted in G. O. Cumpston (George A. Touche and Company) to M. A. Lagen, Victoria, 22 September 1945; Hammill named in "Minutes of a Special Meeting of the Board of Directors of Consolidated Whaling Corporation, Limited," undated copy typed September 1945; UW, Lagen Papers.

4. Cumpston to Lagen, 22 September 1945.

5. "Minutes of A Special Meeting of the Board of Directors."

6. Ottawa loan request noted in G. O. Cumpston to M. A. Lagen, 22 September 1945; U.S. application see RFC Card Index to Business Loans, American Pacific Whaling Company. NARS, RG 234.

7. [William Schupp] to Werner G. Smith Company, [Bellevue, Washington], 1 May 1946. UW, Lagen Papers.

8. G. A. Yardley (Collector of Customs and Excise) to Deputy Minister of National Revenue, Customs, and Excise, Victoria, 16 May 1948, 1. PAC, RG 23, Vol. 764, File 716–14–3 [6].1128.28, fo. 1 (124).

9. "Marc A. Lagen, Whaler, Dies," Marine Digest (20 April 1946), 5.

10. RFC Card Index; the initial request was declined on 8 May 1946. An additional loan of $250,000 was also declined, on 19 February 1947.

11. [Schupp] to the Werner G. Smith Company, 1 May 1946.

12. Ibid..

13. J. K. Nesbitt, "The Whalers Get Ready," News-Herald, 8 May 1946, unidentified news clipping. VMM.

14. G. O. Cumpston to Department of Marine and Fisheries, Victoria, 14 December 1946. PAC, RG 23, Vol. 764, File 716–14–3 [6], fo. 1.

15. Daniel Steen (Norwegian Legation, Montreal) to Secretary of State for External Affairs, Montreal, 16 January 1947. PAC, RG 23, Vol. 764, File 716–14–3 [6], fo. 1.

16. Ibid..

17. F. Warne (Acting Chief Supervisor of Fisheries, Vancouver) to Deputy Minister, Department of Fisheries, Vancouver, 19 March 1947. PAC, RG 23, Vol. 764, File 716–14–3 [6], fo. 1.

18. "Victoria Whaling Industry Ends With Sale of Ships for $1,465," Colonist, 1 April 1947, 17.

19. Ibid..

20. The Orion was a virtual hulk by 1947, and was soon scrapped; see "Only One Original Left," 12. The Sebastian first served as a towboat under its own name, later as the Saanich. Some sources report incorrectly that the Saanich was formerly the Germania.

21. Madison, "British Columbia Whaling Venture," 22.

22. Gordon Newell, ed., The H. W. McCurdy Marine History of the Pacific Northwest (Seattle: Superior Publishing Company, 1966), 543.

23. Madison, "British Columbia Whaling Venture," 22.

24. See "B.C. Whaling Reviving," Pacific Fisherman Year Book (1948), 333; also Norman Hacking, "Whaling Industry Begins Operations in Two Weeks," Province, 27 April 1948.

25. Gordon Gibson later claimed that the Gibson Brothers purchased the stations and four ex-Consolidated vessels, the Blue, White, Brown, and Gray in September 1947; Gordon Gibson, Bull of the Woods: The Gordon Gibson Story, with Carol Renison (Vancouver: Douglas and McIntyre, 1982), 193. These purchases cannot be substantiated, since the chaser boats had already been sold to the scrap dealers, and the Gray awaited its fate on the Oyster Bay breakwater.

26. Gibson, Bull of the Woods; 200–1.

27. "B.C. Whale Catch Biggest Since 1936," *Trade News* 3 (November 1950): 8. The *Tow-mac* was used for this purpose.
28. Pike, "Whaling on the Coast British Columbia," 78–79.
29. D. Creed (Western Whaling Corporation Limited) to A. J. Whitmore (Chief Supervisor of Fisheries, Vancouver), Vancouver, 28 April 1948. PAC, RG 23, Vol. 764, File 716–14–3 [6], fo. 1; also Newell, *H. W. McCurdy Marine History of the Pacific Northwest*, 551.
30. Hacking, "Whaling Industry Begins Operations in Two Weeks"; the *Nahmint* was completed in 1943; it measured 98' x 21.3' x 9.4'.
31. Newell, *H.W. McCurdy Marine History of the Pacific Northwest*, 551.
32. F. Warne to Deputy Minister, Department of Fisheries, Vancouver, 26 August 1948. PAC, RG 23, Vol. 764, File 716–14–3 [6], fo. 1.
33. Creed to Whitmore, 28 April 1948.
34. Bradley, "International Whaling Commission," 80.
35. *Ibid.*.
36. The UFAWU successfully affiliated with the Trade and Labour Congress of Canada until its suspension for alleged Communist activities; see "United Fishermen and Allied Workers Union: A General Introduction," 21. UBC Spec. Coll. HR CD 3647/V3 A25/1978, 6: 20–22.
37. *Ibid.*.
38. Group 1 included finback whales fifty to fifty-nine feet long and humpbacks of thirty-five to thirty-nine feet; each earned a mate $10.50, a first engineer $5.78, a second engineer $4.73, and a deckhand, fireman, or cook $3.15. Group 2 included larger humpbacks to forty-four feet; mates received $21.00, first engineers $10.50, second engineers $9.45, and others $3.30. Group 3—sperm whales forty feet and longer and humpbacks exceeding forty-four feet—paid, respectively, $42.00, $23.10, $18.50, and $12.60. Sei whales earned one-half the Group 1 rate, being smaller and less desirable, while the huge blue whales were worth twice the amount paid on Group 3; "B.C. Whalers Active," *Canadian Fisherman* 36 (June 1949): 43. Whale Group 1 proved to be a total loss. The value of their product did not offset the high cost of catching and rendering them; "Minutes: Negotiating Meeting—Whaling," 28 April 1949, 1. UBC, United Fishermen and Allied Workers Union Records (hereafter UFAWU Records), Vols. 24–25.
39. Gordon Pike recorded a catch of 182, including 113 humpbacks (64 female, 49 male) and 39 finbacks (20 female, 19 male), plus 28 sperm and 2 sei; Canada, Fisheries Research Board, "Whaling Investigation," by Gordon C. Pike. In *Progress Reports of the Pacific Coast Stations* 79 (July 1949): 30–31. The Committee for Whaling Statistics also indicate 182, but tallied 115 humpbacks, 37 finbacks, 28 sperm whales and 2 sei; see *International Whaling Statistics* (1950), 23: 21. A larger total of 184, including 116 humpbacks, 38 finbacks, 28 sperm and 2 sei, was reported in "Whaling," *Trade News* 1 (December 1948): 4. The sperm whales were all males and most of them large, mature animals. Pike noted that the sei whale catch was no indicator of availability, only a factor of the minimal commercial stimulus to catch them, owing to low oil yield and small crew bonuses. Only one blue whale was seen during the 1948 season; none was caught; Canada, Fisheries Research Board, "Whaling Investigation," 31.
40. *International Whaling Statistics*, 23: 21.
41. "Whalers," typescript crew list attached to "Minutes of Whaling Crews of 'Kimsquit,' 'Nahmint,' and 'Tahsis Chief,'" 28 April 1949. UBC, UFAWU Records, Vols. 24–25. The *Saanich*, ex-*Sebastian*, was sold for scrap and destroyed in 1950.
42. *Ibid.*.

43. "Whale Steaks Prove Very Popular on First Appearance in City Stores," *Colonist*, 9 July 1948, 3.
44. "Whale Steaks Coarse, Fishy But Improve With Cooking," *Colonist*, 9 July 1948, 3.
45. "Canning of Whale Meat Studied," *Trade News* 1 (April 1949), 4.
46. Obituary, (unidentified news clipping). UW, W.P. Schupp file.
47. J. L. Dehuff (assistant manager, Reconstruction Finance Corporation) to W. L. Grill, Seattle, 18 April 1949, fo. 1. UW, Lagen Papers.
48. Quoted in "Minutes of Special Meeting of Board of Directors of American Pacific Whaling Company," prepared by E. A. Niemeier, (n.d.), fo. 1. UW, Lagen Papers.
49. William S. Lagen, 17 August 1983. Grill's marriage to Emily Lagen is documented in McDonald, "Whaling on the Washington Coast," 7.
50. Finding Aid to the Consolidated Whaling Corporation Limited Papers. PABC.
51. Gibson, *Bull of the Woods*; 204–5.
52. "News and Notes from British Columbia," *Canadian Fisherman* 36 (April 1949): 16.
53. "British Columbia Whaling," *Pacific Fisherman Year Book* (1951), 317. See also "Fisheries Notes from British Columbia," *Canadian Fisherman* 36 (August 1949): 27–28.
54. The vessel was constructed in Smith's yards for the fleet of A/S Sevilla of Tønsberg, and in 1932 was sold to the well-known Antarctic whaling concern, Salvesen's. The chaser was 116' x 24' x 12'. See *Register over Norske, Svenske, Danske, Finske og Islandske Skibe* (Hovedkontor: Carl C. Werner and Company, 1933), no. 235; also Graeme Somner, *From 70 North to 70 South: A History of the Christian Salvesen Fleet* (Edinburgh: Christian Salvesen, 1984), 83.
55. "B.C. Whale Catch Biggest Since 1936," 8.
56. "Electric Whaling," *Discovery*, 13 (May 1952): 143. Also J. H. Jupe, "The Electrocution of Whales," *New Zealand Electrical Journal* (25 July 1952); 564–65, abstracted in Edward Mitchell, Randall R. Reeves, Anne Evely, and Micheline Stawski, *Whale Killing Methods: An Annotated Bibliography*, press proof (1983). KWM.
57. E. E. Summers (Gulf of Georgia Fish and Curing Company, Nanaimo) to Ralph Smith, M.P., Montreal, 20 July 1904. PAC, RG 23, Vol. 242, File 1536, [1], fo. 1.
58. "Electric Whaling," 143–44.
59. Lorne Hume, 4 August 1983.
60. Don Grady, *The Perano Whalers of Cook Strait 1911–1964* (Sydney: A. H. & A. W. Reed, 1982), 103.
61. See Harry R. Lillie, "Whaling and Its Antarctic Problems Today," *Canadian Geographical Journal* (March 1949); cited in Robotti, *Whaling and Old Salem*, 1st ed., 147.
62. For details of electrical problems, see "Technical Committee Working Group on Humane Killing," Annex 4 to Canada, Fisheries and Oceans, *Canadian Delegation Report*, Thirty-first Annual Meeting of International Whaling Commission (September 1979), 3; also Grady, *Perano Whalers*, 103.
63. "Electro-Harpoon Kills Quickly," *Financial Post* (20 October 1962): 36.
64. "British Columbia Whaling," *Trade News* (July 1951): 4.
65. "B.C. Whale Catch Biggest Since 1936," 8. Statistics submitted to the international record-keeping body in Norway indicate a catch of 317 whales in 1937 from B.C. stations, and 328 whales in 1941; Table M in Committee for Whaling Statistics, *International Whaling Statistics*, 23: 21
66. Sperms earned $100, humpbacks $80, seis $25; see "1951 Whaling Agreement," and "Schedule A" (attachment). Both UBC, UFAWU Records, Vol. 115, File 5. Skippers, independently employed and not part of the union bargaining, negotiated individually with management before commencement of the season.

67. For example, in 1956 management put $40 into the mates' pool and $153 in the crew pool for each blue whale taken; "1956 Whaling Agreement." UBC, UFAWU Records, Vol. 115, File 5.

68. Canada, Fisheries Research Board, "Whale Fishery Study," by Gordon C. Pike. In *Progress Reports of the Pacific Coast Stations* 75 (July 1948): 47.

69. Canada, Fisheries Research Board, "Whaling Investigation," 30—31.

70. Canada, Fisheries Research Board, "Whale Identification," by Gordon C. Pike. In *Progress Reports of the Pacific Coast Stations* 81 (December 1949): 86.

71. "Tests on Whale Carcasses," *Trade News* 8 (February 1956): 12.

72. In field trials, the balloons collapsed on the surface as the whale dived, so the idea was abandoned; Lorne Hume, 4 August 1983.

73. Dale W. Rice and Allen A. Wolman, *The Life History and Ecology of the Gray Whale* (Eschrichtius robustus), Special Publication 3 (American Society of Mammalogists, 1971), 125. The gunners made several "mistakes" during 1951; one of the endangered right whales was also taken in error during that year; see Gilmore, "Rare Right Whale Visits California," 22.

74. British Columbia Packers Limited: Whaling Station, Daily Reports, nos. 1—150, 1 April 1953 – 26 September 1953, H. M. Cowie, manager, keeper. VMM.

75. "B.C. Whalers Start 1951 Campaign Early," *NHT* 40 (August 1951): 396. Victoria Machinery Depot and Letson and Burpee, for example, constructed one additional cooker for the 1951 season.

76. See Charles M. Defieux, "3 Old Whaling Veterans Heading for Scrap Heap," *Vancouver Sun*, 13 February 1965, 20; and "Whaling From the Shore Station Coal Harbour, Vancouver Island, British Columbia, *NHT* 41 (March 1952): 119.

77. Ian MacAskie, *The Long Beaches: A Voyage in Search of the North Pacific Fur Seal* (Victoria: Sono Nis Press, 1979), 96—97.

78. Gordon Newell, ed., *The H. W. McCurdy Marine History of the Pacific Northwest 1966 to 1976* (Seattle: Superior Publishing Company, 1977), xlv.

79. "Special Supplement to the 1956 Whaling Agreement." UBC, UFAWU Records, Vol. 115, File 5.

80. "Minutes of Meeting of Whaling Station Crew," 16 September 1956, 2. UBC, UFAWU Records, Vols. 24—25.

81. Frischauer, *Onassis*, 120. The ship was lengthened 41 feet to 565 feet, nearly 18,650 deadweight tons, and equipped to carry 325 men. A fleet of twelve former corvettes was acquired for the chaser boats; Paul D. Green, *Think* Magazine (March 1950); cited in Robotti, *Whaling and Old Salem*, 1st ed., 145.

82. For details of Norwegian complaints, see "Olympic Challenger," *NHT* 39 (November 1950): 505. Mystery surrounds Lars Andersen; the great Antarctic gunner, nicknamed "Lars the Devil" because of his "reckless" reputation, is supposed to have died in Trondheim during the German occupation of Norway; see "Old Man Whale Comes Back," *Science Digest* (November 1945); cited in Robotti, *Whaling and Old Salem*, 1st ed., 146. A synopsis of the Onassis whaling era may be found in Tønnessen and Johnsen, *History of Modern Whaling*, 535—37, 552—56.

83. See Frischauer, *Onassis*, 129.

84. Lorne Hume, 4 August 1983.

85. See Frischauer, *Onassis*, 129.

86. See " 'Olympic Challenger' Has Not Observed the Provisions of the International Whaling Convention," *NHT* 45 (January 1956): 7. Signatory nations agreed to abide by a 22 December 1950 opening date, but Onassis is said to have begun whaling under the non-member Panamanian flag on 6 December and to have continued more than two

weeks after 9 March 1951, when the IWC total quota was accounted for and member fleets were forced to discontinue the hunt. The *Olympic Challenger* simply sailed into Peruvian territorial waters and continued taking sperm whales.

87. *Ibid.*, 8–9.

88. The Permanent Commission for the Exploitation and Conservation of the Marine Resources of the South Pacific (1952).

89. It operated under Japanese flag until 1970, whaling in the North Pacific during the 1966–68 whaling seasons; Tønnessen and Johnsen, *History of Modern Whaling*, 556.

90. Lorne Hume, 4 August 1983.

91. *Ibid.*. Hume thought that the company had "negligible" success in training Canadian gunners.

92. *Ibid.*.

93. British Columbia Packers Limited: Whaling Station, Daily Report no. 106, 13 September 1956, H. M. Cowie, keeper. VMM.

94. "Minutes of Meeting, Whalers Continuations Committee," 2 January 1957. UBC, UFAWU Records, Vols. 24–25.

95. Captains, being supernumeraries under the new system, were no longer required on the boats; see "Minutes of Meeting, Whaling Station Negotiating Committee," 7 January 1957, 1. UBC, UFAWU Records, Vols. 24–25.

96. British Columbia Packers Limited: Whaling Station, Daily Report no. 76, 2 July 1957, H. M. Cowie, keeper. VMM.

97. Personal communication, Les S. Wilson, Vancouver, September 1983.

98. "Ship Back from First Hunt with Five Whales in Bows," *Vancouver Sun*, 27 June 1962, Section C, 33.

99. *Ibid.*.

100. In 1963, aboard the *Westwhale* 8 out of Coal Harbour; see "Whaler Skipper Sets World Record," *Marine Digest*, 6 July 1963. Clipping, Cadieux Coll, VCA; also Newell, *H. W. McCurdy Marine History of the Pacific Northwest*, 678.

101. Lorne Hume, 4 August 1983.

102. *Ibid.*. The story is variously told; in one version, collected by Allan Reese, the manager (Lorne Hume) was at home, hosting an officer from a large fishing company, when a tourist pulled the trigger of the harpoon-cannon. Arne Borgen is said to have walked up the hill and shouted to the shell-shocked administrators, "Do you want to cut the line or should I haul it in?" In 1985 Reese was shown some hangar repairs which were allegedly made after the accident; personal communication, Allan K. Reese, Sr., Sharon, Massachusetts, 18 October 1987.

103. "Whaling Expands," *Pacific Fisherman Year Book* (January 1957): 223.

104. Canada, Fisheries Research Board, *Annual Report of the Biological Station, Nanaimo, B.C. for 1957–58* (Nanaimo, 1958), 44. Gordon Pike prepared the statistics. The question of immature whales, raised by Pike after the initial whaling season of 1948, was substantiated in further studies. He eventually determined that approximately 43 per cent of all finback and humpback whales taken during the first six years of Western Whaling's operation were immature; so were about 1 in 5 blue whales. Sperm whales were almost all mature bulls; only 11 immature males (under thirty-six feet) were caught in the six-year study period. Of interest to researchers was a relatively large number of females—27—among the 192 sperm whales taken in 1957, attributed to the above-average water temperature at the northern end of their natural range in 40° North; Pike, "Whaling on the Coast British Columbia," 69–79.

105. "Minutes of Meeting, Whaler Crew Men," 19 March 1957, 1. UBC, UFAWU Records, Vols. 24–25.

106. Personal communication to the Author.

107. "Minutes of Meeting, Whaler Crewmen," 27 March 1958, 1. UBC, UFAWU Records, Vols. 24—25.

108. A. L. Gordon to all UFAWU members, form letter, 16 January 1959, 2. UBC, UFAWU Records, Box 166.

109. Newfoundland vote in Wallace Anstey to A. L. Gordon (telegram), Grand Bank, New-foundland, 26 January 1959. B.C. vote noted in Alex Gordon to Slim Farrington, Rod-ney Moss, and Wallace Anstey, [Vancouver], 6 February 1959. Both UBC, UFAWU Records, Box 166.

110. Gordon to Farrington, et al., 6 February 1959; 2; and A. L. Gordon to Rodney Moss and Wallace Anstey, telegram, [Vancouver], 17 February 1959. UBC, UFAWU Re-cords, Box 166.

111. H. M. Cowie to whaling employees, form letter, Vancouver, 17 February 1959. UBC, UFAWU Records, Box 166.

112. United Fishermen and Allied Workers Union, press release, 4 March 1959; agreement of both parties noted in A. L. Gordon to L[orne] Hume, [Vancouver], 16 March 1959, fo. 3. Both UBC, UFAWU Records, Box 166.

113. Johan (or John) Carlsen Borgen had also commanded chaser boats for Aristotle Onassis's Olympic Challenger fleet. He was a huge man, about 6'4" tall; in 1952, when still work-ing for Onassis, he was said to have already killed more than 5,100 whales in twenty-one years of shooting; see D. Richard Statile, "I Sailed the Antarctic Killer Boats," Argosy 335 (November 1952): 63.

114. Gordon to Hume, 16 March 1959, fo. 1.

115. Norman Hacking, "Whalers Won't Sail This Year," Province, 9 March 1960.

116. [Richard Carruthers], Hvalfangst-Oregon 1961 [Warrenton, Ore.: Bioproducts, 1961].

117. Tønnessen and Johnsen, History of Modern Whaling, 649. See also "Ore. Whaling Sta-tion Goes Modern," unidentified news clipping, 31 April 1962. KWM.

118. The Hayashi Kane Shoten from prewar days rapidly expanded into the modern Taiyo Gyogyo K.K.; Kazuhiro Mizue, "Factory Ship Whaling Around Bonin Islands in 1948," Scientific Reports of the Whales Research Institute 3 (February 1950), 106. By 1960 Taiyo was well on its way to becoming the largest privately owned fishing company in the world. One critic has claimed that Taiyo is "virtually one and the same as the Japan Whaling Association," a group which lobbies consistently and effectively against proposals to ban commercial whaling; see Craig Van Note, Outlaw Whalers: An Exposé of Unregulated Whaling Around the World (Washington, D.C.: Whale Protection Fund, 1979), 4—6.

119. Japan's first pelagic whaling fleet sailed in 1934. In 1940 a 10,000-ton factory ship and four chaser boats sailed to the Arctic Sea and shot 681 whales in eighty days, and repeated their effort the following year by taking 590 whales in seventy days off Kam-chatka; Japan, Fisheries Agency, Japanese Whaling Industry, 29. Most Japanese whaling ships were sunk during World War II, but Japan was encouraged by the American oc-cupation government to rebuild; factory-ship operations resumed in 1947—48 both in Antarctic waters and in the eastern North Pacific, around the Aleutian and Komodor Is-lands and the Kamchatka Peninsula; Waldon C. Winston, "The Largest Whale Ever Weighed," Natural History 59 (November 1950): 393—99; and Japanese Whaling Indus-try, 30; see also Takahisa Nemoto, "Foods of Baleen Whale in the Northern Pacific," Scientific Reports of the Whales Research Institute 12 (June 1957): 38.

120. Personal communication, John Lyon, Vancouver, 8 August 1983.

121. Lorne Hume, 4 August 1983.

122. Pike, "Whaling on the Coast British Columbia," 78—79.

123. Les S. Wilson, September 1983. Much canned pet food included whale meat among its contents before the advent of marine mammal protection acts; some of the cans may have been shipped to mink farmers in California.

124. "Letters of Understanding Between Western Canada Whaling Co. Ltd. and United Fishermen And Allied Workers' Union," typescript [December 1961]. UBC, UFAWU Records, Vol. 91, File 18.

125. Bonuses were to be increased by 12.5 per cent for the second (1963) season. Crews would additionally receive a seasonal bonus of $165.50 plus 4-per cent vacation pay, with a further pooled bonus of 65 cents per thousand pounds of meat processed, provided that the total exceeded 4.5 million pounds; "Whalers Sign Two-Year Pact," *Province*, 13 December 1961.

126. List of boat crews, typescript, Vancouver, 5 April 1962. UBC, UFAWU Records, Box 167, File 3.

127. Lorne Hume, 4 August 1983. According to Lorne Hume, then manager at Coal Harbour, the electronic tracking gear did not work as well as expected; the underwater sounds panicked the whales and caused them to swim to exhaustion. The extra adrenalin in their systems ruined the meat for human consumption.

128. "Notes of Meeting with Captain-Gunners Held at 8:30 A.M., April 10, 1962 on board 'Westwhale 2'," fo. 1. UBC, UFAWU Records, Vol. 6.

129. *Ibid.*.

130. Lorne Hume, 4 August 1983.

131. *Ibid.*.

132. Personal communication, Cora Veaudry, Coal Harbour, B.C., 21 July 1983.

133. Les S. Wilson, September 1983.

134. John Lyon, 8 August 1983.

135. Personal communication to the Author.

136. *Ibid.*.

137. *Ibid.*.

138. Mike James to Alex [Gordon], M.V. *Westwhale* 6, Coal Harbour, 22 May 1962; fo. 8. UBC, UFAWU Records, Vol. 6.

139. John Lyon, 8 August 1983.

140. This building was later moved to serve as the Community Hall in Holberg, B.C..

141. Personal communication to the Author.

142. John Lyon, 8 August 1983.

143. Pat Carney, "A Whale of a Problem," *Province*, 24 August 1963, 18. The *Gray*, of course, had exported meat to the United States during World War I; the Pacific Whaling Company exported meat to Japan "in pickle" as early as 1906–07.

144. Pat J. Malone to Alex [Gordon], at sea, *Westwhale* 3, 7 July [1962], 2. UBC, UFAWU Records, Box 167, File 3.

145. Norman Hacking, "Russian Whaling Fleet Shooting off B.C. Coast," *Province*, 31 May 1963; reported by the crew of the *San Juan No. II*.

146. Lorne Hume, 4 August 1983; see also Bob McMurray, "4 Soviet Ships Get into Island Whale Chase," *Province*, 20 July 1963.

147. "Old Ship Steam Engine to Get Movie Role," *Marine Digest*, 17 July 1965, unidentified clipping, VCA, Cadieux Coll., Vol. 240.

148. See Defieux, "3 Old Whaling Veterans Heading for Scrap Heap," 20. The *Westwhale* 2 (ex-*Bouvet III*) was scrapped by Acme Trading Company of Vancouver; see Somner, *From 70 North to 70 South*, 83.

149. See Canada, Fisheries Research Board, *Annual Report of the Biological Station, Nanaimo, B.C., for* 1965 [n.p., 1965].

NOTES TO PP. 284–87

150. See Endangered Species Productions, *A Brief History of the International Whaling Commission (IWC)* (Boston: Endangered Species Productions, [ca. 1975]), 9.

151. M.V. *Westwhale* 7, 1 April 1967 - 25 April 1967. VCA, Add MSS 335, Vol. 434. The chaser was in port for two of the twenty-three days.

152. Norman Hacking, "Japanese Whale Boat Added to B.C. Fleet," *Province*, 28 March 1967. It was 167 feet long, with a 2,000 horsepower diesel.

153. See "West Coast Whaling Fleet to be Laid up During 1968," *Province*, 12 December 1967, 30; and Tønnessen and Johnsen, *History of Modern Whaling*, 649.

154. Resolution, Sixteenth Annual Convention, transmitted to Prime Minister John Diefenbaker and Minister of Fisheries Angus MacLean, 14 March 1960. UBC, UFAWU Records, Box 241.

155. B. Madden (McKenzie Barge & Derrick Company) to Charles M. Defieux, Vancouver, 25 January 1971. VMM.

156. Newell, *H. W. McCurdy Marine History of the Pacific Northwest 1966 to 1976*, 191.

157. See Canada, International Whaling Commission Delegation, "The Canadian Whaling Ban: Its Implementation and Impact," by W. R. Martin. Appendix 5 in *Delegation Report*, Thirty-second Annual Meeting (Ottawa, 1980), 1.

158. *Ibid.*, 3.

159. *Ibid.*; also Endangered Species Productions, *Brief History of the International Whaling Commission*, 11.

160. The ban was announced on 22 December 1972; see Martin, "Canadian Whaling Ban"; and Canada, International Whaling Commission Delegation, Joint Consultative Committee, "Report of the Nova Scotia Whaling Ban." Appendix 2 in *Delegation Report* (Ottawa, 1980), 1–6.

161. The Marine Mammal Protection Act of 1972, 16 U.S.C. 1361–1407; and the Endangered Species Act of 1973, Public Law 93–205, 87 Stat. 884.

162. Chile, the People's Republic of China, the Philippines, South Africa and Switzerland abstained; see Canada, International Whaling Commission Delegation, *Delegation Report*, Thirty-fourth Annual Meeting (Ottawa, 1982); see particularly Appendix 2, "Vote on Moratorium."

163. In 1928, Agnes Laut wrote: "I doubt if there will ever again be a real Indian whale hunt when the whole tribe goes out off the coast of the Pacific Western states. The Indians can make more money salmon fishing and selling timber limits. . . . Whoever wants to see an old-time Indian whale hunt must hie him to these Indians soon, for in another generation there will be no Indian whale hunts." Agnes C. Laut, "Who Wants a Whale Steak?: The Exciting Enterprise of Catching a Whale," *Mentor* (September 1928), 33– 35. The 1940 cultural revival is noted in Jon Van Arsdell, "B.C. Whaling: The Indians," *Raincoast Chronicles First Five* (Madeira Park, B.C.: Harbour Publishing, 1976), 28.

164. Beatrice D. Miller, "Neah Bay: The Makah in Transition," *Pacific Northwest Quarterly*, 42 (October 1952): 266.

BIBLIOGRAPHY

PUBLISHED SOURCES

Books, Reports, Monographs, and Contributions to Books

Allen, Edward Weber. *North Pacific: Japan, Siberia, Alaska, Canada*. New York: Professional and Technical Press, 1936.

Andrews, Ralph W.; and A. K. Larssen. *Fish and Ships*. New York: Bonanza Books, 1959.

Andrews, Roy Chapman. *All About Whales*. New York: Random House, 1954.

——. *Ends of the Earth*. New York: G. P. Putnam's Sons, 1929.

——. *Whale Hunting with Gun and Camera: A Naturalist's Account of the Modern Shore-Whaling Industry, of Whales and Their Habits, and of Hunting Experiences in Various Parts of the World*. New York: D. Appleton and Company, 1928.

Anthony, Joseph R. *Life in New Bedford A Hundred Years Ago: A Chronicle of the Social, Religious and Commercial History of the Period as Recorded in a Diary Kept by Joseph R. Anthony*. Edited by Zephaniah W. Pease. New Bedford: Old Dartmouth Historical Society, 1925.

Arnold, James. "Report on Whale Fishery," in *Proceedings of the National Convention for the Protection of American Interests*, 39–41. New York: Greeley and McElrath; Thaddeus B. Wakeman, 1842.

Bancroft, Hubert Howe. *History of Alaska 1730–1885*. The Works of Hubert Howe Bancroft, Vol. 33. San Francisco: A. L. Bancroft and Company, 1886.

Francis Bannerman Sons. *Illustrated and Descriptive Catalogue of Military Goods*. New York: Francis Bannerman Sons, 1929.

Barbeau, Marius. *Totem Poles of Gitksan, Upper Skeena River, British Columbia*. National Museum of Canada Bulletin 61. Ottawa: F. A. Acland, 1929; facsimile ed. Ottawa: National Museum of Man, 1973.

Barrett-Lennard, C. E. *Travels in British Columbia, with the Narrative of a Yacht Voyage Round Vancouver's Island*. London: Hurst and Blackett, 1862.

Barthelmess, Klaus. *Das Bild des Wals in Fünf Jahrhunderten*. Exhibition catalogue. Cologne: d.m.e. Verlag, 1982.

Beane, J[oshua] F[illebrown]. *From Forecastle to Cabin: The Story of a Cruise in Many*

Seas, Taken From a Journal Kept Each Day, Wherein was Recorded the Happenings of a Voyage Around the World in Pursuit of Whales. New York: Editor Publishing Company, 1905.

Bennett, Frederick Debell. *Narrative of a Whaling Voyage Round the Globe, From the Year 1833 to 1836: Comprising Sketches of Polynesia, California, the Indian Archipelago, Etc.; with an Account of Southern Whales, the Sperm Whale Fishery, and the Natural History of the Climates Visited*. 2 vols. London: Richard Bentley, 1840.

Berman, Bruce D. *Encyclopedia of American Shipwrecks*. Boston: Mariners Press, 1972.

Bernier, J[oseph] E[lzear]. *Report on the Dominion of Canada Government Expedition to the Arctic Islands and Hudson Strait on Board the D.G.S. "Arctic."* Ottawa: Government Printing Bureau, 1910.

Bill, Erastus. *Citizen: An American Boy's Early Manhood Aboard a Sag Harbor Whale-Ship Chasing Delirium and Death Around the World, 1843–1849, Being the Story of Erastus Bill Who Lived to Tell It*. Notes by Robert Wesley Bills. Anchorage: O. W. Frost, 1978.

Birkeland, Knut B. *The Whalers of Akutan: An Account of Modern Whaling in the Aleutian Islands*. New Haven: Yale University Press, 1926.

Boas, Franz. "Tsimshian Mythology." In *Thirty-First Annual Report of the Bureau of American Ethnology*, 1909–1910 (Smithsonian Institution. Bureau of American Ethnology), 27–1037. Washington, D.C.: Government Printing Office, 1916.

Bockstoce, John R. *Steam Whaling in the Western Arctic*. New Bedford: Old Dartmouth Historical Society, 1977.

————. *Whales, Ice, and Men: The History of Whaling in the Western Arctic*. Seattle: University of Washington Press, 1986.

Bogen, Hans. "Norwegian Whaling." In *Norway's Export Trade: A National Publication*, 134–42. Oslo: Belix Publishing Company, [ca. 1939].

Bogen, Hans S.I. *70 år Lars Christensen og hans samtid*. Oslo: Johan Grundt Tanum, 1955.

Bowsfield, Hartwell, ed. *Fort Victoria Letters 1846–1851*. Hudson's Bay Record Society 32. Winnipeg: Hudson's Bay Record Society, 1979.

Brandligt, C. *Geschiedkundige Beschouwing van de Walvisch-Visscherij*. Amsterdam: T. Nusteeg, Jr., 1843.

Brandt, Karl. *Whale Oil: An Economic Analysis*. Fats and Oils Study 7. Palo Alto, Calif.: Food Research Institute, Stanford University, 1940.

[Brewington, Marion V.] *Full Fadom Five*. Sharon, Mass.: Priceless Pearl Press, 1968.

British Columbia, Department of Lands, Geographic Division. *Geographical Gazetteer of British Columbia*. Victoria: Charles F. Banfield, 1930.

————, Provincial Museum of Natural History. *Report of the Provincial Museum of Natural History for the Year 1920*. 11 Geo. 5. Victoria: William H. Cullin, 1921.

————, Provincial Museum of Natural History. *Report of the Provincial Museum of Natural History for the Year 1921*. 12 Geo. 5. Victoria: William H. Cullin, 1922.

————, Provincial Secretary. *Papers Relating to the Protest Against the Paris Award*

Regulations Affecting the Seal Fisheries on the Eastern Side of the North Pacific Ocean.
61 Vict. 1281. Victoria, 1897.

————, Provincial Secretary. *Return to a Respectful Address Presented to His Honour the Lieutenant Governor, Praying Him to Cause to be Laid Before the House a Copy of the Order in Council Relative to the Grievances of the Sealers Referred to in the Answer of the Honourable the Attorney-General on the 12th Day of February Last.* 60 Vict. 965a. Victoria: Richard Wolfenden, 1897.

————, Royal Commission. *Claims by Pelagic Sealers Arising out of the Washington Treaty, 7th July 1911 and the Regulations Made under the Paris Award Which Came into Force 1894.* White Claims, Vol. 13. Victoria: Government Printer, 1914.

British Columbia: Official Centennial Record. Vancouver: Evergreen Press, 1957.

Bronson, G. W. *Glimpses of the Whaleman's Cabin.* Boston: Damrell and Moore, 1855.

Brower, Kenneth. *The Starship and the Canoe.* New York: Holt, Rinehart and Winston, 1978; reprint ed. New York: Harper Colophon Books, 1983.

Brown, James Temple[man]. *The Whale Fishery and its Appliances.* Great International Fisheries Exhibition (London) Publication E. Washington, D.C.: Government Printing Office, 1883.

Brown, Robert. *The Countries of the World: Being a Description of the Various Continents, Islands, Rivers, Seas, and Peoples of the Globe.* London: Cassell and Company, [ca. 1877].

Browne, J. Ross. *Etchings of a Whaling Cruise, with Notes of a Sojourn on the Island of Zanzibar: To Which is Appended a Brief History of the Whale Fishery, Its Past and Present Condition.* New York: Harper and Brothers, 1846; facsimile ed. Cambridge, Mass.: Harvard University Press, 1968.

Budker, Paul. *Baleines et Baleiniers.* Paris: Horizons de France, 1957; English ed. (*Whales and Whaling*) London: George G. Harrap and Company, 1958.

Bull, H. J. *The Cruise of the "Antarctic" to the South Polar Regions.* London: Edward Arnold, 1896.

Bullard, John M. *Captain Edmund Gardner of Nantucket and New Bedford: His Journal and His Family.* New Bedford: published by the author, 1958.

Busch, Briton Cooper. *Alta California 1840–1842: The Journal and Observations of William Dane Phelps, Master of the Ship "Alert."* Glendale, Calif.: Arthur H. Clark Company, 1983.

————. *The War Against the Seals: A History of the North American Seal Fishery.* Kingston, Ontario: McGill-Queen's University Press, 1985.

California Sea Products Company. *California Sea Products Company.* Prospectus. San Francisco, [ca. 1917].

Calkins, Raymond. *The Life and Times of Alexander McKenzie.* Cambridge, Mass.: Harvard University Press, 1935.

Campbell, Archibald. *A Voyage Round the World, from 1806 to 1812; in Which Japan, Kamschatka, the Aleutian Islands, and the Sandwich Islands Were Visited; Including a Narrative of the Author's Shipwreck on the Island of Sannack, and His Subsequent Wreck in the Ship's Long-Boat; with an Account of the Present State of the Sandwich Islands, and a Vocabulary of Their Language*. 4th American ed. Roxbury, Mass.: Allen and Watts, 1825.

Canada, Biological Board. *Annual Announcement of the Biological Board of Canada for 1927*. Ottawa: F. A. Acland, 1927.

————, Department of Fisheries. *Fourteenth Annual Report of the Department of Fisheries: Seventy-Seventh Annual Fisheries Report of the Dominion for the Year 1943–44*. Ottawa: Edmund Cloutier, 1945.

————, Department of Marine and Fisheries. "Annual Report of the Fisheries of British Columbia for the Year 1888," by Thomas Mowat. In *Fifth Annual Report of the Deputy Minister of Fisheries for the Year 1888*. Appendix 8, 233–243. Ottawa, [ca. 1888].

————, Department of Marine and Fisheries. *Forty-Fifth Annual Report of the Department of Marine and Fisheries 1911–12*. 3 Geo. V, 1913, Sess. Paper 22. Ottawa: C. H. Parmelee, 1912.

————, Department of Fisheries and Oceans. "Cetaceans of Canada," by Randall R. Reeves and Edward Mitchell. *Underwater World* no. 59. Ottawa, 1987.

————, Department of Marine and Fisheries. *Forty-First Annual Report of the Department of Marine and Fisheries 1907–8*. 8–9 Edw. VII, 1909, Sess. Paper 22. Ottawa: S. E. Dawson, 1908–9.

————, Department of Marine and Fisheries. "The Fur Sealing Industry of the North Pacific Ocean as Affected by the Behring Sea Award and Consequent Legislation," by R.N. Venning. In *Thirtieth Annual Report of the Department of Marine and Fisheries 1897*. Appendix 13, 325–365. 61 Vict. Sess. Papers 11. Ottawa: S. E. Dawson, 1898.

————, Fisheries Research Board. *Annual Report of the Biological Station, Nanaimo, B.C. for 1957–58*. Nanaimo, 1958.

————, Fisheries Research Board. *Annual Report of the Biological Station, Nanaimo, B.C., for 1965*. n.p., 1965.

————, Fisheries Research Board. *The Chemistry and Technology of Marine Animal Oils with Particular Reference to Those of Canada*. Edited by H. N. Brocklesby. Bulletin 59. Ottawa, 1941.

————, Fisheries Research Board. *Marine Oils: With Particular Reference to Those of Canada*. Edited by B.E. Bailey, N.M. Carter, and L.A. Swain. Bulletin 89. Ottawa, 1952.

————, International Whaling Commission Delegation. *Delegation Report*, Thirty-fourth Annual Meeting. Ottawa, 1982.

————, International Whaling Commission Delegation. "Report of the Nova Scotia Whaling Ban," by the Joint Consultative Committee. Appendix 2, 1–6; and "The Canadian Whaling Ban: Its Implementation and Impact," by W. R. Martin. Appendix 5, 1–14. In *Delegation Report*, Thirty-second Annual Meeting. Ottawa, 1980.

————, International Whaling Commission Delegation. "Technical Committee Working Group on Humane Killing." Annex 4. In *Canadian Delgation Report*, Thirty-first Annual Meeting. Ottawa, 1979.

————, Permanent Commission on Geographical Names. *Gazetteer of Canada: Brit-*

ish Columbia. 2nd ed. Ottawa, 1966.

Carrothers, W. A. *The British Columbia Fisheries*. Toronto: University of Toronto Press, 1941.

[Carruthers, Richard]. *Hvalfangst-Oregon* 1961. [Warrenton, Ore.: Bioproducts, 1961].

Catalogue of Nantucket Whalers, and Their Voyages from 1815 to 1870. Nantucket: Hussey and Robinson, 1876.

Chatterton, E. Keble. *Whalers and Whaling: The Story of Whaling Ships up to the Present Day*. London: T. Fisher Unwin, 1925.

Cheever, Henry T. *The Whale and His Captors: Or, The Whaleman's Adventures and the Whale's Biography, as Gathered on the Homeward Cruise of the "Commodore Preble."* New York: Harper and Brothers, 1850; London: Thomas Nelson and Sons, 1851.

[————.] *The Whaleman's Adventures in the Southern Ocean; as Gathered by the Rev. Henry T. Cheever, on the Homeward Cruise of the "Commodore Preble."* Edited by W[illiam] Scoresby, [Jr.] London: Sampson Low and David Bogue, 1850.

Church, Albert Cook. *Whale Ships and Whaling*. New York: W. W. Norton Company, 1938; reprint ed. New York: Bonanza Books, 1974.

Clark, Joseph G. *Lights and Shadows of Sailor Life, as Exemplified in Fifteen Years' Experience, Including the More Thrilling Events of the U.S. Exploring Expedition, and Reminiscences of an Eventful Life on the "Mountain Wave."* Boston: John Putnam, 1847.

Cleland, Robert G. "Asiatic Trade and American Occupation of the Pacific Coast." In *Annual Report of the American Historical Association for the Year* 1914. Vol. 1, 281–89. Washington, D.C.: Government Printing Office, 1916.

Colby, Barnard L. *New London Whaling Captains*. Mystic, Conn.: Marine Historical Association, 1936.

Colnett, James. *A Voyage to the South Atlantic and Round Cape Horn into the Pacific Ocean, for the Purpose of Extending the Spermaceti Whale Fisheries, and Other Objects of Commerce, by Ascertaining the Ports, Bays, Harbours, and Anchoring Births, in Certain Islands and Coasts in Those Seas, at Which the Ships of the British Merchants Might be Refitted*. London: W. Bennett, 1798.

Colwell, Max. *Ships and Seafarers in Australian Waters*. Melbourne: Lansdowne Press, 1973.

Cook, James; and James King. *A Voyage to the Pacific Ocean: Undertaken, by the Command of His Majesty, for Making Discoveries in the Northern Hemisphere; Performed Under the Direction of Captains Cook, Clerke, and Gore, in His Majesty's Ships the* Resolution *and* Discovery; *in the Years* 1776, 1777, 1778, 1779, *and* 1780. 3rd ed. 3 vols. London: H. Hughs, 1785.

Cook, John A. *Pursuing the Whale: A Quarter-Century of Whaling in the Arctic*. Boston: Houghton Mifflin Company, 1926.

Cook, Warren L. *Flood Tide of Empire: Spain and the Pacific Northwest, 1543–1819*. New Haven: Yale University Press, 1973.

Corney, Peter. *Voyages in the Northern Pacific: Narrative of Several Trading Voyages from 1813 to 1818, between the Northwest Coast of America, the Hawaiian Islands and*

China, with a Description of the Russian Establishments on the Northwest Coast. Honolulu: Thomas G. Thrum, 1896.

Crapo, Henry H. *The New Bedford Directory*. 5th ed. New Bedford: Benjamin Lindsey, 1845.

Crawfurd, Frederik. *Captain Frederik Crawfurds Dagbok: En Norsk Hvalfangstferd 1843–1846 og Andre Europeiske Landes Deltagelse i Stillehavsfangsten 1800–1860*. Oslo: J. W. Cappelen, 1953.

Crile, George. *George Crile: An Autobiography*. Edited by Grace Crile. 2 vols. Philadelphia: J. B. Lippincott Company, 1947.

Cristiani, R. S. *A Technical Treatise on Soap and Candles; With a Glance at the Industry of Fats and Oils*. Philadelphia: Henry Carey Baird and Company, 1881.

Cumpston, J[ohn] S. *Shipping Arrivals & Departures: Sydney, 1788–1825*. Vol. 1. Roebuck Society Publication 22. Canberra: Roebuck Society, 1977.

Curtis, Edward S. *The Nootka and Haida. The North American Indian*, Vol. II. New York: Johnson Reprint Corporation, 1916.

Cushing, Caleb. *The Treaty of Washington: Its Negotiation, Execution, and the Discussions Relating Thereto*. New York: Harper and Brothers, 1873.

Dakin, William John. *Whalemen Adventurers: The Story of Whaling in Australian Waters and Other Southern Seas Related Thereto, from the Days of Sails to Modern Times*. Sydney: Angus and Robertson, 1934.

Dalzell, Kathleen E. *The Queen Charlotte Islands 1774–1966*. Terrace, B.C.: C. M. Adam, 1968.

Damon, Ethel M. *Samuel Chenery Damon*. Honolulu: Hawaiian Mission Children's Society, 1966.

Dana, Samuel L.; Linus Child; Alex[ander] Wright; and James B. Francis. *A Report on Rosin Oil, by a Committee Appointed by the Agents of the Manufacturing Companies at Lowell, Massachusetts*. 2nd ed. Lowell, Mass.: S. J. Varney, 1852.

Davidson, George. *The Alaska Boundary*. San Francisco: Alaska Packers Association, 1903.

Davis, William Heath. *Seventy-Five Years in California: A History of Events and Life in California: Personal, Political and Military; Under the Mexican Regime; During the Quasi-Military Government of the Territory by the United States, and after the Admission of the State to the Union*. San Francisco: John Howell, 1929.

Dawson, George M. *Report on the Queen Charlotte Islands 1878*. Montreal: Dawson Brothers, 1880.

[Debrett, J.] *The Errors of the British Minister, in the Negotiation with the Court of Spain*. London: printed for J. Debrett, 1790.

Densmore, Frances. *Nootka and Quileute Music*. Smithsonian Institution Bureau of American Ethnology Bulletin 124. Washington, D.C.: Government Printing Office, 1939.

Dexter, Elisha. *Narrative of the Loss of Whaling Brig William and Joseph, of Martha's Vineyard, and the Sufferings of Her Crew for Seven Days, a Part of the Time on a Raft in the Atlantic Ocean; with an Appendix, Containing Some Remarks on the Whaling Business, and Descriptions of the Mode of Killing and Taking Care of Whales*. 2nd ed. Boston: Charles C. Mead, 1848.

Dixon, George. *Remarks on the Voyages of John Meares, Esq. in a Letter to That Gentleman*. London: printed for the author, 1790.

Doig, Ivan. *Winter Brothers: A Season at the Edge of America*. New York: Harcourt Brace Jovanovich, 1980.

Dorion-Robitaille, Yolande. *Captain J. E. Bernier's Contribution to Canadian Sovereignty in the Arctic*. Ottawa: Indian and Northern Affairs Canada, 1978.

Drucker, Philip. *Cultures of the North Pacific Coast*. San Francisco: Chandler Publishing Company, 1965.

———. *The Northern and Central Nootkan Tribes*. Smithsonian Institution, Bureau of American Ethnology Bulletin 144. Washington, D.C.: Government Printing Office, 1951.

Duboc, Gustave. *Les Nuées Magellaniques: Voyages au Chili, au Pérou et en Californie a la Pêche de la Baleine*. Part 1. Paris: Amyot, 1853.

Duflot de Mofras, [Eugène]. *Duflot de Mofras' Travels on the Pacific Coast*, translated and edited by Marguerite Eyer Wilbur. 2 vols. Santa Ana, Calif.: Fine Arts Press, 1937.

———. *Exploration du Territoire de l'Orégon, des Californies et de la Mer Vermille, Exécutée Pendant les Années 1840, 1841 et 1842*. Paris: Arthus Bertrand, 1844.

D'Wolf, John. *A Voyage to the North Pacific*. Cambridge, Mass.: Welch, Bigelow and Company, 1861; facsimile ed. Fairfield, Wash.: Ye Galleon Press, 1968.

———. *A Voyage to the North Pacific and a Journey Through Siberia More Than Half a Century Ago*. Bristol, R.I.: Rulon-Miller Books, 1983.

Edward Eberstadt and Sons. *The Northwest Coast: A Century of Personal Narratives of Discovery, Conquest & Exploration from Bering's Landfall to Wilkes' Surveys 1741–1841*. New York: Edward Eberstadt and Sons, [1941].

Ehrman, John. *The Younger Pitt: The Years of Acclaim*. London: Constable and Company, 1969.

Elliott, Henry W. *Our Arctic Province: Alaska and the Seal Islands*. New York: Charles Scribner's Sons, 1886.

Elsdon, G. D. *The Chemistry and Examination of Edible Oils and Fats, Their Substitutes and Adulterants*. London: Ernest Benn, 1926.

Endangered Species Productions. *A Brief History of the International Whaling Commission (IWC)*. Boston: Endangered Species Productions, [ca. 1975].

Enderby, Charles. *Proposal for Re-Establishing the British Southern Whale Fishery, Through the Medium of a Chartered Company, and in Combination with the Colonisation of the Auckland Islands, as the Site of the Company's Whaling Station*. London: Effingham Wilson, 1847.

Erskine, Wilson Fiske. *White Water: An Alaskan Adventure*. London: Abelard-Schuman, 1960.

Espinosa y Tello, José. *A Spanish Voyage to Vancouver and the Northwest Coast of America*. Translated by Cecil Jane. London: Argonaut Press, 1930.

Evans, Robley D. *A Sailor's Log: Recollections of Forty Years of Naval Life*. New York: D. Appleton and Company, 1901.

Fahey, James C. *The Ships and Aircraft of the United States Fleet*. Victory ed. New York: Ships and Aircraft, 1945.

Fairbank, John K., Edwin O. Reischauer, and Albert M. Craig. *East Asia: The Modern Transformation*. Boston: Houghton Mifflin Company, 1965.

Fairburn, William Armstrong. *Merchant Sail*. 6 vols. Center Lovell, Maine: Fairburn Marine Educational Foundation, 1945–55.

Findlay, Alexander. *A Directory for the Navigation of the North Pacific Ocean: With Descriptions of its Coasts, Islands, Etc., from Panama to Behring Strait and Japan, its Winds, Currents, and Passages*. 2nd ed. London: Richard Holmes Laurie, [ca. 1870].

Fingard, Judith. *Jack in Port: Sailortowns of Eastern Canada*. Toronto: University of Toronto Press, 1982.

Forester, Joseph E., and Anne D. Forester. *Fishing: British Columbia's Commercial Fishing Industry*. Saanichton, B.C.: Hancock House Publishers, 1975.

Forster, Honore. *The South Sea Whaler: An Annotated Bibliography of Published Historical, Literary and Art Material Relating to Whaling in the Pacific Ocean in the Nineteenth Century*. Sharon, Mass.: Kendall Whaling Museum; Fairhaven, Mass.: Edward J. Lefkowicz, 1985.

Frank, Stuart M. " 'Vast Address and Boldness': The Rise and Fall of the American Whale Fishery." In *The Spirit of Massachusetts: Our Maritime Heritage*, edited by George S. Perry, Jr. and Helen Richmond, 20–29. [Boston]: New England Historic Seaport and Massachusetts Department of Education, 1985.

Frischauer, Willi. *Onassis*. London: Book of the Month Club, 1968.

Frouin, Charles. *Journal de Bord 1852–1856: Charles Frouin, Chirurgien du Baleinier "L'Espadon."* Paris: Editions France-Empire, 1978.

Gelett, Charles Wetherby. *A Life on the Ocean: Autobiography of Captain Charles Wetherby Gelett, a Retired Sea Captain, Whose Life Trail Crossed and Recrossed Hawaii Repeatedly*. Advertiser Historical Series 3. Honolulu: Hawaiian Gazette Company, 1917.

Gibson, Gordon, with Carol Renison. *Bull of the Woods: The Gordon Gibson Story*. Vancouver: Douglas and McIntyre, 1982.

Gibson, James R. *Imperial Russia in Frontier America: The Changing Geography of Supply of Russian America 1784–1867*. New York: Oxford University Press, 1976.

Gilkerson, William. *The Scrimshander*. San Francisco: Troubador Press, 1975.

Goddard, Joan. "North to the Station." In *Tales from the Queen Charlotte Islands, Book 2* (Regional Council, Senior Citizens of the Queen Charlotte Islands), 6–13. Masset, B.C.: Regional Council, 1982.

Gough, Barry M. *Distant Dominion: Britain and the Northwest Coast of North America, 1579–1809*. Vancouver: University of British Columbia Press, 1980.

————. *The Royal Navy and the Northwest Coast of North America, 1810–1914: A Study of British Maritime Ascendancy*. Vancouver: University of British Columbia Press, 1971.

Grady, Don. *The Perano Whalers of Cook Strait 1911–1964*. Sydney: A. H. and A. W. Reed, 1982.

Great Fisheries of the World, The. London: T. Nelson and Sons, [ca. 1878].

Greenhow, Robert. *The History of Oregon and California, and the Other Territories on the*

North-West Coast of North America. 2nd ed. Boston: Charles C. Little and James Brown, 1845.

Hagelund, W[illiam] A. *Flying the Chase Flag: The Last Cruise of the West Coast Whalers*. Toronto: Ryerson Press, 1961.

[Hamilton, James H.] *Western Shores: Narratives of the Pacific Coast, By "Captain Kettle."* Vancouver: Progress Publishing Company, 1933.

Hare, Lloyd C. M. *Salted Tories: The Story of the Whaling Fleets of San Francisco*. Mystic, Conn.: Marine Historical Association, 1960.

Hartwig, G. *The Sea and Its Living Wonders: A Popular Account of the Marvels of the Deep and of the Progress of Maritime Discovery from the Earliest Ages to the Present Time*. 4th ed. London: Longmans, Green and Company, 1873.

Hegarty, Reginald B. *Addendum to "Starbuck" and "Whaling Masters."* New Bedford: New Bedford Free Public Library, 1964.

———. *Returns of Whaling Vessels Sailing from American Ports 1876–1928*. New Bedford: Old Dartmouth Historical Society and Whaling Museum, 1959.

Heizer, Robert F. *Aconite Poison Whaling in Asia and America: An Aleutian Transfer to the New World*. Smithsonian Institution, Bureau of American Ethnology, Anthropological Papers 24. Washington, D.C.: Government Printing Office, 1943.

Henderson, David A. *Men & Whales at Scammon's Lagoon*. Baja California Travels Series 29. Los Angeles: Dawson's Book Shop, 1972.

Henderson, David S. *Fishing for the Whale: A Guide-Catalogue to the Collection of Whaling Relics in Dundee Museum*. Dundee, Scotland: Dundee Museum, 1972.

Henry, John Frazier. *Early Maritime Artists of the Pacific Northwest Coast, 1741–1841*. Seattle: University of Washington Press, 1984.

Hohman, Elmo P. *The American Whaleman: A Study of Life and Labor in the Whaling Industry*. New York: Longmans, Green and Company, 1928; reprint ed. Clifton, N.J.: Augustus M. Kelley, 1972.

Holm, Bill. *Northwest Coast Indian Art: An Analysis of Form*. Thomas Burke Memorial Washington State Museum Monograph 1. Seattle: University of Washington Press, 1965.

Hopkins, Caspar T., *et al. Report on Port Charges, Shipping and Ship-Building to the Manufacturers' Association, the Board of Trade, and the Chamber of Commerce, of San Francisco*. San Francisco: H. S. Crocker, 1885.

Howay, Frederic W., ed. *Voyages of the "Columbia" to the Northwest Coast 1787–1790 and 1790–1793*. Boston: Massachusetts Historical Society, 1941.

———; W. N. Sage; and H. F. Angus. *British Columbia and the United States: The North Pacific Slope from Fur Trade to Aviation*. Toronto: Ryerson Press; New Haven: Yale University Press; London: Humphrey Milford, Oxford University Press, 1942.

Huff, Boyd. *El Puerto de los Balleneros: Annals of the Sausalito Whaling Anchorage*. Early California Travel Series 42. Los Angeles: Glen Dawson, 1957.

Hunter, Robert. *Warriors of the Rainbow: A Chronicle of the Greenpeace Movement*. New York: Holt, Rinehart and Winston, 1979.

———; and Rex Weyler. *To Save a Whale: The Voyages of Greenpeace*. Vancouver: Douglas and McIntyre, 1978.

Jackson, Gordon. *The British Whaling Trade*. London: Adam and Charles Black, 1978.

Janssen, Albrecht. *Tausend Jahre Deutscher Walfang*. Leipzig: F. A. Brodhaus, 1937.

Japan, Ministry of Agriculture and Forestry, Fisheries Agency. *Japanese Whaling Industry*. Tokyo: Japan Whaling Association, 1954.

Jenkins, J[ames] T[ravis]. *A History of the Whale Fisheries: From the Basque Fisheries of the Tenth Century to the Hunting of the Finner Whale at the Present Date*. London: H. F. and G. Witherby, 1921; reprint ed. Port Washington, N.Y.: Kennikat Press, 1971.

———. *Whales and Modern Whaling*. London: H. F. and G. Witherby, 1932.

Jenkins, Thomas H. *Bark Kathleen Sunk By a Whale, as Related by the Captain, Thomas H. Jenkins, to Which is Added an Account of Two Like Occurrences, the Loss of Ships Ann Alexander and Essex*. New Bedford: H. S. Hutchinson and Company, 1902.

Jewitt, John R. *Narrative of the Adventures and Sufferings of John R. Jewitt; Only Survivor of the Crew of the Ship Boston, During a Captivity of Nearly Three Years Among the Savages of Nootka Sound: With an Account of the Manners, Mode of Living, and Religious Opinions of the Natives*. New York: printed for the publisher, [ca. 1815].

Johnsen, Arne Odd. *Norwegian Patents Relating to Whaling and the Whaling Industry: A Statistical and Historical Analysis*. Kommandør Chr. Christensens Hvalfangstmuseum Publikation 16. Oslo: A. W. Brøggers, 1947.

Jones, A. G. E. *Ships Employed in the South Seas Trade 1775–1861 (Parts I and II) and Registrar General of Shipping and Seamen: Transcripts of Registers of Shipping 1787–1862 (Part III)*. Roebuck Society Publication 36. Canberra: Roebuck Society, 1986.

Jones, Mary Lou; Steven L. Swartz; and Stephen Leatherwood, eds. *The Gray Whale* (Eschrichtius robustus). Orlando, Florida: Academic Press, 1984.

Jore, Léonce. *L'Océan Pacifique au Temps de la Restauration et de la Monarchie de Juillet (1815–1848)*. 2 vols. Paris: Editions Besson et Chantemerle, 1959.

Judd, Bernice. *Voyages to Hawaii Before 1860*. Honolulu: Hawaiian Mission Children's Society, 1929.

Jupp, Ursula, ed. *Home Port: Victoria*. Victoria: published by the author, 1967.

Jürgens, Hans Peter. *Abenteuer Walfang: Wale, Männer, und das Meer*. Herford: Koehlers, 1977.

Keithahn, Edward L. *Monuments in Cedar*. Ketchikan, Alaska: Roy Anderson, 1945.

Kemble, John H. *San Francisco Bay: A Pictorial Maritime History*. Cambridge, Maryland: Cornell Maritime Press, 1957.

Kingsford, Maurice Rooke. *The Life, Work and Influence of William Henry Giles Kingston*. Toronto: Ryerson Press, 1947.

Kingston, William H. G. *Peter the Whaler: His Early Life and Adventures in the Arctic Regions*. London: Blackie and Son, n.d.

Kirk, Ruth; and Richard P. Daugherty. *Hunters of the Whale: An Adventure in Northwest Coast Archeology*. New York: William Morrow and Company, 1974.

Kommerchesko-Tekhnischeskoye Otdeliniye. Kniga 2. Moscow: Moskovskago Kommercheskago Instituta, 1914.

Krasheninnikov, S. *The History of Kamtschatka and the Kurilski Islands, with the Coun-*

tries Adjacent. Translated by James Grieve. Gloucester, England: T. Jefferys, 1764.

La Croix, Louis. *Les Derniers Baleiniers Français: Un Demi-Siècle d'Histoire de la Grande Pêche Baleinière en France de 1817 à 1867*. Nantes: Aux Portes du Large, 1947.

Langdon, Robert, ed. *American Whalers and Traders in the Pacific: A Guide to Records on Microfilm*. Canberra: Pacific Manuscripts Bureau, Research School of Pacific Studies, Australian National University, 1978.

Langevin, H. L. *Report of the Hon. H.L. Langevin, C.B., Minister of Public Works*. Ottawa: I. B. Taylor, 1872.

Larkin, P. A.; and W. E. Ricker. *Canada's Pacific Marine Fisheries: Past Performance and Future Prospects*. Fifteenth British Columbia Natural Resources Conference. Vancouver, 1962.

Larkin, Thomas O. *The Larkin Papers: Personal, Business, and Official Correspondence of Thomas Oliver Larkin, Merchant and United States Consul in California*. Edited by George P. Hammond. 10 vols. Berkeley and Los Angeles: University of California Press, 1951–64.

Lawson, Will. *Blue Gum Clippers and Whale Ships of Tasmania*. Melbourne: Georgian House, 1949.

Lecomte, Jules. *Pratique de la Pêche de la Baleine dans les Mers du Sud*. Paris: Lecointe et Pougin, 1833.

Leonard, L. Larry. *International Regulation of Fisheries*. Carnegie Endowment for International Peace Monograph 7. Washington, D.C.: Carnegie Endowment for International Peace, Division of International Law, 1944.

Lindeman, Moritz. *Die Arktische Fischerei der Deutschen Seestädte 1620–1868*. Gotha: Justus Perthes, 1869.

Lower, J. Arthur. *Ocean of Destiny: A Concise History of the North Pacific, 1500–1978*. Vancouver: University of British Columbia Press, 1978.

Lubbock, Basil. *The Arctic Whalers*. Glasgow: Brown, Son and Ferguson, 1937; reprint ed. 1955.

Lyman, Chester S[mith]. *Around the Horn to the Sandwich Islands and California 1845–1850*. Edited by Frederick J. Teggart. New Haven: Yale University Press, 1924.

Lytle, Thomas G. *Harpoons and Other Whalecraft*. New Bedford: Old Dartmouth Historical Society Whaling Museum, 1984.

MacAskie, Ian. *The Long Beaches: A Voyage in Search of the North Pacific Fur Seal*. Victoria: Sono Nis Press, 1979.

Macfie, Matthew. *Vancouver Island and British Columbia: Their History, Resources, and Prospects*. London: Longman, Green, Longman, Roberts, and Green, 1865; facsimile ed. Toronto: Coles Publishing Company, 1972.

Malham, John. *The Naval Gazetteer: or, Seaman's Complete Guide*. 2 vols. Boston: for W. Spotswood and J. Nancrede, 1797.

Manguin, André. *Trois Ans de Pêche de la Baleine: D'après le Journal de Pêche du Capitaine Dufour (1843–1846)*. Paris: J. Peyronnet et Cie., 1938.

Marchand, Etienne. *Voyage Autour du Monde Pendant les Années 1790–92*. 3 vols. Paris, [1798–1800].

Marshall, W. P. *Afloat on the Pacific, or Notes of Three Years Life at Sea: Comprising Sketches of People, Places, and Things Along the Pacific Coast and Among the Islands of Polynesia, Visited During Several Voyages of the U.S.S. Lancaster and Saranac.* Zanesville, Ohio: Sullivan and Parsons, 1876.

Maurelle, Francisco Antonio. "Journal of a Voyage in 1775, to Explore the Coast of America, Northward of California, By the Second Pilot of the Fleet, Don Francisco Antonio Maurelle, in the King's Schooner, called the Sonora, and commanded by Don Juan Francisco de la Bodega." In *Miscellanies*, translated and edited by Daines Barrington, 470–534. London: J. Nichols, 1781.

Maury, M[atthew] F[ontaine]. *Explanations and Sailing Directions to Accompany the Wind and Current Charts.* 3rd ed. Washington, D.C.: G. Alexander, 1851; 6th ed. Philadelphia: E.C. and J. Biddle, 1854.

McCann, Leonard G. *The Honourable Company's Beaver.* Vancouver: Vancouver Museums and Planetarium Association, 1980.

McCourt, Ben. *The Story of the Whale.* n.p.: Pacific Whaling Company, n.d.

Meares, John. *An Answer to Mr. George Dixon, Late Commander of the Queen Charlotte, in the Service of Messrs. Etches and Company; in Which the Remarks of Mr. Dixon on the Voyages to the North West Coast of America, &c, Lately Published, are Fully Considered and Refuted.* London: Logographic Press, 1791.

———. *Voyages Made in the Years 1788 and 1789, from China to the North West Coast of America: To Which are Prefixed, an Introductory Narrative of a Voyage Performed in 1786, from Bengal, in the Ship Nootka; Observations on the Probable Existence of a North West Passage; and Some Account of the Trade Between the North West Coast of America and China; and the Latter Country and Great Britain.* 1st ed. London: Logographic Press, 1790.

———. *Voyages Made in the Years 1788 and 1789, from China to N.W. Coast of America: With an Introductory Narrative of a Voyage Performed in 1786, from Bengal, in the Ship Nootka to Which are Annexed Observations on the Probable Existence of a North West Passage and Some Account of the Trade Between the North West of America and China; and the Latter Country and Great Britain.* 2nd ed. 2 vols. London: Logographic Press, 1791.

Melville, Herman. *Mardi: And a Voyage Thither.* London: Richard Bentley, 1849; New York: Harper and Brothers, 1849; reprint ed. Evanston, Illinois: Northwestern University Press and the Newberry Library, 1970.

———. *Moby-Dick: Or, the Whale.* New York: Harper and Brothers, 1851.

Michelet, Jules. *The Sea.* Translated by W. H. Davenport Adams. London: T. Nelson and Sons, 1875.

Middleton, Lynn. *Place Names of the Pacific Northwest Coast.* Victoria: Elldee Publishing Company, 1969.

Millais, J. G. *Newfoundland and Its Untrodden Ways.* London: Longmans, Green and Company, 1907.

Miller, Polly; and Leon Gordon Miller. *The Lost Heritage of Alaska: The Adventure and Art of the Alaskan Coastal Indians.* New York: Bonanza Books, 1967.

Minasian, Stanley M.; Kenneth C. Balcomb III; and Larry Foster. *The World's*

Whales: The Complete Illustrated Guide. Washington, D.C.: Smithsonian Books, 1984.

Mitchell, Edward. *Magnitude of Early Catch of East Pacific Gray Whale* (Eschrichtius robustus). Washington, D.C.: Center for Environmental Education, 1981.

————; Georgina Blaylock; and V.M. Kozicki. *Modifiers of Effort in Whaling Operations: With a Survey of Anecdotal Sources on Searching Tactics and Use of ASDIC in the Chase*. Washington, D.C.: Center for Environmental Education, 1981.

Morfit, Campbell. *A Treatise on Chemistry Applied to the Manufacture of Soap and Candles; Being a Thorough Exposition, in All Their Minutiae, of the Principles and Practice of the Trade, Based Upon the Most Recent Discoveries in Science and Art*. Philadelphia: Parry and McMillan, 1856.

Morley, F. V.; and J. S. Hodgson. *Whaling North and South*. London: Methuen and Company, 1927.

Moziño, José Mariano. *Noticias de Nutka: An Account of Nootka Sound in 1792*. Translated and edited by Iris Higbie Wilson. Toronto: McClelland and Stewart, 1970.

Murdoch, W. G. Burn. *Modern Whaling & Bear-Hunting*. Philadelphia: J. B. Lippincott Company, 1917.

Naval Medicine in the Early Nineteenth Century. Boston: USS Constitution Museum, 1981.

Nesbitt, Henrietta. *White House Diary*. Garden City, N.Y.: Doubleday and Company, 1948.

Newell, Gordon, ed. *The H.W. McCurdy Marine History of the Pacific Northwest*. Seattle: Superior Publishing Company, 1966.

————. *The H.W. McCurdy Marine History of the Pacific Northwest 1966 to 1976*. Seattle: Superior Publishing Company, 1977.

Niblack, Albert P. "The Coast Indians of Southern Alaska and Northern British Columbia." In *Annual Report of the Board of Regents of the Smithsonian Institution . . . And Report of the U.S. National Museum for the Year Ending June 30, 1888*, 225–386. Washington, D.C.: Government Printing Office, 1890.

Nicol, John. *The Life and Adventures of John Nicol, Mariner*. Edinburgh: William Blackwood; London: T. Cadell, 1822.

Oesau, Wanda. *Die Deutsche Südseefischerei auf Wale im 19. Jahrhundert*. Glückstadt: J. J. Augustin, 1939.

Ogden, Adele. *The California Sea Otter Trade, 1784–1848*. Berkeley and Los Angeles: University of California Press, 1941.

[Oliver, Simeon], with Alden Hatch. *Son of the Smoky Sea, by "Nutchuk."* New York: Julian Messner, 1941.

Olson, Ronald L. *The Quinault Indians*. University of Washington Publications in Anthropology 6. Seattle: University of Washington Press, 1936; reprint ed. 1967.

O'May, Harry. *Wooden Hookers of Hobart Town*. [Hobart Town, Tasmania]: L. G. Shea, n.d..

Ormsby, Margaret A. *British Columbia: A History*. Toronto: Macmillan of Canada, 1964.

Packard, Winthrop. *The Young Ice Whalers*. Boston: Houghton Mifflin and Company, 1903.

Pasquier, Thierry du. *Les Baleiniers Français au XIXe Siècle (1814–1868)*. Grenoble: Terre et Mer, 1982.

Pease, Zephaniah W., ed. *History of New Bedford*. 3 vols. New York: Lewis Historical Publishing Company, 1918.

Pennant, Thomas. *Introduction to the Arctic Zoology*. 2nd ed. London: Robert Faulder, 1792.

Perkins, Edward T. *Na Motu: Or Reef-Rovings in the South Seas*. New York: Pudney and Russell, 1854.

Perkins, John T. *John T. Perkins' Journal at Sea: 1845*. Mystic, Conn.: Marine Historical Association, 1934.

Pérouse, Jean F. G. de la. *A Voyage Round the World Performed in the Years 1785, 1786, 1787 and 1788 by the Boussole and Astrolabe*. 3 vols. London: A. Hamilton, 1799; facsimile ed. Amsterdam: N. Israel; New York: Da Capo Press, 1968.

Petersen, Kaare. *The Saga of Norwegian Shipping*. Oslo: Dreyers, 1955.

Pethick, Derek. *The Nootka Connection: Europe & the Northwest Coast 1790–1795*. Vancouver: Douglas and McIntyre, 1980.

Petit-Thouars, Abel du. *Voyages autour du Monde sur la Frégate La Vénus, Pendant les Années 1836–1839*. 10 vols. + 4 atlases. Paris: Gide, 1840–46, 1855.

Pierce, Richard A. *Alaskan Shipping, 1867–1878: Arrivals and Departures at the Port of Sitka*. Materials for the Study of Alaskan History 1. Kingston, Ontario: Limestone Press, 1972.

Portlock, Nathaniel. *A Voyage Round the World; But More Particularly to the North West Coast of America: Performed in 1785, 86, 87 and 88 in the King George & Queen Charlotte, Captain Portlock and Dixon*. London: printed for John Stockdale and George Goulding, 1789.

Proceedings of the National Conference for the Protection of American Interests. New York: Greeley and McElrath; Thaddeus B. Wakeman, 1842.

Reeves, John. *A History of the Law of Shipping and Navigation*. Dublin: Thomas Burnside, 1792.

Regehr, T. D. *The Canadian Northern Railway: Pioneer Road of the Northern Prairies 1895–1918*. Toronto: Macmillan Company of Canada, 1976.

Reilly, Robin. *Pitt the Younger*. Toronto: Cassell, 1978.

Reynolds, J. N. *Address, on the Subject of a Surveying and Exploring Expedition to the Pacific Ocean and South Seas*. New York: Harper and Brothers, 1836.

Rice, Dale W.; and Allen A. Wolman. *The Life History and Ecology of the Gray Whale* (Eschrichtius robustus). Special Publication 3. n.p.: American Society of Mammologists, [1974].

Rich, E. E., ed. *The Letters of John McLoughlin from Fort Vancouver to the Governor and Committee: Second Series, 1839–44*. Hudson's Bay Record Society 6. London: Hudson's Bay Record Society, 1943.

————. *The Letters of John McLoughlin from Fort Vancouver to the Governor and Committee: Third Series, 1844–46*. Hudson's Bay Record Society 8. London: Hudson's Bay Record Society, 1944.

Ricketson, Daniel. *The History of New Bedford, Bristol County, Massachusetts: Including a History of the Old Township of Dartmouth, and the Present Townships of Westport, Dartmouth and Fairhaven, from Their Settlement to the Present Time*. New Bedford: published by the author, 1858.

Risting, Sigurd. *Av Hvalfangstens Historie*. Kommandør Chr. Christensens Hvalfangstmuseum Publikation 2. Christiania: J. W. Cappelens, 1922.

Robertson, Elsie. *Where the Sea-Flag Floats: A Story of the Seal Fisheries*. Elgin, Ill.: David C. Cook Publishing Company, 1912.

Robotti, Frances Diane. *Whaling and Old Salem: A Chronicle of the Sea*. 1st ed. Salem, Mass.: Newcomb and Gauss, 1950; 2nd ed. New York: Fountainhead Publishers, 1962.

Rolt-Wheeler, Francis. *The Boy with the U.S. Fisheries*. Boston: Lothrop, Lee and Shepard Company, 1912.

Roquefeuil, Camille de. *Journal d'un Voyage autour du Monde*. Paris: Ponthieu, Lesage and Gide, 1823; reprint ed. *Voyage Around the World, and Trading for Sea Otter Fur on the Northwest Coast of America*. Fairfield, Wash.: Ye Galleon Press, 1981.

Roscoe, Theodore. *United States Submarine Operations in World War II*. Annapolis: United States Naval Institute, 1949.

Ruby, Robert H.; and John A. Brown. *Indians of the Pacific Northwest: A History*. Norman: University of Oklahoma Press, 1981.

Ruffner, James A. "Two Problems in Fuel Technology." In *History of Technology, Third Annual Volume 1978*, edited by A. Rupert Hall and Norman Smith, 123–61. London: Mansell, 1978.

Sampson, William R., ed. *John McLoughlin's Business Correspondence, 1847–48*. Seattle: University of Washington Press, 1973.

Sapir, Edward; and Morris Swadesh. *Nootka Texts: Tales and Ethnological Narratives*. Philadelphia: Linguistic Society of America, University of Pennsylvania, 1939.

Sarytschew, Gawrila. *Account of a Voyage of Discovery to the North-East of Siberia, the Frozen Ocean, and the North-East Sea*. 2 vols. London: J. G. Barnard, 1807.

Sauer, Martin. *An Account of a Geographical and Astronomical Expedition to the Northern Parts of Russia, for Ascertaining the Degrees of Latitude and Longitude of the Mouth of the River Kovima; of the Whole Coast of the Tshutski, to the East Cape; and of the Islands in the Eastern Ocean, Stretching to the American Coast; Performed . . . by Commodore Joseph Billings, in the Years 1785, Etc. to 1794*. London: A. Strahan, 1802.

Scammon, Charles Melville. *Journal Aboard the Bark* Ocean Bird *on a Whaling Voyage to Scammon's Lagoon, Winter of 1858–1859*. Edited by David A. Henderson. Baja California Travel Series 21. Los Angeles: Dawson's Book Shop, 1970.

———. *The Marine Mammals of the North-Western Coast of North America, Together with An Account of the American Whale-Fishery*. San Francisco: John H. Carmany and Company, 1874; reprint ed. New York: Dover Publications, 1968.

Schmitt, Frederick P. *Mark Well the Whale!: Long Island Ships to Distant Seas*. Port Washington, N.Y.: Kennikat Press, 1971.

———; Cornelis de Jong; and Frank H. Winter. *Thomas Welcome Roys: America's Pioneer of Modern Whaling*. Charlottesville: University Press of Virginia, 1980.

Scott, R. Bruce. *Barkley Sound: A History of the Pacific Rim National Park Area*. n.p.: published by the author, 1972.

Seed, Alice, ed. *Toothed Whales in Eastern North Pacific and Arctic Waters*. 2nd ed. Seattle: Pacific Search Press, 1971.

Seward, William H. *Commerce in the Pacific Ocean: Speech of William H. Seward, in the Senate of the United States, July 29, 1852*. Washington, D.C.: Buell and Blanchard, 1852.

Shortt, Adam; and Arthur G. Doughty, eds. *Canada and Its Provinces: A History of the Canadian People and Their Institutions by One Hundred Associates*. Vol. 22. Toronto: Publishers' Association of Canada, 1914.

Simmonds, P. L. *The Commercial Products of the Sea; or, Marine Contributions to Food, Industry, and Art*. New York: D. Appleton and Company, 1879.

Simpson, George. *Narrative of a Voyage to California Ports in 1841–42, together with Voyages to Sitka, the Sandwich Islands & Okhotsk, to Which are Added Sketches of Journeys Across America, Asia, & Europe* (1847). Edited by Thomas C. Russell. San Francisco: Thomas C. Russell, 1930.

————. *An Overland Journey Round the World, During the Years 1841 and 1842*. Philadelphia: Lea and Blanchard, 1847.

Smith, C. Fox. *Sailor Town Days*. Boston: Houghton Mifflin Company, 1923.

Somner, Graeme. *From 70 North to 70 South: A History of the Christian Salvesen Fleet*. Edinburgh: Christian Salvesen, 1984.

Spence, Bill. *Harpooned: The Story of Whaling*. New York: Crescent Books, 1980.

Stackpole, Edouard A. *The Sea-Hunters: The New England Whalemen During Two Centuries 1635–1835*. New York: J. B. Lippincott Company, 1953.

————. *Whales & Destiny: The Rivalry Between America, France, and Britain for Control of the Southern Whale Fishery, 1785–1825*. [Amherst]: University of Massachusetts Press, 1972.

Starbuck, Alexander. *The History of Nantucket: County, Island and Town*. Boston: C. E. Goodspeed and Company, 1924.

Stewart, Hilary. *Looking at Indian Art of the Northwest Coast*. Vancouver: Douglas and McIntyre, 1979.

Strange, James. *Records of Fort St. George: James Strange's Narrative of the Commercial Expedition from Bombay to the North-West Coast of America*. Introduction by A. V. Venkatarama Ayyar. Madras: Superintendent, Government Press, 1929.

Sullivan, Josephine. *A History of C. Brewer & Company Limited: One Hundred Years in the Hawaiian Islands 1826–1926*. Edited by K. C. Leebrick. Boston: Walton Advertising and Printing Company, 1926.

Swan, James G[ilchrist]. "The Indians of Cape Flattery, at the Entrance to the Strait of Fuca, Washington Territory." Article 8 in *Smithsonian Contributions to Knowledge* 16. Washington, D.C.: Smithsonian Institution, 1870.

————. *The Northwest Coast: Or, Three Years' Residence in Washington Territory*. New York: Harper and Brothers, 1857; facsimile ed. Seattle: University of Washington Press, 1972.

Swanton, John R. *Haida Texts and Myths: Skidegate Dialect*. Smithsonian Institution,

Bureau of American Ethnology Bulletin 29. Washington, D.C.: Government Printing Office, 1905.

Thiercelin, [Louis]. *Journal d'un Baleinier: Voyages en Océanie*. 2 vols. Paris: L. Hachette et Cie., 1866.

Thomas, Philip J. *Songs of the Pacific Northwest*. Saanichton, B.C.: Hancock House Publishers, 1979.

Tikhmenev, P. A. *A History of the Russian-American Company*. Translated and edited by Richard A. Pierce and Alton S. Donnelly. Seattle: University of Washington Press, 1978.

Tønnessen, J. N., and Arne Odd Johnsen, *The History of Modern Whaling*. Translated by R. I. Christophersen. Berkeley and Los Angeles: University of California Press; London: C. Hurst and Company; Canberra: Australian National University Press, 1982.

―――. *Den Moderne Hvalfangsts Historie: Opprinnelse og Utvikling*. 4 vols. Oslo: H. Aschehoug and Company, 1959; Sandefjord: Norges Hvalfangstforbund, 1967–70.

Tower, Walter S. *A History of the American Whale Fishery*. Series in Political Economy and Public Law 20. Philadelphia: University of Pennsylvania, 1907.

Townsend, Charles Wendell. *Along the Labrador Coast*. Boston: Dana Estes and Company, 1907.

True, Frederick W. "The Whalebone Whales of the Western North Atlantic: Compared with Those Occurring in European Waters with Some Observations on the Species of the North Pacific." In *Smithsonian Contributions to Knowledge* 33. Washington, D.C.: Smithsonian Institution, 1904.

United States, Coast Survey. *Coast Pilot of Alaska (First Part), from Southern Boundary to Cook's Inlet*, by George Davidson. Washington, D.C.: Government Printing Office, 1869.

―――, Commission of Fish and Fisheries. "Aquatic Products in Arts and Industries: Fish Oils, Fats, and Waxes; Fertilizers from Aquatic Products," by Charles H. Stevenson. In *Report of the Commissioner for the Year Ending June 30, 1902*. Part 28, 177–279. Washington, D.C.: Government Printing Office, 1904.

―――, Commission of Fish and Fisheries. "Bringing Whale Oil from the Pacific to New York," by Frederick Habershaw. In *Bulletin of the U.S. Fish Commission, Vol. III, for 1882*, 215. Washington, D.C.: Government Printing Office, 1883.

―――, Commission of Fish and Fisheries. "Commercial Fisheries of the Hawaiian Islands," by John N. Cobb. In *Report of the Commissioner for the Year Ending June 30, 1901*. Part 27, 381–499. Washington, D.C.: Government Printing Office, 1902.

―――, Commission of Fish and Fisheries. *The Fisheries and Fishery Industries of the United States*, by George Brown Goode. 7 vols. Congress 47:1 [124]. Washington, D.C.: Government Printing Office, 1884–87.

―――, Commission of Fish and Fisheries. "The Fisheries of the Pacific Coast," by William A. Wilcox. In *Report of the Commissioner for the Year Ending June 30, 1893*. Part 19, 143–304. Washington, D.C.: Government Printing Office, 1895.

————, Commission of Fish and Fisheries. "History of the American Whale Fishery: From its Earliest Inception to the Year 1876," by Alexander Starbuck. In *Report of the Commissioner for* 1875–1876. Part 4, Appendix A, 1–782. Washington, D.C.: Government Printing Office, 1876; reprint ed. New York: Argosy-Antiquarian, 1964.

————, Commission of Fish and Fisheries. "Report Submitted to the Department of the Interior on the Practical and Scientific Investigations of the Finmark Capelan-Fisheries, Made During the Spring of the Year 1879," by G.O. Sars, translated by Herman Jacobson. In *Report of the Commissioner for* 1880. Part 8, 167–87. Washington, D.C.: Government Printing Office, 1883.

————, Commission of Fish and Fisheries. "Notes Upon the History of the American Whale Fishery," by F. C. Sanford. In *Report of the Commissioner for* 1882, 205–91. Washington, D.C.: Government Printing Office, 1884.

————, Commission of Fish and Fisheries. "Report of the Division of Statistics and Methods of the Fisheries," by Hugh M. Smith. In *Report of the Commissioner for the Year Ending June* 30, 1893. Part 19, 52–77. Washington, D.C.: Government Printing Office, 1895.

————, Commission of Fish and Fisheries. "Report on the Work of the Fish Commission Steamer Albatross, for the Year Ending June 30, 1893," by Z.L. Tanner. In *Report of the Commissioner for the Year Ending June* 30, 1893. Part 19, 305–41. Washington, D.C.: Government Printing Office, 1895.

————, Commission of Fish and Fisheries. "Svend Foyn's Whaling Establishment." *Deutsche Fischerei-Zeitung* 6 (1883), translated in *Report of the Commissioner for* 1883. Part 11, 337–40. Washington, D.C.: Government Printing Office, 1885.

————, Department of Commerce, Bureau of Fisheries. "Alaska Fisheries and Fur Industries in 1916," by Ward T. Bower and Henry D. Aller. Appendix 2 in *Report of the United States Commissioner of Fisheries for* 1916. Document 838. Washington, D.C.: Government Printing Office, 1917.

————, Department of Commerce, Bureau of Fisheries. "Alaska Fisheries and Fur Industries in 1917," by Ward T. Bower and Henry D. Aller. Appendix 2 in *Report of the United States Commissioner of Fisheries for* 1917. Document 847. Washington, D.C.: Government Printing Office, 1918.

————, Department of Commerce, Bureau of Fisheries. "Alaska Fisheries and Fur Industries in 1918," by Ward T. Bower. Appendix 7 in *Report of the United States Commissioner of Fisheries for* 1918. Document 872. Washington, D.C.: Government Printing Office, 1919.

————, Department of Commerce, Bureau of Fisheries. "Alaska Fisheries and Fur Industries in 1919," by Ward T. Bower. Appendix 9 in *Report of the United States Commissioner of Fisheries for* 1919. Document 891. Washington, D.C.: Government Printing Office, 1920.

————, House of Representatives. *Letter from the Acting Secretary of the Treasury, Transmitting the Papers in the Claim of the North American Commercial Company for Transportation and Subsistence Afforded the Surviving Officers and Crew of the Whaling*

Bark Sea Ranger, Wrecked at Kayak Island May 26, 1893. Congress 53:2 [186]. Washington, D.C.: Government Printing Office, 1894.

————, House of Representatives. *Letter from the Secretary of War, in Relation to the Territory of Alaska*, by William W. Belknap. Congress 42:1 [5]. Washington, D.C.: Government Printing Office, 1871.

————, House of Representatives. *Military Posts —Council Bluffs to the Pacific Ocean*, by [Nathaniel G.] Pendleton. Congress 27:2 [830]. [Washington, D.C., 1842].

————, House of Representatives. "Report on the Population, Industries, and Resources of Alaska," by Ivan Petroff. In *Seal and Salmon Fisheries and General Resources of Alaska.* 4 vols., Vol. 4, 167–452. Congress 55:1 [92]. Washington, D.C.: Government Printing Office, 1898.

————, Department of the Interior. "Tribes of Western Washington and Northwestern Oregon," by George Gibbs. Part 2 in *Contributions to North American Ethnology*, Vol. 1, 157–361. Washington, D.C.: Government Printing Office, 1877.

————, Department of the Interior, Fish and Wildlife Service. *Fur Seal Industry of the Pribilof Islands*, by Francis Riley. Bureau of Commercial Fisheries Circular 275. Washington, D.C.: Government Printing Office, 1967.

————, Department of the Interior, Fish and Wildlife Service. *The Northern Fur Seal*, by Ralph C. Baker, Ford Wilke, and C. Howard Baltzo. Bureau of Commercial Fisheries Circular 169. Washington, D.C.: Government Printing Office, 1963.

————, Department of the Interior, Geological Survey. *Dictionary of Alaska Place Names*, by Donald J. Orth. Professional Paper 567. Washington, D.C.: Government Printing Office, 1971.

————, Department of the Navy, Naval History Division. *Civil War Naval Chronology 1861–1865.* Vol. 1. Washington, D.C.: Government Printing Office, 1971.

————, President. *Message from the President of the United States, Transmitting a Copy of the Convention Between the United States and the Emperor of Russia. Concluded at St. Petersburg on the 5th of April Last*, by James Monroe. Congress 18:2, 21 January 1825. Washington, D.C.: Gales and Seaton, 1825.

————, President. *Message from the President of the United States, Transmitting the Correspondence with the British Government, in Relation to the Boundary of the United States on the Pacific Ocean*, by John Quincy Adams. Congress 19:1 [65]. Washington, D.C.: Gales and Seaton, 1826.

————, Select Committee. *Report of the Select Committee, Appointed on the 29th of December Last, with Instructions to Inquire into the Expediency of Occupying the Mouth of the Columbia River*, by Thomas S. Jesup. Congress 18:1 [110]. Washington, D.C., 1824.

————, Senate. "Historical Review of the Formulation of the Russian-American Company, and Their Proceedings Up to the Present Time," by P. Tikhmenieff (St. Petersburg, 1863). Translated in *Proceedings of the Tribunal of Arbitration, Convened at Paris Under the Treaty Between the United States of America and Great*

Britain Concluded at Washington February 29, 1892, for the Determination of Questions Between the Two Governments Concerning the Jurisdictional Rights of the United States in the Waters of Bering Sea. 16 vols., Vol. 2, 130–39. Senate 53:2 [177]. Washington, D.C.: Government Printing Office, 1895.

————, Senate. *Memoir, Geographical, Political, and Commercial, on the Present State, Productive Resources and Capabilities for Commerce, of Siberia, Manchuria, and the Asiatic Islands of the Northern Pacific Ocean; and on the Importance of Opening Commercial Intercourse with Those Countries, &c,* by Aaron H. Palmer. Senate 30:1 [80]. Washington, D.C.: Tippin and Streeper, 1848.

————, Senate. *Memoir, Historical and Political, on the Northwest Coast of North America, and the Adjacent Territories,* by Robert Greenhow. Senate 26:1 [174]. Washington, D.C.: Blair and Rives, 1840.

————, Senate. *Message from the President of the United States in Response to Senate Resolution of January 8, 1895, Transmitting Information Relating to the Enforcement of the Regulations Respecting Fur Seals.* Senate 53:3 [67]. Washington, D.C.: Government Printing Office, 1895.

————, Senate. *Proceedings of the Tribunal of Arbitration, Convened at Paris Under the Treaty Between the United States of America and Great Britain Concluded at Washington February 29, 1892, for the Determination of Questions Between the Two Governments Concerning the Jurisdictional Rights of the United States in the Waters of Bering Sea.* 16 vols. Senate 53:2 [177]. Washington, D.C.: Government Printing Office, 1895.

————, Department of the Treasury. *The Fur Seals and Fur-Seal Islands of the North Pacific Ocean,* by David Starr Jordan. Document 2017. 4 parts. Washington, D.C.: Government Printing Office, 1898–99.

————, Works Progress Administration. *Whaling Masters.* New Bedford: Old Dartmouth Historical Society, 1938.

Vamplew, Wray. *Salvesen of Leith.* Edinburgh and London: Scottish Academic Press, 1974.

Van Arsdell, Jon. "B.C. Whaling: The Indians." In *Raincoast Chronicles First Five,* 20–28. Madeira Park, B.C.: Harbour Publishing, 1976.

Van Note, Craig. *Outlaw Whalers: An Exposé of Unregulated Whaling Around the World.* Washington, D.C.: Whale Protection Fund, 1979.

Vancouver, George. *A Voyage of Discovery to the North Pacific Ocean and Round the World 1791–1795.* Edited by W. Kaye Lamb. 4 vols. London: Hakluyt Society, 1984.

Victoria, Corporation of the City of. *Victoria Illustrated: Containing a General Description of the Province of British Columbia, and a Review of the Resources, Terminal Advantages, General Industries, and Climate of Victoria, the "Queen City," and its Tributary Country.* Victoria: Ellis and Company, 1891.

Villiers, A. J. *Whaling in the Frozen South: Being the Story of the 1923–24 Norwegian Expedition to the Antarctic.* Indianapolis: Bobbs-Merrill Company, 1925.

Walbran, John T. *British Columbia Place Names 1592–1906, to Which are Added a Few Names in Adjacent United States Territory; Their Origin and History.* Vancouver: J. J. Douglas, 1971.

Wallace, Frederick William. *In the Wake of the Wind-Ships: Notes, Records and Biographies Pertaining to the Square-Rigged Merchant Marine of British North America*. New York: George Sully and Company, 1927.

[Ward, James]. *Perils, Pastimes, and Pleasures of an Emigrant in Australia, Vancouver's Island and California*. London: Thomas Cautley Newby, 1849.

Wardman, George. *A Trip to Alaska: A Narrative of What Was Seen and Heard During a Summer Cruise in Alaskan Waters*. San Francisco: Samuel Carson and Company, 1884.

Waterman, T. T. *The Whaling Equipment of the Makah Indians*. University of Washington Publications in Anthropology 1. Seattle: University of Washington Press, 1920; reprint ed. 1955.

Watson, Lyall. *Sea Guide to Whales of the World*. New York: E.P. Dutton, 1981.

Webb, Robert Lloyd. "The American Whaleman as a Deterrent to Pacific War, 1845–47." In *Proceedings* of the International Commission for Maritime History/ North American Society for Oceanic History Joint Meeting (1987) (forthcoming).

————. *West Whaling: A Brief History of Whale-Hunting in the Pacific Northwest*. Vancouver: Vancouver Maritime Museum, 1984.

Webster, Peter S. *As Far as I Know: Reminiscences of an Ahousat Elder*. Campbell River, B.C.: Campbell River Museum and Archives, 1983.

Weiss, Harry B.; Howard R. Kemble; and Millicent T. Carre. *Whaling in New Jersey*. Trenton: New Jersey Agricultural Society, 1974.

Whipple, A. B. C. *The Whalers*. Alexandria, Va.: Time-Life Books, 1979.

Wilkes, Charles. *Narrative of the United States Exploring Expedition During the Years 1838, 1839, 1840, 1841, 1842*. 5 vols. Philadelphia: Lea and Blanchard, 1845.

Williams, E. C. *Life in the South Seas: History of the Whale Fishery; Habits of the Whale; Perils of the Chase and Methods of Capture*. New York: Polhemus and DeVries, 1860.

Wise, Terence. *To Catch a Whale*. London: Geoffrey Bles, 1970.

Wright, E.W., ed. *Lewis & Dryden's Marine History of the Pacific Northwest*. [Portland, Ore.]: Lewis and Dryden Printing Company, 1895; facsimile ed. New York: Antiquarian Press, 1961.

Articles and Contributions to Serials

Anderson, Stuart. "British Threats and the Settlement of the Oregon Boundary Dispute." *Pacific Northwest Quarterly* 66 (October 1975): 153–60.

Andrews, Clarence L. "Alaska Whaling." *Washington Historical Quarterly* 9 (January 1918): 3–10.

Andrews, Roy Chapman. "American Museum Whale Collection." *American Museum Journal* 14 (December 1914): 275–94.

————. "Shore-Whaling: A World Industry." *National Geographic* 22 (May 1911): 411–42.

————. "Whale-Hunting As It Is Now Done." *World's Work* (December 1908): 11031–47.

Arestad, Sverre. "The Norwegians in the Pacific Coast Fisheries." *Pacific Northwest Quarterly* 34 (January 1943): 3–17.

Bach, John. "The Imperial Defense of the Pacific Ocean in the Mid-Nineteenth Century: Ships and Bases." *American Neptune* 32 (October 1972): 233–46.

Barthelmess, Klaus. "Deutsche Walfanggesellschaften in Wilhelminischer Zeit: Germania AG und Sturmvogel GmbH." *Deutsches Schiffahrtsarchiv* 9 (1986): 227–50.

————. "Julius Tadsen—Manager des Walfanges: Die 'Germania Walfang-und Fischindustrie Aktiengesellschaft' zu Hamburg." *Nordfriesland Tageblatt* (5 October 1981).

Bean, Margaret. "Whale Hunters of the Olympic Peninsula." *Travel* (June 1931): 23–25.

Bettum, Frithjof. "Position of the Whaling Industry at the Turn of the Year." *Norsk Hvalfangst-Tidende* 46 (January 1957): 14–19.

Bigg, Michael A.; and Allen A. Wolman. "Live-Capture Killer Whale (*Orcinus orca*) Fishery, British Columbia and Washington, 1962–73." *Journal of the Fisheries Research Board of Canada* 32 (July 1975): 1213–21.

Biggins, Patricia. "Doughnuts in the Tryworks: A Child's Life Aboard the *Charles W. Morgan*." *Log of Mystic Seaport* 27 (May 1975): 8–16.

Blue, George Vern. "France and the Oregon Question." *Oregon Historical Quarterly* 34 (March 1933): 39–59.

Bockstoce, John R. "History of Commercial Whaling in Arctic Alaska." *Alaska Geographic* 5 (1978): 17–26.

Bonaparte, R. "Whale Fishery off the Coast of Norway." *Scientific American* 52 (16 May 1885): 1.

Boone, Andrew R. "Wings for Whale Shooters." *Popular Mechanics Magazine* 52 (September 1929): 353–55.

Boutilier, Helen R. "Vancouver's Earliest Days." *British Columbia Historical Quarterly* 10 (1946): 151–86.

Brett, Oswald L. "Charles Robert Patterson: The Sailorman's Painter." *Sea History* 30 (Winter 1983): 12–17.

Brown, Beriah, Jr. "Quinaielt Hunters, Part 2," *Youth's Companion*, no. 3572 (7 November 1895): 569–70.

Buehler, Jack. "The Whaling Epoch in Humboldt County, Part II." *Ship's Log* (Humboldt Bay Maritime Museum) 8 (May-June 1986): 6–8.

Burgess, Alan. "Now the Whales Die Quickly." *True* 33 (November 1953): 50–53.

"Byzantium." *Underwater Archaeological Society of B.C. Newsletter* (January 1986): 3–4.

Canada, Fisheries Research Board. "Whale Fishery Study." By Gordon C. Pike. In *Progress Reports of the Pacific Coast Stations* 75 (July 1948): 47.

————. "Whale Identification." By Gordon C. Pike. In *Progress Reports of the Pacific Coast Stations* 81 (December 1949): 84–86.

————. "Whaling Investigation." By Gordon C. Pike. In *Progress Reports of the Pacific Coast Stations* 79 (July 1949): 30–31.

"Capture of a Fifty Foot Whale in Puget Sound, Washington." *Scientific American* 76 (30 January 1897): 73–74.

Chandler, Charles Lyon. "List of United States Vessels in Brazil, 1792–1805, Inclusive." *Hispanic American Historical Review* 26 (November 1946): 599–617.

Cobb, John N. "Overcoming Waste in the Fishing Industry." *Pacific Fisherman* 21 (May 1923): 15.

————. "Pacific Coast Fishing Methods." *Pacific Fisherman Year Book* (January 1916): 19–33.

Cocks, Alfred Heneage. "Additional Notes on the Finwhale Fishery on the North European Coast." *Zoologist* 10 (April 1885): 1–10.

————. "The Finwhale Fishery of 1885 on the North European Coast." *Zoologist* 11 (April 1886): 1–16.

————. "The Finwhale Fishery of 1886 on the Lapland Coast." *Zoologist* 12 (June 1887): 1–16.

————. "The Finwhale Fishery on the Coast of Finmarken." *Zoologist* 9 (1884): 1–22.

"Commercial Legislation of France." *Fisher's National Magazine and Industrial Record* (New York) 3 (October 1846): 397–413.

Crawford, Richard W. "The Whalemen of San Diego Bay." *Mains'l Haul* (Maritime Museum Association of San Diego) 19 (Winter 1983): 3.

Davidson, Donald C. "Relations of the Hudson's Bay Company with the Russian American Company on the Northwest Coast, 1829–1867." *British Columbia Historical Quarterly* 5 (1941): 33–88.

Efrat, Barbara S. "The Hesquiat Project: Research in Native Indian Aural History." *Sound Heritage* 4, no. 3–4 (1976): 6–11.

"Electric Whaling." *Discovery* 13 (May 1952): 143–44.

Freidel, Frank. "A Whaler in Pacific Ports, 1841–42." *Pacific Historical Review* 12 (December 1943): 380–90.

Galbraith, John S. "James Edward Fitzgerald Versus the Hudson's Bay Company: The Founding of Vancouver Island." *British Columbia Historical Quarterly* 16 (1952): 191–207.

Geiling, E. M. K.; and G. B. Wislocki. "The Anatomy of the Hypophysics of Whales." *Anatomical Records* 66 (1936), 17–41.

Gibson, James R. "Russian America in 1833: The Survey of Kirill Khlebnikov." *Pacific Northwest Quarterly* 63 (January 1972): 1–13.

Gilbert, Benjamin F. "Rumours of Confederate Privateers Operating in Victoria, Vancouver Island." *British Columbia Historical Quarterly* 18 (July-October 1954): 239–55.

Gilmore, Raymond M. "Census and Migration of the California Gray Whale." *Norsk Hvalfangst-Tidende* 49 (September 1960): 409–31.

————. "Rare Right Whale Visits California." *Pacific Discovery* 9 (July-August 1956): 20–25.

Godden, R. R. "Sealing Voyage—1897." *Bulletin* (Maritime Museum of British Columbia) 44 (Summer 1979): 21–23.

"Gordon C. Pike, 1922–1968." *Journal of Mammalogy* 51 (May 1970): 434–35.

Granberg, W. J. "There Go Flukes!" *Beaver* 281 (March 1951): 38–41.

Gunther, Erna. "Reminiscences of a Whaler's Wife." *Pacific Northwest Quarterly* 33 (January 1942): 65–69.

Harmer, Sidney F. "The History of Whaling." *Proceedings of the Linnean Society of London*, 140th Session (1928): 51–95.

"Have a Whale?" *Business Week* (10 July 1943): 48.

Heater, William. "Victoria's Whaling Fleet." *Canadian Fisherman* 22 (November 1935): 22.

Hines, Clarence. "Adams, Russia and Northwest Trade, 1824." *Oregon Historical Quarterly* 36 (December 1935): 348–58.

Hirata, Moriso. "Experimental Investigation on Flattened Head Harpoon: An Attempt for Restraining Ricochet." *Scientific Reports of The Whales Research Institute* (Tokyo) 6 (December 1951): 199–207.

Hjort, Johan. "Human Activities and the Study of Life in the Sea." *Geographical Review* 25 (October 1935): 529–64.

———. "The Story of Whaling: A Parable of Sociology." *Scientific Monthly* 45 (July 1937): 19–34.

———; Gunnar Jahn; and Per Ottestad. "The Optimum Catch." *Hvalrådets Skrifter: Scientific Results of Marine Biological Research* 7 (1933): 92–127.

———; J. Lie; and Johan T. Ruud. "Norwegian Pelagic Whaling in the Antarctic, VI: The Season 1935–1936." *Hvalrådets Skrifter: Scientific Results of Marine Biological Research* 14 (1937): 5–45.

Howay, Frederic W. "Early Relations Between the Hawaiian Islands and the Northwest Coast." *Publications of the Archives of Hawaii* 5 (1930): 11–38.

———; and T. C. Elliott, eds. "Reprint of Boit's Log of the *Columbia* 1790–1793: Remarks on the Ship *Columbia*'s voyage from Boston (on a Voyage Round the Globe)." *Oregon Historical Quarterly* 22 (September 1921): 265–351.

Hustwick, Alfred. "The War on the Whale." *British Columbia Magazine* 7 (December 1911): 1246–54; 8 (January 1912): 7–12.

Illerbrun, W. J. "A Selective Survey of Canadian-Hawaiian Relations." *Pacific Northwest Quarterly* 63 (July 1972): 87–103.

"Den Japanske Hvalfangst-industri." *Norsk Hvalfangst-Tidende* 43 (November 1954): 385–91.

Johnsen, Arne Odd. "Causation Problems of Modern Whaling." *Norsk Hvalfangst-Tidende* 36 (August 1947): 281–94.

———. "Granatharpunen: En Kort Utredning om Hvordan Svend Foyn Løste Projektil-problemet." *Norsk Hvalfangst-Tidende* 29 (September 1940): 222–41.

Jones, A.G.E. "The British Southern Whale and Seal Fisheries; Part II: The Principal Operators." *Great Circle* 3 (October 1981): 90–102.

Kean, A.D. "Whale Hunting off the Coast of British Columbia." *Industrial Progress and Commercial Record* (Vancouver) 4 (November 1916): 125–28.

Kirk, Ruth. "The Pompeii of the Northwest." *Historic Preservation* 32 (March/April 1980): 2–9.

Knaplund, Paul. "James Stephen on Granting Vancouver Island to the Hudson's Bay Company, 1846–1848." *British Columbia Historical Quarterly* 9 (1945): 259–62.

Kool, Richard. "Northwest Coast Indian Whaling: New Considerations." *Canadian Journal of Anthropology* 3 (Fall 1982): 31–44.

Lagen, William S. "Lake Washington Remembered." *Puget Soundings* (November 1977): 22–23.

Lamb, W. Kaye. "British Columbia Official Records: The Crown Colony Period." *Pacific Northwest Quarterly* 29 (January 1938): 17–25.

————. "Diary of Robert Melrose, Part 1," *British Columbia Historical Quarterly* 7 (1943): 119–34.

————. "Five Letters of Charles Ross, 1842–44." *British Columbia Historical Quarterly* 7 (1943): 103–18.

————. "Founding of Fort Victoria." *British Columbia Historical Quarterly* 7 (1943): 71–92.

"Last Gam, The: The Old Whale Hunter Passes." *National Fisherman Yearbook* 57 (1977): 114–16.

Latham, Charles L. "Whaling in Southern Chile." *Norsk Hvalfangst-Tidende* 2 (22 November 1913): 139.

Laut, Agnes C. "Who Wants a Whale Steak?: The Exciting Enterprise of Catching a Whale." *Mentor* (September 1928): 33–35.

Litten, Jane. "Greenhand Hero or Mutineer?: Mutiny Aboard the Whaleship *Meteor* 1846." *Log of Mystic Seaport* 39 (Summer 1987): 54–64.

McDonald, Lucile. "Whaling on the Washington Coast." *Sea Chest* 6 (September 1972): 3–12.

M'Gonigle, R. Michael. "The 'Economizing' of Ecology: Why Big, Rare Whales Still Die." *Ecology Law Quarterly* 9 (1980): 120–237.

McGrath, P. T. "Wonderful Whale-Hunting by Steam." *Cosmopolitan* 37 (May 1904): 49–56.

McKelvie, B. A. "When Washington Irving Slept." *Beaver* 289 (Summer 1958): 3–8.

Macpherson, Donald B. "Discovery: 1917." *British Columbia Historical News* 16, no. 4: 14–17.

Madison, Peter. "British Columbia Whaling Venture." *Canadian Fisherman* 35 (September 1948): 22.

Merilees, Bill. "Humpbacks in Our Strait." *Waters* (Vancouver Aquarium), 8 (1985): 7–24.

Merriam, Paul G. "Riding the Wind: Cape Horn Passage to Oregon, 1840s-1850s." *Oregon Historical Quarterly* 77 (March 1976): 37–60.

Miller, Beatrice D. "Neah Bay: The Makah in Transition." *Pacific Northwest Quarterly* 42 (October 1952): 262–72.

Mitchell, Edward. "The Status of the World's Whales." *Nature Canada* 2 (October-December 1973): 9–27.

Mizue, Kazuhiro. "Factory Ship Whaling Around Bonin Islands in 1948." *Scientific Reports of The Whales Research Institute* (Tokyo) 3 (February 1950): 106–18.

Moment, David. "The Business of Whaling in America in the 1850's." *Business History Review* 31 (Autumn 1957): 259–91.

Monty, T. J. "Norway's Share of the Whaling Industry." *Commercial Intelligence Journal* (30 September 1939): 627.

Morgan, Lael. "Modern Shore-Based Whaling." *Alaska Geographic* 5 (1978): 34–43.

Nemoto, Takahisa. "Foods of Baleen Whale in the Northern Pacific." *Scientific Reports of The Whales Research Institute* (Tokyo) 12 (June 1957): 33–89.

Nicol, C.W. "The Whaling Controversy." *Whaling Review* (London), no. 20 (ca. 1983): 1–8.

Nishiwaki, Masaharu. "One Eyed-Monster of Fin Whale." *Scientific Reports of The Whales Research Institute* (Tokyo) 12 (June 1957): 193–95.

"Nylands Verksted." *Norsk Hvalfangst-Tidende* 26 (June 1937): 185–90.

"Nylands Verksted 1854–1954." *Norsk Hvalfangst-Tidende* 43 (August 1954): 295–97.

Old Dartmouth Historical Society. "A Lad Before the Wind." *Bulletin from Johnny Cake Hill* (Fall 1987): 2.

Orr, Robert T. "The Distribution of the More Important Marine Mammals of the Pacific Ocean, as it Affects Their Conservation." *Proceedings of the Sixth Pacific Science Conference* [1939]: 217–22.

Pike, Gordon C. "Whaling on the British Columbia Coast." *Proceedings of the Seventh Pacific Science Congress* (1949), 4 (1953); reprinted in *Studies* 1954 (Canada, Fisheries Research Board): 370–72.

———. "Whaling on the Coast British Columbia." *Norsk Hvalfangst-Tidende* 43 (March 1954): 69–79.

Plutte, Will. "The Whaling Imperative: Why Norway Whales." *Oceans* 17 (March-April 1984): 24–26.

"Prodigious Panorama, The." *American Heritage* 12 (December 1960): 55–62.

Quimby, George I. "James Swan Among the Indians: The Influence of a Pioneer from New England on Coastal Indian Art." *Pacific Northwest Quarterly* 61 (October 1970): 212–16.

Rabot, Charles. "The Whale Fisheries of the World." *La Nature* (14 September 1912), translated in *Annual Report of the Board of Regents of the Smithsonian Institution... for the Year Ending June 30, 1913.* Washington, D.C.: Government Printing Office, 1914: 481–89.

Reagan, Albert B. "Archaeological Notes on Western Washington and Adjacent British Columbia." *Proceedings of the California Academy of Sciences* 7 (18 July 1917): 1–31.

Reid, R. L. "The Whatcom Trails to the Fraser River Mines in 1858, Part 2," *Washington Historical Quarterly* 17 (October 1926): 271–76.

Rice, Dale W.; and Clifford H. Fiscus. "Right Whales in the Southeastern North Pacific." *Norsk Hvalfangst-Tidende* 57 (October 1968): 105–07.

Risting, Sigurd. "Den Norske Hvalfangst: Gjennem 50 Aar." *Norsk Hvalfangst-Tidende* Supplement (1914): 63–73.

Roe, Michael. "Australia's Place in 'The Swing to the East,' 1788–1810." *Historical Studies Australia and New Zealand* 8 (May 1958): 202–13.

Sapir, Edward. "The Rival Whalers: A Nitinat Story." *International Journal of American Linguistics* 3 (1924): 76–102.

Schafer, Joseph. "Letters of Sir George Simpson, 1841–1843." *American Historical Review* 14 (October 1908): 70–94.

733 [pseud.] "Whaling—and Some Whalers." *Canadian Merchant Service Yearbook* (1926): 6.

Statile, D. Richard. "I Sailed the Antarctic Killer Boats." *Argosy* 335 (November 1952): 24–25, 63.

Stevenson, Charles H. "Whale Oil." *Scientific American Supplement* 57 (5 March 1904): 23549–52.

Taylor, Herbert; and James Bosch. "Makah Whalers." *Carnivore* 2, Part 3 Special Supplement (November 1979): 10–15.

Townsend, Charles Haskins. "The Distribution of Certain Whales as Shown by Logbook Records of American Whaleships." *Zoologica* 19 (3 April 1935): 1–50.

———. "Twentieth Century Whaling." *Bulletin of the New York Zoological Society* 33 (January-February 1930): 1–31.

Wallace, Frederick William. "The Old Time Whalers of British North America." *Canadian Fisherman* 19 (June 1932): 17–20; 19 (July 1932): 17–20.

Webb, Robert Lloyd. "Connecticut Yankees in Queen Victoria's Fort: New London Whalemen on the Northwest Coast." *Log of Mystic Seaport* 39 (Summer 1987): 43–53.

———. "A Feast of Whales." *Horizon Canada* 6 (1986): 1472–77; published simultaneously as "Les Géants de la Mer."

———. "Invented Too Late: The Introduction of Steam to the Arctic Whaling Fleet." *American Neptune* 44 (Winter 1984): 11–21.

———. "Whale Hunters of the Northwest Coast." *Whalewatcher* (American Cetacean Society) 16 (Fall 1982): 3–5.

"Whale Fisheries of Today, The." *Nation* 80 (23 March 1905): 226.

"Whale Fishery on the Coast of Japan." *Ballou's Pictorial Drawing Room Companion* 9 (1 December 1855): 340.

"Whale Meat as a Food." *Scientific American* 105 (15 July 1911): 56–57.

"Whale Stocks at Coal Harbour, The." *Norsk Hvalfangst-Tidende* 44 (October 1955): 596–98.

"Whaling, 1925." *Canadian Merchant Service Guild Yearbook* (1926): 23.

"Whaling on the North Pacific," *Canadian Merchant Service Guild Yearbook* (1923): 23.

Winston, Waldon C. "The Largest Whale Ever Weighed." *Natural History* 59 (November 1950): 393–99.

Winter, Frank H.; and Frederick P. Schmitt. "Captain Thomas Welcome Roys: America's First Scientific Whaler." *Oceans* 8 (May 1975): 34–39.

————; and Mitchell R. Sharpe. "The California Whaling Rocket and the Men Behind It." *California Historical Quarterly* 50 (December 1971): 349–62.

MANUSCRIPT LOGBOOKS AND JOURNALS

Sailing Whaleships

Abigail, ship of New Bedford, William H. Reynard, master and keeper; 24 October 1835 - 22 October 1838. MSM (96).

Abigail, ship of New Bedford, David Barnard, master and keeper; 27 November 1843 - 26 July 1847. KWM (360).

Acushnet, ship of Fairhaven, Valentine Pease, master; 1841–45, abstract of voyage only. Maury 12, no. 13.

Adele, ship of Havre, --- Luhrs, master; 1842–43, abstract of whaling cruises. FNA CC⁵611.

Ajax, ship of Havre, Joseph Marie Letellier, master; 1838–40, abstract of whaling cruises. FNA CC⁵611.

Alexander, ship of Nantucket, George B. Chase, master; 18 August 1821 - 3 May 1824. KWM (10).

Alexander Barclay, ship of Bremen, Clement Norton, master; Levi C. Fisher (second officer), keeper; 3 January 1840 - 25 December 1840. KWM (11).

American, ship of Nantucket, David Baker, master and keeper; 4 July 1838 - 25 July 1841. NHA.

Andrew Hicks, bark of New Bedford, Charles A. Church, master; Charlotte E. Church, keeper; 6 December 1905 - 1 June 1906. KWM (607).

Angelina, ship of Havre, Edouard Sebastian L'hynne, master; 1841–42, abstract of whaling cruises. FNA CC⁵611.

Armata, ship of New London, J. Fitch, master; 24 August 1846 - 6 June 1848. KWM (21).

Beaver, bark of Hudson, --- Rogers, master; 31 August 1840 - 15 March 1842. RP Wh/B386/18401.

Benjamin Tucker, ship of New Bedford, Albert A. Barber, master; Sina Stevens, keeper; 28 February 1857 - 22 August 1858. MSM MR 90.

Braganza, ship of New Bedford, Charles C. Waterman, master; Robert Kline, keeper; 1 August 1843 - 6 May 1846. ODHS (550).

Cabinet, ship of Stonington, --- Noyes, master; Theophilus Brown, keeper; 21 April 1843 - 21 February 1845. KWM (566).

Cachalot, ship of Havre, Louis Frederic Mauger, master; 1841–43, abstract of whaling cruises. FNA CC⁵611.

Callao, ship of New Bedford, James A. Norton, master; 1 November 1842 - 16 June 1845. KWM (41).

Canton, ship of New Bedford, Abraham Gardner, master; David Wordell, keeper; 25 October 1834 - 20 May 1838. KWM (252).

Caroline, ship of New Bedford, Daniel McKenzie, master; George W.R. Bailey (green hand), keeper; 17 December 1842 - 22 February 1844. KWM (596).

Charles, ship of New Bedford, Barzilla Morselander, master; George Snow Cleveland, keeper; 1 December 1837 - 4 August 1839. KWM (339).

Charles Phelps, ship of Stonington, Palmer Hall, master; Charles W. Austin (boatsteerer), keeper, 27 August 1842 - 24 May 1843. ODHS (899).

Charles Phelps, ship of Stonington, Palmer Hall, master; 29 August 1842 - 1 April 1844. ODHS (1041C).

Charles Phelps, ship of Stonington, Gilbert Pendleton, Jr., master; 25 June 1844 - 15 April 1847. KWM (368).

Citizen, ship of Sag Harbor, --- Lansing, master; 1844–46, 1846–49, abstracts of voyages only. Maury 78, no. 25.

Columbus, ship of Fairhaven, Frederick Fish, master; William Wyse (boatsteerer), keeper; 23 November 1843 - 27 August 1844. PMS 656/1844C.

Congress, bark of Groton [Mystic], Austin M. Lester, master; 12 October 1844 - 7 December 1846. ODHS (1068B).

Constance, ship of Havre, Narcisse Chaudière, master; 25 October 1832 - 5 August 1834. FNA 5JJ354.

Contest, ship of New Bedford, Jeremiah Ludlow, master; James A. Stubbs, keeper; 24 October 1856 - 13 October 1860. MSM (169).

Coral, bark of San Francisco, Rodolphus Delano Wicks, master and keeper; 14 April 1889 - 6 October 1889. KWM (470).

Cowper, ship of New Bedford, John K. Hathaway, master and keeper; 3 June 1845 - 24 September 1848. KWM (64).

Dartmouth, ship of New Bedford, Abraham W. Pierce, master and keeper; 1 June 1848 - 20 March 1851. KWM (549).

Dartmouth, ship of New Bedford, Nathan B. Heath, master; Ephraim A. Chapman, keeper; 19 November 1854 - 22 August 1858. KWM (572).

Drymo, bark of New Bedford, John S. Taber, master and keeper; 28 August 1844 - 21 September 1845. KWM (71).

Edward, ship of New Bedford, John S. Barker and Nathan B. Heath, masters; 15 July 1845 - 5 April 1849. ODHS (551).

Elbe, ship of Poughkeepsie, Josiah B. Whippey, master; 6 September 1833 - 26 March 1837. IMA (A105).

Elisa, ship of Havre, --- Malherbe, master; 1841–42, abstract of whaling cruises. FNA CC⁵611.

Eliza Adams, ship of New Bedford, Francis C. Smith, master; John Jones and Clothier Peirce, Jr., keepers; 1 January 1852 - 22 November 1854. KWM (319).

Emily Morgan, ship of New Bedford, Benjamin Dexter, master; 8 September 1866 - 13 June 1868. NBFPL.

Emma F. Herriman, bark of San Francisco, Eliel T. Fish, master; A. Judson James, keeper; 4 December 1890 - 24 October 1891. KWM (603).

Endeavour, ship of New Bedford, Ebenezer J. Stetson, master; Edward Collins, keeper; 13 May 1835 - 25 April 1837. KWM (81).

Endeavour, ship of New Bedford, Ebenezer J. Stetson, master; Willie A. Hersey, keeper; 11 November 1835 - 26 April 1837. KWM (271).

Eugene, ship of Stonington, Gurdon Pendleton, master; John A. States, keeper; 15 July 1844 - 29 September 1845. MSM (69).

Europa, ship of Bremen, Ezra Fitch, master and keeper; 30 May 1842 - 30 April 1845. KWM (333).

Europa, ship of Edgartown, Thomas Mellen, master; George Wilbur Piper, keeper; manuscript notebook probably kept during the voyage of 29 August 1866 - 17 August 1872. KWM (A-194).

Fabius, ship of New Bedford, Lyman Wing, master; [27 July 1854 - 17 March 1857], crew list only. KWM.

Faune, ship of Havre, O. de Grandsaigne, master; 1841–43, abstract of whaling cruises. FNA CC⁵611.

Favorite, bark of Fairhaven, Henry T. Smith, master; John E. Perkins, keeper; 11 November 1857 - 4 June 1861. KWM (512).

Fortune, bark of New Bedford, Thomas S. Bailey, master; Hiram King, keeper; 17 November 1844 - 17 March 1847. KWM (375).

Gange, ship of Havre, Narcisse Chaudière, master; 5 September 1834 - 7 April 1837. FNA 5JJ355.

Gange, ship of Havre, Pierre Francis Phobin, master; 1839–40, abstract of whaling cruises. FNA CC⁵611.

Général Teste, ship of Havre, Louis Jean Baptiste Morin, master; 1847–48, abstract of whaling cruises. FNA CC⁵611.

Gideon Howland, ship of New Bedford, Arthur Cox, master and keeper; 26 October 1842 - 10 November 1844. KWM (231).

Golconda, ship of New Bedford, Francis Dougherty, master; James S. Taber (first officer), H. Johnson (first officer), and Charles W. Pratt (first officer), keepers; 31 July 1851 - 11 April 1855. KWM (244).

Good Return, ship of New Bedford, John S. Taber, master and keeper; 21 October 1841 - 31 December 1843. KWM (97).

Gratitude, ship of New Bedford, Peleg H. Stetson, master; Joseph E. Eayrs, keeper; 25 April 1841 - 7 April 1845. KWM (433).

Hibernia, ship of New Bedford, N. P. Simmons, master; Henry Eldredge, keeper; 15 June 1844 - 19 May 1846. PMS 656/1844M.

India, ship of New Bedford, Charles W. Gelett, master; 1840–43, abstract of voyage only. Maury 71, no. 5.

Israel, ship of New Bedford, James Finch, master; Daniel Kimball Ritchie (second mate), keeper; 5 December 1843 - 21 February 1845. KWM (478).

Java, ship of Fairhaven, Jarvis Wood, master; 16 May 1854 - 20 September 1856. SAG.

Java, ship of New Bedford, William Shockley, master; [10 June 1841 - 22 April 1843], list of whales seen and taken only. KWM.

Josephine, bark of San Francisco, John McInnes, master; Mrs. John McInnes, keeper; 16 December 1891 - 22 June 1892. KWM (253).

Julian, ship of New Bedford, Cyrus Taber, master; Frederick Cady (green hand), keeper; 20 June 1847 - 30 October 1850. KWM (404).

Julian, ship of New Bedford, Samuel P. Winegar, master and keeper; 11 October 1858 - 2 January 1860. Yale.

Kutusoff, ship of New Bedford, William H. Cox, master; 1842—45, abstract of voyage only. Maury 12, no 1.

Leonidas, ship of Bristol, Nelson Waldron, master; Benjamin L. West, keeper; 12 November 1843 - 13 February [1846]. KWM (624).

Logan, ship of New Bedford, Luther J. Briggs, master; 1838—41, abstract of voyage only. Maury 71, no. 7.

Lucy Ann, ship of Wilmington, Henry King, master; John F. Martin, keeper; 28 November 1841 - 15 June 1844. KWM (434).

Luminary, ship of Warren, Henry Cleveland, master; Edwin W. Athearn, unidentified second man, and L. A. Baker (surgeon), keepers; 22 May 1844 - 2 September 1847. KWM (463).

Magnet, ship of Warren, Samuel S. Munro, master and keeper; 6 June 1843 -12 April 1845. KWM (294).

Magnolia, ship of New Bedford, David Barnard, master and keeper; 3 December 1838 - 3 August 1842. KWM (359).

Majestic, ship of New Bedford, Worthen Hall, master; 1 November 1848 - 25 April 1851. PMS 656/1848M.

Maria, ship of Nantucket, Benjamin Ray, master; Thomas S. Worth, keeper; 1 January 1829 - 13 March 1830. IMA (426).

Mary, ship of Edgartown, Henry Pease 2d, master; 1 December 1844 - 9 October 1847. Dukes.

Merlin, bark of New Bedford, David E. Allen, master; [23 June 1868 - 13 April 1872], manuscript memoir of the voyage, "When I Was Seven, or Sea Memories," by Helen C. [Allen] Bradford. KWM (402-A).

Milo, ship of New Bedford, Francis M. Gardner, master; Charles Goodall, keeper; 19 April 1843 - 19 May 1846. Huntington (HM 26538).

Monmouth, bark of Cold Spring Harbor, Hiram B. Hedges, master and keeper; 12 October 1843 - 2 January 1845. KWM (586).

Montpelier, ship of New Bedford, Henry T. [Ellery Tompkins] Taber, master; 21 September 1844 - 26 July 1847. Kelly.

Montpelier, ship of New Bedford, Ellery Tompkins Taber, master; 2 September 1844 - 30 July 1847. Yale (HM21, v. 2).

Morning Light, bark of New Bedford, Hervey E. Luce, master; 22 September 1859 - 19 May 1862. KWM (383).

Morrison, ship of New London, Samuel Green, Jr., master; Rev. Thomas Douglass, keeper; 16 September 1844 - 23 May 1846 (account of voyage ends 24 March 1845, at Lahaina). MSM (343).

Morrison, ship of New London, Samuel Green, Jr., master. "Ship Morrison Book of Accounts," kept by Captain Green, 1845—47. MSM Misc. Vol. 318.

Moss, ship of New Bedford, William Austin, master; Melzar J. Comery, keeper; 24 December 1840 - 18 January 1844. KWM (456).

Mount Vernon, ship of New Bedford, George A. Covell, master; Israel G. Halstead, keeper; 1 December 1843 - 21 May 1846. KWM (241).

Nancy, ship of Havre, Thomas Jay, master; 1841−43, abstract of whaling cruises. FNA CC⁵611.

Narwal, ship of Havre, Antoine Gustave Radou, master; 5 June 1844 - 13 April 1847. KWM (559).

Navy, ship of New Bedford, George F. Bauldry, master; 7 October 1869 - 14 September 1871. KWM (156).

Navy, ship of New Bedford, George F. Bauldry, master and keeper; 8 October 1869 - 2 August 1871. de Lucia.

Neptune, ship of New London, Samuel Green, Jr., master and keeper; 20 June 1842 - 27 September 1843. KWM (157).

Neptune, ship of Sag Harbor, W. Pierson, master; Henry Rogers, keeper; 9 July 1843 - 7 May 1845. KWM (158).

Nil, ship of Havre, Gilbert Smith, master; 1841−43, abstract of whaling cruises. FNA CC⁵611.

Ocean, ship of New Haven, W. W. Clark, master; 16 August 1860 - 31 October 1861. KWM (161).

Oregon, ship of Fairhaven, Charles Tobey, master; 8 October 1857 - 25 May 1861. KWM (163).

Orozimbo, ship of New Bedford, David H. Bartlett, master; 16 November 1840 - 6 April 1843. Harvard F6870.65F.

Pantheon, bark of Fall River, --- Borden, master; Frederick L. Crapser, keeper; 26 November 1842 - 25 May 1845. MSM (769).

Philippe Delanoye, ship of Fairhaven, David G. Pierce, master; William Childs, keeper; 6 September 1852 - 13 November 1854. KWM (51).

Phoenix, ship of New London, --- Allen and James M. Fitch, masters; 11 November 1837 - 2 April 1841. KWM (21).

Portsmouth, ship of Warren, Samuel S. Munro, master and keeper; 5 February 1846 - 12 May 1849. KWM (294).

Robin Hood, ship of Mystic, William Pendleton, master; 8 October 1845 - 25 June 1848. MSM (48).

Roman, ship of New Bedford, Humphrey A. Shockley, master; Joseph Ammons, keeper; 11 May 1845 - 27 April 1847. MSM (792).

Sapphire, ship of Salem, Alexander Cartwright, master; John Crowell, keeper; 1 December 1839 - 24 June 1841. Essex M.656−1839S2.

Sapphire, ship of Salem, Alexander Cartwright, master; John Crowell, keeper; 25 June 1841 - 17 December 1842. Essex M.656−1841S.

Saratoga, ship of New Bedford, Frederick Slocum, master; Washington Fosdick, keeper; 23 April 1857 - 12 December 1858. KWM (180).

Sea Breeze, bark of New Bedford, Thomas J. Smith, then Benjamin Cushman, and George Duffy, masters; 1 September 1853 - 30 June 1856. KWM (472).

Sea Ranger, bark of New Bedford, James T. Sherman, master and keeper; 22 November 1888 - 7 May 1889. KWM (262).

Shepherdess, bark of Mystic, Hiram Clift, master and keeper; 30 April 1842 - 11 July 1844. MSM, Coll. 65, Vol. 12.

South America, bark of New Bedford, Washington T. Walker, master; Benjamin

Franklin Gallup (foremast hand), keeper; [3 October 1855 - 5 May 1859]. KWM (555).

South Carolina, ship of New Bedford, Ansel N. Stewart, master; Richard S. Sears, keeper; 30 June 1842 - 18 March 1844. KWM (445).

South Carolina, ship of New Bedford, Ansel N. Stewart, master; 1 July 1842 - 4 April 1844. KWM (564).

Splendid, ship of Edgartown, John S. Smith, master; 2 November 1854 - 18 May 1858. KWM (187).

Stonington, ship of New London, George W. Hamley, master and keeper; 17 August 1843 - 29 August 1847. MSM (335).

Superior, ship of New London, [Albert McLane, master]; [29 September 1840 - 3 July 1842], abstract of whales only. NLCHS .B495m 1914.04.78B.

Syren, ship of London, Frederick Coffin, master; 8 August 1819 - 18 April 1822. CRMM 1982.8.30.

Tiger, ship of Stonington, William E. Brewster, master; Mrs. William E. Brewster, keeper; 4 December 1845 - 8 March 1848. MSM (38).

Timoleon, ship of New Bedford, Joshua Bunker, master; 13 November 1835 - 24 June 1839. RP Wh/T585/18351.

Triton, ship of New Bedford, Reuben Chase, master and keeper; 2 August 1828 - 25 February 1831. KWM (576).

Virginia, ship of New Bedford, Joseph T. Chace, master; John Francis Akin, keeper; 7 November 1843 - 5 June 1847. KWM (407).

Wade, bark of New Bedford, George W. Downs, master; William H. Chappell, keeper; 19 July 1840 - 10 April 1844. KWM (288).

Wade, bark of New Bedford, George W. Downs, master; William H. Hardwick (chief officer), keeper; 28 June 1844 - 29 December 1846. KWM (201).

William Hamilton, ship of New Bedford, Lorenzo Fisher, master; Abraham Wilcox Pierce, keeper; 2 September 1845 - 14 January 1848. KWM (548).

Powered Vessels

Blue, of [Liverpool], pilot house log book, 14 June - 12 October [n.d.]. PABC Add MSS 21, Vol. 18.

Blue, of [Liverpool], chief officer's log book, 10 May 1935 - 18 September 1935. PABC Add MSS 21, Vol. 18.

Brown, of Liverpool, chief officer's log book, 1 May 1930 - 20 September 1930. PABC Add MSS 21, Vol. 14.

Brown, of [Liverpool], chief engineer's log book, H. D. Hornibrook, keeper, 21 April 1934 - 24 September 1934. PABC Add MSS 21, Vol. 14.

Brown, of [Liverpool], chief engineer's log book, 2 May 1935 - 21 September 1935. PABC Add MSS 21, Vol. 14.

Brown, of [Liverpool], pilot house log book, 20 April 1938 - 16 September 1938. PABC Add MSS 21, Vol. 18.

Brown, of Liverpool, log book, 5 May 1941 - 18 September 1941. PABC Add MSS 21, Vol. 18.

Green, of [Liverpool], pilot house log book, 17 March - 15 October [n.d.]. PABC Add MSS 21, Vol. 16.

Green, of [Liverpool], pilot house log book, 4 June - 8 October [n.d.]. PABC Add MSS 21, Vol. 16.

Green, of [Liverpool], engineer's log book, 23 April 1936 - 16 September 1936. PABC Add MSS 21, Vol. 16.

Westwhale 6, of [Victoria], whale catcher's log, [K. Kitayama, keeper], August 1963. MMBC (L3399b).

Westwhale 7, of [Victoria], whale catcher's log, 1 April 1967 - 25 April 1967. VCA Add MSS 335, Vol. 434.

White, of [Liverpool], gunner's report, 23–30 May 1942. PABC Add MSS 21, Vol. 11.

W[illiam] Grant, of Victoria, pilot house log book, 2 June 1933 - 29 August 1933. PABC Add MSS 21, Vol. 18.

W[illiam] Grant, of Victoria, chief officer's log book, 18 April 1937 - 17 September 1937. PABC Add MSS 21, Vol. 18.

W[illiam] Grant, of Victoria, pilot house log book, 27 April 1940 - 2 September 1940. PABC Add MSS 21, Vol. 18.

UNPUBLISHED DOCUMENTS

Public Archives

Many archives in Canada and the United States provided manuscript material and other unpublished documents. Eighteenth-century British interest in the Pacific Northwest is documented in various collections scattered through the holdings of the Public Record Office at Kew Gardens, London, and can be identified in the Notes accompanying Chapter 1. The primary sources on French whaling are the records of the French National Archives, stored on microfilm (particularly file CC^{5}611) at the Old Dartmouth Historical Society Whaling Museum in New Bedford, Massachusetts. These contain the official abstracts of the voyages of French whaling ships, as well as the abstracted logbook of the whaleship *Gange*, whose crew may have been first to take right whales north of 49° North in the eastern Pacific Ocean.

Considerable American whaling material of the nineteenth century survives in New England and elsewhere, and includes logbooks, journals of voyages, and various types of account books kept by shipowners, agents, and outfitters. The present work draws heavily upon collections in the Kendall Whaling Museum in Sharon, Massachusetts; the Old Dartmouth Historical Society Whaling Museum in New Bedford; the G.W. Blunt White Library of Mystic Seaport Museum in Mystic, Connecticut; the Melville Whaling Room of the New Bedford Free Public Library in New Bedford; the New London County Historical Society in New London, Connecticut; the Nicholson Whaling Collection at the Providence, Rhode Island, Free

Public Library; the Peabody Museum of Salem, Massachusetts; and the Essex Institute in that same city.

Particular attention need be drawn to two unpublished records of American whaling. The first, kept by Dennis Wood of New Bedford during the period 1840–70, tracks the reports of whaleships spoken during their voyages around the world. In many cases, Wood's accounts are the most complete and most accurate extant. His original manuscripts are housed in the New Bedford Free Public Library, with copies in both the Kendall Whaling Museum and the Old Dartmouth Historical Society. The other records, prepared by Matthew Fontaine Maury to support his study of the world's winds and currents, constitute another invaluable source for the itineraries and events of American whaling. Microfilms of the Maury Abstracts were consulted in the Old Dartmouth Historical Society Whaling Museum.

Old Dartmouth holds also the microfilm records of the now-defunct International Marine Archives of Nantucket. This project attempted to film all existing whaling records, and its international scope makes it invaluable for the study of whaling in the nineteenth century. The Frederick C. Sanford Papers, included among the IMA microfilms, proved to be of particular usefulness, as was a copybook of Sanford's letters, held by the Nantucket Historical Association.

Material relating to the activities of the Hudson's Bay Company and its relationship with whaling is documented in the Hudson's Bay Company Archives at the Provincial Archives of Manitoba in Winnipeg; supplementary material may be found in the Lyman Allyn Papers at the Old Dartmouth Historical Society, and among the Records of the Collector of Customs of the District of New London, Connecticut (RG 36), housed at the Federal Records Center in Waltham, Massachusetts.

Very little documentation of the Dawson-Roys period of shore whaling in British Columbia exists, save for newspaper accounts surviving in the microfilm collections of university libraries in British Columbia. Additional information was gleaned from lists of vessels entering and clearing the Port of Victoria (Add MSS 1073) in the Provincial Archives of British Columbia; and from the Shipping Registers of the Port of Victoria, on microfilm roll B2527 at the same venue.

The most significant public collection covering the development of modern whaling in British Columbia is the correspondence file of the Ministry of Marine and Fisheries, RG 23, Vol. 242, File 1536, in the Public Archives of Canada at Ottawa. Most of this material is also available on microfilm at the University of British Columbia Library (AW1 R5474, Vol. 41). The Public Archives of Canada holds other Fisheries files of relevance, including RG 23, Vols. 764, 1081, and 1082. Much of this material pertains to the activities of the Pacific Whaling Company (1904–09), and the Canadian North Pacific Fisheries Limited (1910–15), although some continues through World War II. Users of File 1536 must be aware that it is paginated in reverse, and is so cited herein.

The only known archive of correspondence and records covering the transitional period of the CNPF and the Victoria Whaling Company (1914–18) survives in the Vancouver Maritime Museum Collections at Vancouver. The activities of William Schupp are documented in two collections: the Consolidated Whaling Company Papers in the Provincial Archives of British Columbia at Victoria (particularly Add

MSS 21); and the William S. Lagen Collection in the Suzallo Library of the University of Washington, Seattle. Additional information on Schupp's dealings with the U.S. Government may be found in the collections of the Franklin D. Roosevelt Library in Hyde Park, New York (see OF 2371, the Henry Morgenthau Papers, and the Eleanor Roosevelt Papers).

The postwar whaling from Coal Harbour is documented in the United Fishermen and Allied Workers Union records in the Special Collections of the University of British Columbia, and additional information may be found in a small collection at the Maritime Museum of British Columbia. The Vancouver Maritime Museum holds many of the Daily Reports from this period.

Two important collections housed in the Vancouver City Archives were also consulted: the papers of the City Archivist, Major James Matthews (notably Add MSS 54); and the Hubert Lindsay Cadieux Collection. These cover the entire modern whaling period in British Columbia. Though not manuscripts, considerable background information was obtained from two newspaper files from the 1908–11 period, one held by the Museum of Northern British Columbia in Prince Rupert; and the other in the "morgue" of the *Prince Rupert Daily News*.

Unpublished Documents

Bradley, Michael Douglas. "The International Whaling Commission: Allocating an International Pelagic Ocean Resource." Ph.D. dissertation, University of Michigan, 1971. University Microfilms. Ann Arbor 71–23,707.

[Consolidated Whaling Corporation]. Whaling station statistics, 1908–23. Unpublished typescript. Parks Canada Anthropological Research Project, Victoria.

Cass, John. "Nanaimo's Whaling Station." VMM.

Clapp, Frank. "Notes on Whalers Out of Victoria." Mss notes from *Colonist* newspapers.

French, Charlie Hunt. "First Mention of Whaling—1840." Unpublished typescript, ca. 1940.

Goddard, Joan. "Reminiscences of Life on the Shore Whaling Stations of British Columbia 1905–1918." Typescript of paper presented to the Society for Historical Archaeology/Conference on Underwater Archaeology annual meeting, Sacramento, California, 9 January 1986.

Huelsbeck, David R. "The Economic Context of Whaling at Ozette." Typescript of paper presented to the Eleventh International Congress of Archaeological and Ethnological Science, Vancouver, 1983.

Messing, Jeffery K. "A Review of Some Historical and Economic Aspects of Hudson River Whaling." Unpublished typescript, [1975]. KWM.

Mitchell, Edward; Randall R. Reeves; Anne Evely; and Micheline Stawski. *Whale Killing Methods: An Annotated Bibliography*, press proofs.

O'Leary, Beth. "Magic and Poison: The Whaling Technologies of Three Northern Cultures." Unpublished paper, University of New Mexico, [n.d]. KWM.

Robbins, Lewis. "Letters Written of Whaling Expeditions 1936, 1937, & 1938."

Typescript of personal correspondence outward. Robbins.

Sanford, Frederick C. [Account of Whaleship Departures from Nantucket, 1815–1868]. Untitled mss. IMA (425).

Scarff, James E. "Historic and Present Distribution of the Right Whale (*Eubalaena glacialis*) in the Eastern North Pacific South of 50° N and East of 180° W." International Whaling Commission *Reports* (1986), press proofs.

Seabury, Humphrey W. "Right Whaling." Unpublished typescript (1868). KWM.

Smith, Sandra Truxtun. "A History of the Whaling Industry in Poughkeepsie, N.Y. (1830–1845)." Undergraduate thesis, Vassar College, 1956. FDR.

[Stafford, Samuel S.] "There She Blows: Journal of a Whaling Voyage Around the World in the Barque 'Laurens,' by 'Uncle Sam.' " Unpublished typescript. KWM.

Swiderski, Richard M. "Playing the Whale's Part: Magic & Technology in Arctic Whaling." Unpublished paper presented at Sixth Annual Whaling Symposium, Kendall Whaling Museum, 17 October 1981.

United States, Works Progress Administration, Survey of Federal Archives. *New Bedford Crew Lists*. Report No. 1610, [n.d.], typescript. KWM.

———, Works Progress Administration. *New Bedford Whalemen's Inward Manifests*. Report No. 836, [n.d.], typescript. KWM.

———, Works Projects Administration. Survey of Federal Archives. *Ship Registers of New Bedford, Massachusetts*. Vol. 1, 1796–1850. Boston: National Archives Project, 1940. Typescript. KWM.

[Van de Venter, A.] Whaling Inspector's Report, Port Hobron, Alaska, 21 May 1937 - 7 August 1937. KWM (466).

Webb, Robert Lloyd. " 'Nootkans,' Norwegians and 'Peter the Whaler': The Evolution of Coastal Whaling in British Columbia." Unpublished paper presented at Sixth Annual Whaling Symposium, Kendall Whaling Museum, 17 October 1981.

Westlake, Paulette. "Whaling in British Columbia." Undergraduate paper, University of British Columbia, December 1973.

NEWSPAPERS AND SERIAL PUBLICATIONS

(Separate articles are not listed)

British Columbia Gazette (Victoria)
Colonist (variously *British Colonist*, *Weekly Colonist*, and *Daily Colonist*) (Victoria)
Daily Standard (Victoria)
Evening Empire (Prince Rupert)
The Friend (Honolulu)
Friend of Virtue (Boston)
International Whaling Statistics (Oslo)
Nanaimo Free Press (Nanaimo)
Norsk Hvalfangst-Tidende [Norwegian Whaling Gazette] (Sandefjord)

Pacific Fisherman and *Pacific Fisherman Yearbook* (Seattle)
Polynesian (Honolulu)
Prince Rupert Journal (Prince Rupert)
Prince Rupert Optimist (Prince Rupert)
Province (or *Vancouver Daily Province*) (Vancouver)
Sailor's Magazine and Naval Journal (New York)
Sheet-Anchor (Boston)
Shipping & Commercial List (New York)
Times (London)
Trade News (Fisheries Department of Canada, Ottawa)
Union (San Diego)
Vancouver Sun (Vancouver)
Victoria Daily Times (Victoria)
Whalemen's Shipping List and Merchants' Transcript (New Bedford)

INDEX

Pagination referring to illustrations appears in *italic*.

University of British Columbia Press
PACIFIC MARITIME STUDIES SERIES

Sept 29 These here from the N & E cruising after Right Whales. the Larbard boat fasten to one & Killed him & got him alongside, the Waist boat fasten to another, the Bow boat Killed him Starbard boat fastened to another and the Jun drew that Bow boat fastened to another and brought him alongside.

saved   saved

Lost   Lost

28

Light heres from the N. cruising after Right Whales the Starbard boat fastened & Killed one the Larbard boat fastened to another & Killed him. they both sunk

sunk   sunk

Lat. 58. 39 N
Lo. 145. 15 W.